The Phylogeny and Classification of the Tetrapods

Proceedings of a Symposium held in London,
UK, March 1987

The Systematics Association
Special Volume No. 35B

The Phylogeny and Classification of the Tetrapods

VOLUME 2
Mammals

Edited by

M. J. BENTON
Department of Geology
The Queen's University of Belfast

Published for the SYSTEMATICS ASSOCIATION by the
CLARENDON PRESS · OXFORD
1988

Oxford University Press, Walton Street, Oxford OX2 6DP

Oxford New York Toronto
Delhi Bombay Calcutta Madras Karachi
Petaling Jaya Singapore Hong Kong Tokyo
Nairobi Dar es Salaam Cape Town
Melbourne Auckland

and associated companies in
Berlin Ibadan

Oxford is a trade mark of Oxford University Press

Published in the United States
by Oxford University Press, New York

British Library Cataloguing in Publication Data

The Phylogeny and classification of the tetrapods.
—(The Systematics Association special volume, ISSN 0309-2593:35)
1. Quadrupeds—Evolution
I. Benton, M. J. II. Systematics Association
596'.038 QL607.5

ISBN 0-19-857705-2 Vol. I
ISBN 0-19-857712-5 Vol. II

Library of Congress Cataloging in Publication Data

The Phylogeny and classification of the tetrapods
edited by M. J. Benton.
Proceedings of an international symposium held Mar. 26–27, 1987, and sponsored by the Systematics Association, the Linnean Society of London, and the Palaeontological Association.
Contents: v. 1. Amphibians, reptiles, birds—v. 2. Mammals.
1. Vertebrates—Evolution—Congresses. 2. Vertebrates—Classification—Congresses. 3. Vertebrates, Fossil—Congresses.
I. Benton, M. J. (Michael J.) II. Systematics Association.
III. Linnean Society of London. IV. Palaeontological Association.
QL607.5.P49 1988 597.6'0438'012—dc19
87-34832 CIP

ISBN 0-19-857705-2 (v. 1)
ISBN 0-19-857712-5 (v. 2)

Typeset by Wyvern Typesetting Ltd, Bristol
Printed in Great Britain by
St Edmundsbury Press
Bury St Edmunds
Suffolk

Preface

The papers in this volume represent part of the proceedings of an international symposium on 'The Phylogeny and Classification of the Tetrapods' held in the rooms of the Linnean Society of London, 26–27 March 1987, and sponsored by the Systematics Association, the Linnean Society, and the Palaeontological Association. A companion volume, *The Phylogeny and Classification of the Tetrapods, Volume 1: Amphibians, Reptiles, Birds*, was published concurrently with this volume. The intention of the meeting was to bring together zoologists and palaeontologists from both sides of the Atlantic who are actively working on the higher-order systematics of tetrapods as a whole, and of the major groups of tetrapods. In the event, the contributors to this volume are about equally split between Europeans (8) and Americans (11). The delegates at the conference were also diverse in their origins (U K, 51; U S A, 25; Germany, 6; Belgium, 5; Italy, 5; France, 3; The Netherlands, 3; Spain, 3; Denmark, 2; Sweden, 2; Eire, 1; Norway, 1; Switzerland, 1) and their backgrounds (biology, geology, anatomy, medicine).

The original intention was to try to present a balanced coverage of the systematics of all major mammal groups. It soon became evident, however, that although some parts of the cladogram appear to be well corroborated and accepted by most workers, there are still several major problems. Some of these areas of controversy are highlighted here: What are the closest living and extinct sister-groups of the Mammalia (Chapter 1)?; What are the basal splits within the Mammalia (Chapter 2)?; What are the relationships within the Eutheria of edentates and pholidotans (Chapter 3), of carnivores, insectivores, tubulidentates, primates, tree shrews, and bats (Chapters 3–6), of rodents, lagomorphs, and macroscelideans (Chapters 3 and 7), and of the various ungulate groups (Chapters 3, 8, 9, 10, and 11)?

Many of the disputed nodes on the cladogram of the Mammalia arise because of incomplete data sets (especially for fossil forms); the lack of convincing synapomorphies (especially when lineages are so highly autapomorphous that potential synapomorphies have been obscured); conflicting, but apparently equally congruent, alternative character distributions; and different data sets. These problems are exemplified by some of the biggest character conflicts over the relationships of the major living eutherian groups. A recent survey of molecular data by Miyamoto and Goodman (1986; *Syst. Zool.* **35**, 230–40) found rather different groupings from those proposed on the basis of morphological data (Chapter 3). For example, molecular data point to a grouping of primates, lagomorphs, and rodents as a clade, while morphological data tend to place the primates closer to the scandentians, dermopterans, and bats. Although there are still many conflicts like this in the large-scale picture of mammalian diversification, the tree has now been resolved to some extent, and it is not simply a 'bush' of 15 to 20 lineages radiating from a single point in the Late Cretaceous to Early Tertiary.

There are also controversies at lower levels. The authors of Chapter 3 (Novacek *et al.*) are not clear about the validity of Ungulata, while the authors of Chapter 8 (Prothero *et al.*) defend this clade. Novacek *et al.*, and Tassy and Shoshani (authors of Chapter 11) see the hydraxes as the sister-group of the tethytheres (elephants plus sirenians), while Prothero *et al.* interpret them as perissodactyls. Gentry and Hooker (authors of Chapter 9) and Janis and Scott (authors of Chapter 10) present different views of ruminant systematics.

This volume should be of interest to a wide range of biologists, palaeontologists, and naturalists. The papers all contain mostly new data, since it has only been in the last few years that a number of large-scale cladistic surveys of the higher-level systematics of mammals have been undertaken. We are privileged, in most cases, to have papers in this volume that represent the first publication of major and continuing projects that have not yet been published substantially elsewhere. The two companion volumes should provide a review of the present state of tetrapod systematics.

I thank Professors R. G. Davies and A. Hallam, Presidents of the Systematics Association, and Mr R. W. Sims, Dr A. R. Stone (deceased), and Dr D. M. John, officers of that body, for their encouragement and help throughout. Professor W. G. Chaloner and Dr D. M. Kermack of the Linnean Society graciously offered the use of the rooms of that body for the meeting, and Commander John Fiddian-Green and his staff ran the domestic aspects most efficiently, despite the large numbers of delegates. I thank Mr W. D. Maxwell and Dr Mary Benton for other assistance with the organization and running of the meeting. The chairmen were Dr R. P. S. Jefferies, Professor A. d'A. Bellairs, Dr W. A. Clemens, and Professor R. J. G. Savage, who controlled the loquacity of the assembled cladists admirably. I thank Mrs Gail Lyons for typing parts of the manuscript and Ms Libby Lawson for designing the meeting logo. The following acted as referees for the various contributions: D. P. Domning, J. J. Flynn, A. W. Gentry, J. J. Hooker, J. A. Hopson, R. Hunt, C. Janis, T. S. Kemp, C. Patterson, D. R. Prothero, R. J. G. Savage, J. H. Wahlert, A. E. Wood, B. Wood, and one anonymous reviewer. Financial assistance was received from the Systematics Association, the Linnean Society, the British Council, and other sources, which enabled a number of overseas speakers to attend.

Belfast M. J. B.
July 1987

Contents

VOLUME 2: MAMMALS

List of contributors ix

1. Interrelationships of the Synapsida
T. S. Kemp 1

2. A note on the Mesozoic mammals, and the origin of therians
T. S. Kemp 23

3. The major groups of eutherian mammals
Michael J. Novacek, Andre R. Wyss, and Malcolm C. McKenna 31

4. Phylogeny of the Carnivora
John J. Flynn, Nancy A. Neff, and Richard H. Tedford 73

5. Phylogeny of the insectivores
P. M. Butler 117

6. A phylogenetic analysis of the Primates
Peter Andrews 143

7. Rodent phylogeny: new data and old problems
J.-J. Jaeger 177

8. The phylogeny of the ungulates
Donald R. Prothero, Earl M. Manning, and Martin Fischer 201

9. The phylogeny of the Artiodactyla
A. W. Gentry and J. J. Hooker 235

10. The phylogeny of the Ruminantia (Artiodactyla, Mammalia)
Christine M. Janis and Kathleen M. Scott 273

11. The Tethytheria: elephants and their relatives
Pascal Tassy and Jeheskel Shoshani 283

Index 317

List of Systematics Association Publications 327

Contents

VOLUME 1: AMPHIBIANS, REPTILES, BIRDS

List of contributors ix

1. The relationships of the earliest tetrapods
A. L. Panchen and T. R. Smithson 1

2. Estimating the interrelationships of tetrapod groups on the basis of molecular sequence data
M. J. Bishop and A. E. Friday 33

3. The relationships and origin of living amphibians
Andrew R. Milner 59

4. The early evolution of the Amniota
Jacques A. Gauthier, Arnold G. Kluge and Timothy Rowe 103

5. A phylogeny of turtles
Eugene S. Gaffney and Peter A. Meylan 157

6. The early history and relationships of the Diapsida
Susan E. Evans 221

7. The classification of the Squamata
Olivier Rieppel 261

8. Archosaur phylogeny and the relationships of the Crocodylia
Michael J. Benton and James Clark 295

9. The major clades of birds
Joel Cracraft 339

Index 363

List of Systematics Association Publications 375

Contributors

P. ANDREWS
Department of Palaeontology, British Museum (Natural History), Cromwell Road, London SW7 5BD, UK.

P. M. BUTLER
Department of Zoology, Royal Holloway and Bedford New College, Englefield Green, Surrey TW20 9TY, UK.

M. FISCHER
Institut für Biologie III, Universität Tübingen, Tübingen, West Germany.

J. J. FLYNN
Department of Geological Sciences, Wright Geological Laboratory, Busch Campus, Rutgers University, New Brunswick, New Jersey 08903, USA.

A. W. GENTRY
Department of Palaeontology, British Museum (Natural History), Cromwell Road, London SW7 5BD, UK.

J. J. HOOKER
Department of Palaeontology, British Museum (Natural History), Cromwell Road, London SW7 5BD, UK.

J. J. JAEGER
Laboratoire de Paléontologie des Vertébrés et de Paléontologie Humaine, Université Pierre et Marie Curie, 4 Place Jussieu (Paris VI), 75252 Paris, Cedex 05, France.

C. M. JANIS
Division of Biology and Medicine, Box G, Brown University, Providence, Rhode Island 02912, USA.

T. S. KEMP
University Museum, Parks Road, Oxford OX1 3PW, UK.

E. M. MANNING
Museum of Geoscience, Louisiana State University, Baton Rouge, Louisiana 70803, USA.

M. C. McKENNA
Department of Vertebrate Paleontology, American Museum of Natural History, Central Park West at 79th Street, New York, New York 10024, USA.

N. A. NEFF
Department of Ecology and Evolutionary Biology, U–43 University of Connecticut, Storrs, Connecticut 06268, USA.

M. J. NOVACEK
Department of Vertebrate Paleontology, American Museum of Natural History, Central Park West at 79th Street, New York, New York 10024, USA.

D. R. PROTHERO
Department of Geology, Occidental College, Los Angeles, California 90041, USA.

K. M. SCOTT
Department of Biological Sciences, Rutgers University, Box 1059, Piscataway, New Jersey 08855, USA.

J. SHOSHANI
Department of Biology, Wayne State University, Detroit, Michigan 48202, USA.

P. TASSY
Laboratoire de Paléontologie des Vertébrés et de Paléontologie Humaine, Université Pierre et Marie Curie (Paris VI), 4 place Jussieu, 75252 Paris, Cedex 05, France, and INSERM, Unité 155, Château de Longchamp, Bois de Boulogne, 75016 Paris, France.

R. H. TEDFORD
Department of Vertebrate Paleontology, American Museum of Natural History, Central Park West at 79th Street, New York, New York 10024, USA.

A. R. WYSS
Department of Vertebrate Paleontology, American Museum of Natural History, Central Park West at 79th Street, New York, New York 10024, USA.

1. Interrelationships of the Synapsida

T. S. KEMP
University Museum and Department of Zoology, Parks Road, Oxford, UK

Abstract

Historically the Synapsida, which strictly includes the mammals, have been subject to excessive creation of paraphyletic groups, both the classical sequence of Pelycosauria and Therapsida, and also series of subgrades within them. The reason for this is the extensiveness of the fossil record coupled with a focus on the question of the origin of mammals. During the last few years several authors have reanalysed parts of the Synapsida cladistically, and their results are reviewed.

A number of controversial areas remain, notably the interrelationships of the main therapsid groups, the classification of the advanced cynodont families, and the precise relationship of the Mammalia within the cynodonts. On the latter point, the hypothesis that the Tritylodontidae form the sister-group of the Mammalia (possibly with the Tritheledontidae) appears to be better supported than the alternative, that the Tritylodontidae are related to the other herbivorous cynodonts.

Introduction

The Synapsida, or 'mammal-like reptiles', have been accepted for over a century as the group from which the mammals evolved. Furthermore, within the Synapsida, several grade-groups have traditionally been given the status of ancestral taxa, principally the Pelycosauria for the most primitive forms, the Therapsida for the more advanced forms, and the Cynodontia for the most mammal-like of the therapsids prior to the mammals themselves. Even within these major paraphyletic group-

The Phylogeny and Classification of the Tetrapods, Volume 2: Mammals (ed. M. J. Benton), Systematics Association Special Volume No. 35B, pp. 1–22. Clarendon Press, Oxford, 1988.

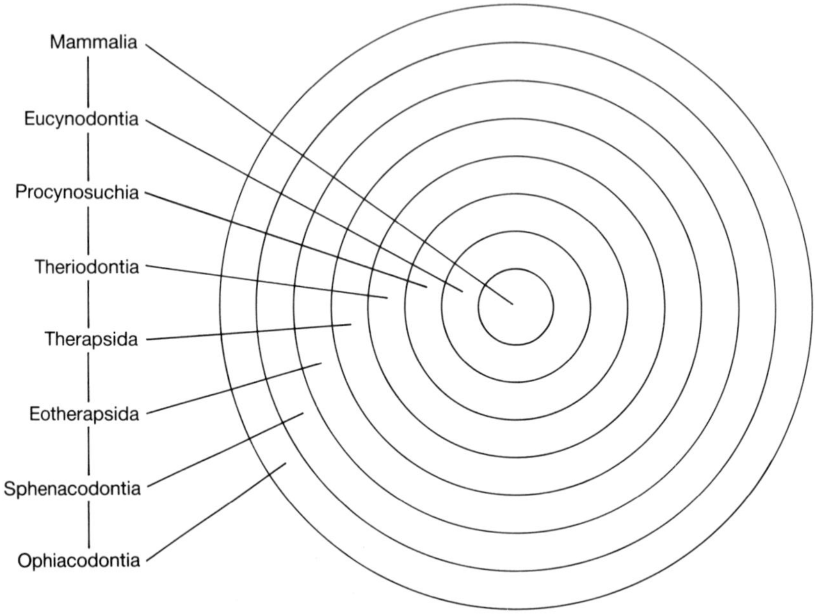

Fig. 1.1. The internested arrangement of paraphyletic groups frequently used in the classification of the Synapsida.

ings, classifications have invariably included ancestral taxa, such as the Ophiacodontia (Romer and Price 1940) for primitive pelycosaurs, the Eotherapsida (Olson 1962, 1986) for the primitive therapsids, and the Procynosuchia for the most primitive cynodonts. Indeed, no major vertebrate group has been subject to a more comprehensively evolutionary mode of classification than the Synapsida, containing as it does so many arbitrarily defined ancestral or paraphyletic groupings (Fig. 1.1).

The reasons for this state of affairs are clear. The fossil record of the synapsids appears to be relatively complete, with the large morphological difference between a primitive amniote, i.e. a pelycosaur, and a mammal spanned by a series of forms of intermediate structure. It is therefore easy to perceive certain lineages of transforming characters, and to label the forms possessing them as a sequence of ancestor-descendent taxa, ignoring those characters (autapomorphies) that do not show such evident transformations. What is more, this sequence of increasingly mammal-like fossil forms corresponds very closely with relative stratigraphic position. To this is added the fortuitous geological circumstance that the larger morphological gaps tend to coincide with biogeographic and stratigraphic discontinuities, which encourages agreement between different authors on taxonomic boundaries.

Thus, the great majority of pelycosaurs are from the Early Permian of North America, while the therapsids have been found in greatest abundance in southern Africa, and to a lesser extent in South America and Russia. Within the therapsids, the more primitive groups are largely of Late Permian age while, apart from a very few primitive forms, the cynodonts are a Triassic group. Finally, there is the legitimate interest in understanding the origin of a group deemed as important as the Mammalia, and in expressing the conclusions reached in the form of a grade-classification of the relevant fossils.

Nevertheless, despite this evolutionarily satisfying classification, the need for a cladistic approach has become increasingly clear in the last few years, both to create a more objective and, consequently, a more testable phylogeny, and also to highlight more clearly a number of problems and areas of dispute about the exact phylogenetic relationships of certain forms. The first attempt at a comprehensive cladogram of the major groups of the synapsids was that of Gaffney (1980), while Reisz (1980) noted the synapomorphies of the main groups of pelycosaurs, and discussed their cladistic interrelationships. Kemp (1982) reviewed all the synapsids and offered cladograms for most of the subgroups.

The purpose of this paper is to review the more important cladistic work on synapsids that has appeared since the publication of Kemp (1982). In 1982, Battail first published a cladogram of the cynodonts. Cluver and King (1983) initiated the cladistic analysis of the dicynodonts, the dominant herbivorous group of therapsids, and King (1988) has extended this work and also included the dinocephalians. Kemp (1983) and Sues (1985) discussed, and disagreed about, the exact relationships of the Mammalia to the various cynodont groups. Brinkman and Eberth (1983) analysed the cladistic relationships of certain pelycosaurs, and Reisz (1986) has reviewed this group in detail. Hopson and Barghusen (1986; although their paper was read at a conference in 1981 and has been available in manuscript to other authors for several years) have reviewed the interrelationships of all the therapsids. Most recently, Rowe (MS) has made a very detailed study of the relationships of the mammals to other synapsid groups.

The cladistic relationships of the Synapsida

Synapsida

To qualify for strictly monophyletic status, the group Synapsida as traditionally conceived must be enlarged to include the mammals. (In fact, Goodrich (1916) long ago created a taxon Theropsida for synapsids plus mammals, distinguishing them from the rest of the reptiles which he termed Sauropsida. Accepting Goodrich's term would leave the term Synapsida redundant, referring only to the non-monophyletic group of Theropsida minus mammals, but it seems preferable to retain the more familiar expression Synapsida.)

The remarkably wide range of forms hypothesized to be synapsids, from exceedingly primitive amniotes to the highly derived mammals, means that there are few synapomorphies defining the whole group, and that sequences of transformational homologies (Kemp 1988) play an important role in establishing membership in the group of the more derived forms.

The most familiar character of the group is the single temporal fenestra, bounded primitively by squamosal and postorbital bones above. Gaffney (1980) and Reisz (1980) both pointed out that the lower temporal fenestra of diapsids has this construction as well, and therefore the synapsid condition might be seen as plesiomorphic for both diapsids and synapsids and not unique to the latter. Reisz (1986) nevertheless lists it as a characteristic of pelycosaurs, and therefore of synapsids as a whole, with the implication that it is convergent in the diapsids. Synapomorphies of synapsids are thus (Reisz 1986):

(1) temporal fenestra primitively bounded above by squamosal and postorbital;

(2) occipital plate broad and tilted forwards; post-temporal fenestra small;

(3) postparietal single and median;

(4) septomaxilla with a broad base and dorsal process;

(5) medial process of the jugal sutured to the pterygoid, but ectopterygoid primitively present.

There are no known characters of the postcranial skeleton of synapsids that are not also to be found in at least some other groups of amniotes (Carroll 1986).

Two challenges to the monophyletic status of the Synapsida have been made. Reig (1967) proposed that the archosaurs evolved from pelycosaurs, a hypothesis based primarily on the common possession of plesiomorphic characters that has been effectively refuted by Romer (1971; see also Benton 1985).

A more recent and more radical challenge has come from Gardiner (1982*a,b*). He proposed that amongst living amniotes, the mammals and the birds are sister-groups, forming a taxon Haemothermia which is in turn the sister-group of the crocodiles. On attempting to fit the various fossil amniote groups into the cladogram of living groups, he claimed that the Synapsida are actually a highly artificial assemblage. Pelycosaurs are the sister-group of all living amniotes, dicynodonts the sister-group of turtles, and cynodonts the sister-group of archosaurs plus haemothermians. Kemp (1988) reconsidered Gardiner's data in detail and concluded that many of the characters used were unreliable, while many other apparently sound characters had been ignored. By far the most parsimonious hypothesis of relationships of living amniotes has birds and crocodiles as sister-groups, and mammals as relatively distant. The traditional non-mammalian synapsid groups continue to fit most congruently as relatives of the mammals.

Pelycosauria

In their seminal review of the group, Romer and Price (1940) classified the pelycosaurs into three major taxa that are still in wide use. One of these, the Ophiacodontia, is clearly paraphyletic for it consists of two families, Eothyrididae and Ophiacodontidae, whose only common characters are primitive pelycosaurian features. The second group, Edaphosauria, is equally clearly polyphyletic, for it mainly consists of the two herbivorous families Caseidae and Edaphosauridae which share only superficially similar adaptations for herbivory, but each of which possesses synapomorphies of quite different respective groups. The third group, Sphenacodontia, is paraphyletic, not only for the trivial reason that it is technically ancestral to the Therapsida, but also because its more primitive constituent family Varanopseidae does not appear to be the sister-group of the more advanced family Sphenacodontidae.

Reisz (1980; followed by Kemp 1982) was the first author to address these problems, and he proposed a cladogram of the families that was based on apparent synapomorphies alone. Brinkman and Eberth (1983) subsequently restudied the better-known genera of pelycosaurs and offered an alternative cladogram that differed in certain fundamental respects from that of Reisz (1980). Most recently, Reisz (1986) has returned to the question of pelycosaur interrelationships and has attempted a compromise classification based on both his and Brinkman and Eberth's (1983) analyses. As the most recent and the most comprehensive, Reisz's (1986) hypothesis of relationships is presented here (Fig. 1.2).

Therapsida

Monophyly of the Therapsida As with the Synapsida, the taxon Therapsida must be taken formally to include the mammals, whereupon there is no serious doubt that it is monophyletic. From Kemp (1982), Hopson

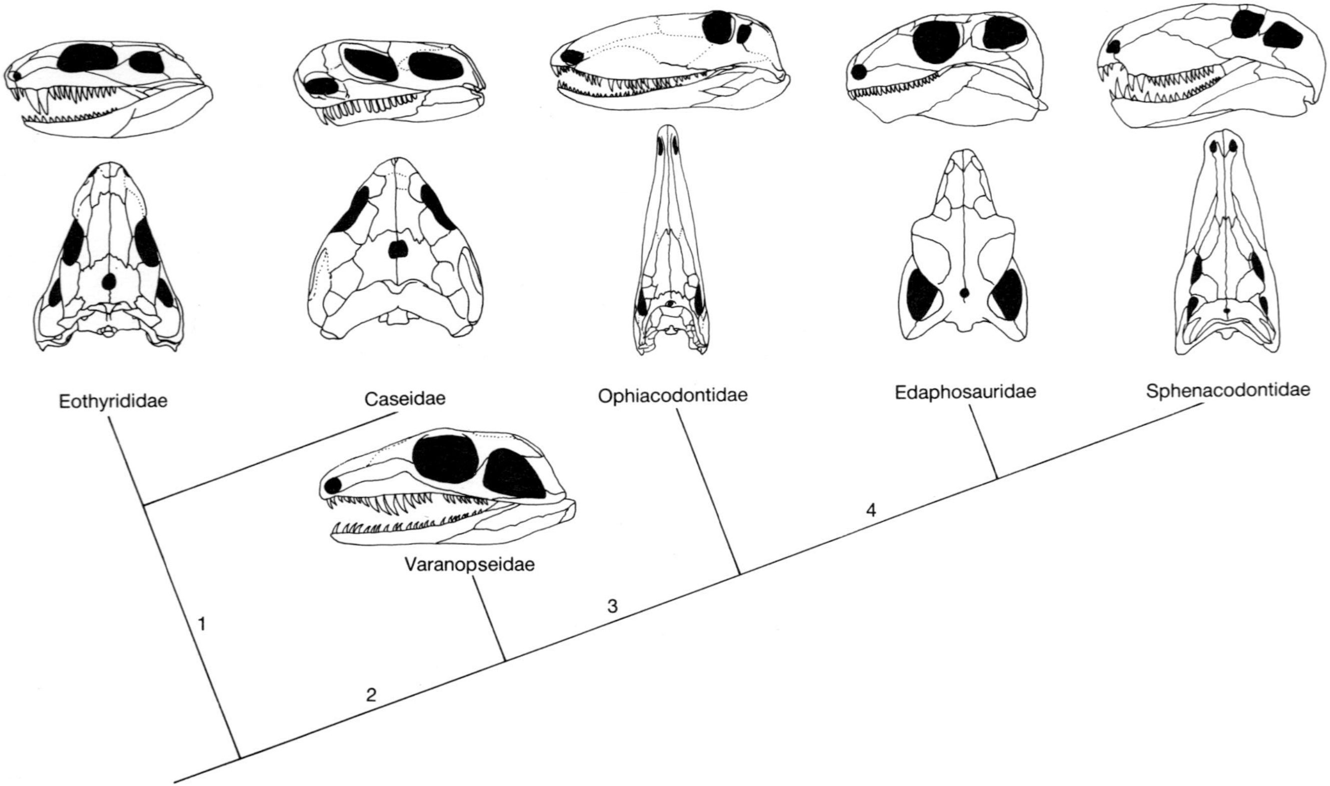

Fig. 1.2. Cladogram of the main pelycosaur families, after Reisz (1986). Synapomorphies of clades as follows: 1, pointed rostrum formed by anteriorly tilted premaxilla; external narial shelf on nasal; (?) ventral orbital margin partly maxilla; anteroposteriorly elongated external narial opening; 2. height of snout greater than width; frontal contributes to dorsal orbital margin; supratemporal long and narrow; pineal foramen in posterior half of parietals; 3, dorsal process of stapes in socket in paroccipital process; angular with ventral keel; margin of cheek concave; (?) nasal longer than parietal in midline; 4, zygomatic arch formed by jugal and squamosal; no shelf between parasphenoid wings; frontal with lateral lappet; quadrate ramus of pterygoid with rounded ventral edge; (?) prearticular twisted; (?) pterygoideus process formed entirely by articular.

and Barghusen (1986), and Rowe (MS) the following synapomorphies of the group are recognized:

(1) lateral temporal fenestra enlarged;

(2) supratemporal lost;

(3) shallow groove ('external auditory meatus') on posterior face of squamosal;

(4) septomaxilla with long posterodorsal process;

(5) pineal foramen on a raised boss;

(6) reflected lamina of the angular greatly enlarged as a thin sheet;

(7) jaw articulation shifted forwards;

(8) loss of mobile basipterygoid articulation;

(9) interpterygoid vacuity reduced;

(10) reduction of palatal teeth;

(11) paroccipital process extending more laterally than ventrally;

(12) vomer transversely widened between internal nares;

(13) transverse process of pterygoid shifted forwards;

(14) no ossified intercentra in trunk;

(15) ossified sternum;

(16) cleithrum reduced, scapular blade narrow, coracoid and procoracoid reduced; loss of the screw-shaped glenoid;

(17) expansion of the ilium both anteriorly and posteriorly, acetabulum circular; loss of fourth trochanter of the femur;

(18) reduction in phalanges (possibly to the mammalian number of 2.3.3.3.3., Rowe and Van den Heever 1986).

The only serious argument against monophyly of the Therapsida has been that of Olson (1962; and much more tentatively, 1986), who discerned certain resemblances between the herbivorous therapsids and the caseid pelycosaurs, as distinct from those characters shared by the carnivorous therapsids and the sphenacodontid pelycosaurs (see below). In the face of the list of apparent synapomorphies of therapsids just presented, such resemblances are more parsimoniously interpreted as convergences for a herbivorous mode of life.

The relationship of the Therapsida to pelycosaurs With the exception of Olson's (1962) doubts, there has been practically universal agreement that the therapsids are most closely related to the sphenacodontid pelycosaurs, the two groups sharing the following derived characters (Romer and Price 1940; Reisz 1980, 1986; Kemp 1982; Hopson and Barghusen 1985):

(1) reflected lamina of the angular present;

(2) retroarticular process of the angular turned downwards;

(3) loss of the zygomatic process of the quadratojugal;

(4) canine tooth enlarged and compressed laterally;

(5) maxilla enlarged, separating lachrymal from nares;

(6) ventrolateral orientation of paroccipital process.

Romer and Price (1940) proposed that the primitive sphenacodontid *Haptodus* is ancestral to the therapsids, but this was based on the non-cladistic argument that the more derived forms (sphenacodontines) such as *Dimetrodon* are too specialized to be ancestral. In fact, *Haptodus* lacks characters 4 and 5 above, and has a smaller reflected lamina of the angular than the sphenacodontines and the therapsids, and is evidently the plesiomorphic sister-group of those two (Kemp 1982; Rowe MS). Nevertheless, Currie (1979) argued the less parsimonious case that *Haptodus* and the sphenacodontines were sister-groups, and that these characters evolved convergently in sphenacodontines and therapsids.

The major monophyletic groups of therapsids (Fig. 1.3) Aside from certain problematical primitive forms, 6 and 7 below, the therapsids fall into five distinctive, uncontroversially monophyletic groups, as follows:

1. *Dinocephalia.* These are large forms, both herbivores and carnivores, which tend towards pachyostosis of the skull bones. They are characterized by: incisor teeth with lingual heels, which intermesh when the jaws are closed; temporal fenestra expanded dorsoventrally; jaw adductor muscles invading the lateral and dorsal faces of the bones of the intertemporal regions of the skull. Most authors now include the Russian brithopodids as fairly primitive members, and also the enigmatic *Estemmenosuchus*. Hopson and Barghusen (1986) and, in greater detail, King (1988) have presented cladograms showing the interrelationships within the group.

2. *Dicynodontia.* These herbivorous forms, from the earlier Late Permian to the Late Triassic, are highly specialized and are characterized by numerous features of the skull and jaws, for example: dorsal displacement of the zygomatic arch; dentaries fused at the symphysis; articular surface of the lower jaw sloping posteroventrally; horn devel-

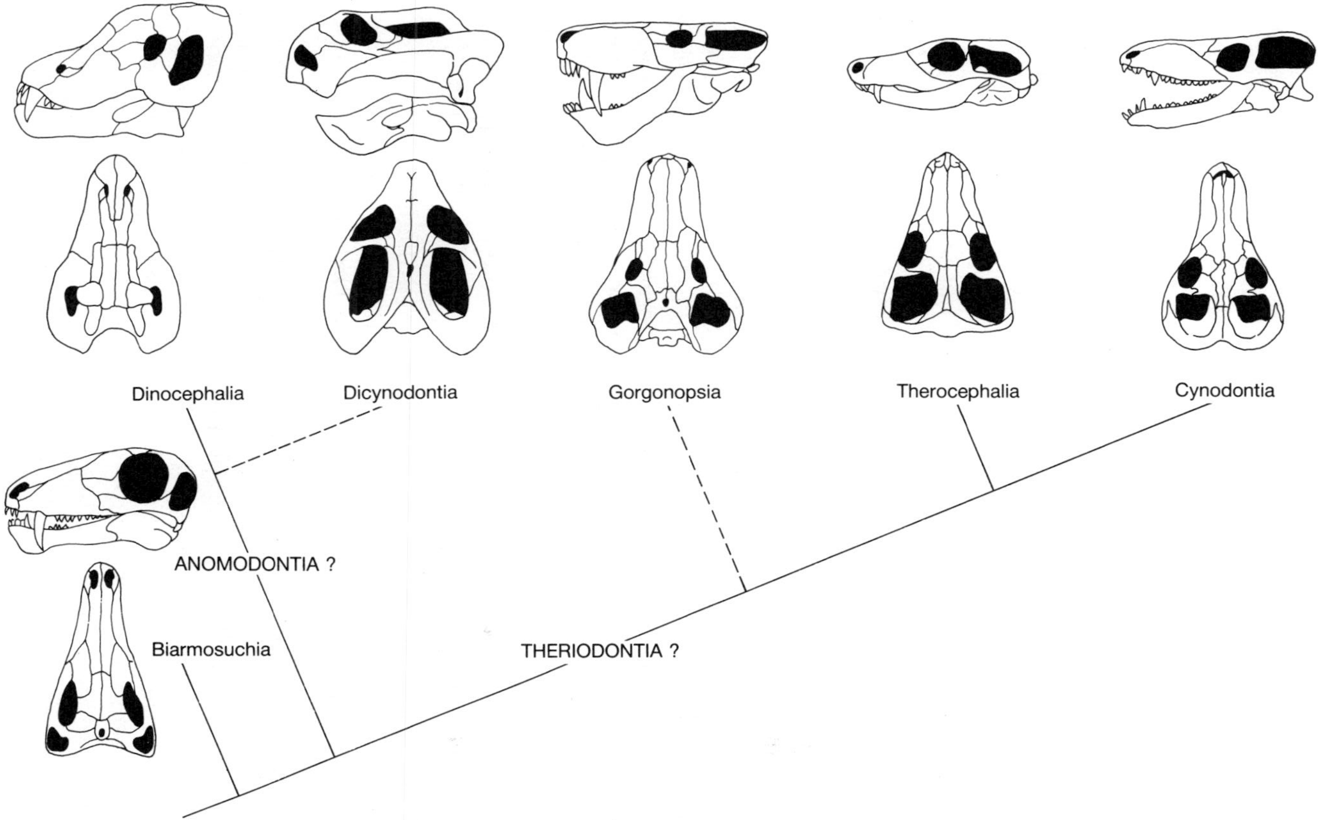

Fig. 1.3. The major groups of the Therapsida. For characters of hypothesized clades, see text.

oped on the jaws; intramandibular fenestra; premaxillary secondary palate. Included as the most primitive members are *Otsheria* and *Venjukovia* of Russia, and the dromasaurs of South Africa. Cluver and King (1983) and King (1988) have analysed their cladistic interrelationships in detail.

3. *Gorgonopsia*. These are relatively primitive carnivores restricted to the Late Permian. They are characterized by numerous features including: reflected lamina of the angular far in front of the jaw articulation, attached dorsally, and with a single vertical ridge on the lateral surface; quadrate bulbous and loosely set in a squamosal socket; articular with dorsal process; palatine bones meet midventrally; lower incisors capable of interdigitating with uppers, canines further enlarged, postcanines reduced in size and number; vomers fused and expanded transversely. The gorgonopsians are a very conservative group and difficult to classify (Sigogneau 1970), and no cladistic analysis has yet been published.

4. *Therocephalia (including bauriamorphs)*. These are carnivorous forms (except perhaps for a few late members), occurring from Late Permian to Middle Triassic, and characterized for example by: reflected lamina large, posteriorly placed, with radiating external ridges; process of the squamosal bounding pterygo-paroccipital foramen; suborbital vacuities in palate; stapes lacking foramen and dorsal process. Earlier authors (e.g. Watson and Romer 1956) recognized a group called Bauriamorpha, separate from the Therocephalia, for the later forms. This makes the Therocephalia paraphyletic (Kemp 1982), and Hopson and Barghusen (1986) have accordingly reduced the bauriamorphs to a superfamily Baurioidea, contained within the Therocephalia. The latter authors have presented a cladogram for the group.

5. *Cynodontia*. These appear in the latest Late Permian and form the dominant therapsid group of the Triassic. They are also the most derived of the non-mammalian therapsids, characterized by a great many features that are essentially mammalian. Hopson and Barghusen (1986) list 28 such characters of the skull, and Kemp (1982) and Rowe (MS) note several postcranial features. Among the more significant characters are: lateral flaring of the zygomatic arches; adductor fossa on lateral surface of the enlarged dentary; reflected lamina of the angular reduced; quadrate with posterior process in slit in squamosal; lateral prootic process contacting epipterygoid; palatal processes of palatine bones approaching midline; occipital condyles double. The further classification of cynodonts, including mammals, is discussed later.

In addition to these well-defined therapsid groups, there are certain very primitive forms from the earlier part of the Late Permian of Russia and South Africa, whose relationships to one another and to other therapsids are contentious.

6. *Biarmosuchia*. The Russian genus *Biarmosuchus* (with the probably synonymous *Biarmosaurus*) is the most primitive, i.e. the most sphenacodontid-like, having a relatively very small temporal fenestra, wide intertemporal roof, and a convex dorsal skull margin (Fig. 1.3). A series of South African genera, included in a family 'Ictidosuchidae' by Sigogneau (1970), is very similar to *Biarmosuchus*. Hopson and Barghusen (1986) erected a group Biarmosuchia for them, and tentatively placed the rather poorly-known Russian form *Phthinosuchus* in this group. Although they were unable to discover any synapomorphies shared by all these forms, which suggests that the Biarmosuchia could be paraphyletic, they nevertheless proposed that the 'Biarmosuchia' form the plesiomorphic sister-group of all other therapsids (Fig. 1.3). This is justified by the belief that all the remaining therapsids form a monophyletic group categorized by a single character, a further increase in the size of the temporal fenestra. However, there are unexpected differences in the detailed form of the temporal fenestra in the different advanced therapsid groups (Kemp 1982). In the Dinocephalia the fenestra is dorsoventrally expanded, in the Gorgonopsia posterolaterally, in the Dicynodontia mainly posteriorly, and in the Therocephalia and Cynodontia mainly medially. This throws doubt on the homology of the large size of the fenestra, and therefore on the proposition that the 'non-biarmosuchian' therapsids are monophyletic. This in turn leaves open the possibility that the 'Biarmosuchia' contains primitive relatives of one or more of the individual later therapsid groups.

7. *Eotitanosuchia*. Another Russian genus, *Eotitanosuchus*, is slightly more advanced because, although still small compared to later therapsids, the temporal fenestra is enlarged over that of the 'biarmosuchians', and shows signs of some invasion of its posteromedial border by jaw musculature. Kemp (1982) took *Eotitanosuchus* to be related to the gorgonopsians, particularly because of similarities of the palate. But Hopson and Barghusen (1986) regard it as the sister-group of all the remaining therapsids. Because of the doubts about the monophyletic status of the advanced therapsids, and about the status of the 'Biarmosuchia', the exact position of *Eotitanosuchus* must remain unresolved.

Interrelationships of the major groups of the Therapsida The cladistic interrelationships of these therapsid groups is a matter of considerable dispute. Indeed, the only relationship upon which all recent authors agree is one between the Therocephalia and the Cynodontia, a taxon referred to as the Therosauria by Kemp (1982), and as the Eutheriodonta by Hopson and Barghusen (1986; followed by Rowe, MS). Many clear-cut synapomorphies support this grouping, including:

(1) narrowed intertemporal roof; postorbital shortened; postfrontal reduced;

(2) parietal expanded posteriorly;

(3) epipterygoid expanded anteroposteriorly;

(4) secondary palate at least rudimentary;

(5) thickened posteroventral part of dentary supporting angular;

(6) laterally directed prootic process in front of paroccipital process;

(7) notch in squamosal behind quadrate;

(8) similar differentiation of the vertebral column into cervical, thoracic, and lumbar regions; lumbar ribs short, horizontal and immovable (Kemp 1986);

(9) very similar atlas-axis complex (Kemp 1986).

Kemp's (1982) claim that the therocephalian family Whaitsiidae is the sister-group of cynodonts, and with it the implication that the Therocephalia as a whole is paraphyletic, has been shown by Hopson and Barghusen (1986) to be less parsimonious than the hypothesis that Therocephalia and Cynodontia are sister-groups.

Apart from the Therosauria (Eutheriodontia), there is little agreement about the interrelationships of the therapsids. Prior to 1982, the most influential classification was that of Romer (1966; largely following Watson and Romer 1956), which recognized a dichotomy of the Therapsida into two branches, the mainly herbivorous Anomodontia (dinocephalians and dicynodonts) and the mainly carnivorous Theriodontia (gorgonopsids, therocephalians, and cynodonts). This scheme was essentially followed by Gaffney (1980). However, Kemp (1982) felt that there were no adequate shared characters to support either Anomodontia or Theriodontia with confidence, and left an unresolved polychotomy at the base of the Therapsida. Nevertheless, Hopson and Barghusen (1986) did recognize the Theriodontia (Fig. 1.3A), on the basis of the following characters:

(1) postdentary bones reduced in height and posterodorsal part of the dentary increased in height so that the latter is left as a free-standing coronoid process;

(2) quadrate and quadratojugal reduced in height;

(3) quadrate and quadratojugal lie in a depression in the anterior face of the squamosal, with no sutural union with the squamosal;

(4) dorsal process of the premaxilla reduced in length;

(5) skull reduced in height.

However, it should be pointed out that the coronoid process (1) is triangular in section in gorgonopsids but parallel-sided in therocephalians, which suggests that the structure is associated with a different muscle pattern and not homologous in the two groups. Also, the quadrate complex (2 and 3) of the gorgonopsids is structurally (and functionally) quite distinct from that of the Therosauria, and therefore the simple character state as quoted appears to reflect a similarity that does not really exist. Finally, the other two characters are also found in later dicynodonts.

King (1988) rather tentatively recognizes the Anomodontia as a monophyletic grouping of the dinocephalians and dicynodonts (Fig. 1.3B), on the basis of:

(1) loss of coronoid bone(s);

(2) non-terminal nostrils and long posterior spur of premaxilla (modified in higher dicynodonts);

(3) grooved or troughed palatal exposure of vomers;

(4) reduction or loss of internal trochanter of femur.

Of these, there may actually be a coronoid bone in the primitive, brithopodid dinocephalians (Hopson, *pers. comm.*). The nature of the premaxilla resembles the condition found in biarmosuchians and *Eotitanosuchus*, which implies that it is the primitive therapsid state (Hopson and Barghusen 1986). But since the exact phylogenetic position of the latter two groups is by no means certain, then it remains perfectly possible that King (1988) is correct in identifying the posteriorly extended premaxilla as a synapomorphy, and that biarmosuchians and *Eotitanosuchus* are also anomodonts. Pelycosaurs are no help in establishing the polarity because they are highly variable in this character.

Rowe (MS) has proposed the most radical alternative, whereby neither Anomodontia nor Theriodontia is accepted. Instead, the Dicynodontia are taken to be the sister-group of the Therosauria. He lists 28 characters in support of this particular relationship, although several of them are also found in the gorgonopsids, several are not present in all members of the proposed group, and others are very vague. On the other hand, some of his proposed synapomorphies look sounder than those proposed for alternative therapsid groupings, e.g.

the dorsal bowing of the zygomatic arch, and the structure of the atlas-axis complex. Detailed consideration of Rowe's (MS) conclusions must await publication, but certainly his case could prove to be better supported than the Anomodontia–Theriodonta dichotomy, and to be preferable to Kemp's (1982) agnosticism.

Cynodontia

Cynodont grades The undisputed monophyletic status of the Cynodontia (if taken to include the mammals) has already been indicated. Particularly on the basis of the relative size of the dentary, three grades of cynodonts are recognizable. The most primitive grade is represented by *Dvinia* and *Procynosuchus*, in which the dentary is little enlarged over that of other therapsids. *Thrinaxodon* and *Galesaurus* represent an intermediate grade, while all other cynodonts have reached the advanced grade of a dentary completely dominating the postdentary bones. Kemp (1979, 1982) challenged the then prevailing classification of cynodonts into a predominantly carnivorous group and a predominantly herbivorous group (e.g. Hopson and Kitching 1972), with its implication that the advanced cynodont condition had been achieved independently in two lineages. Battail (1982) presented a cladogram continuing to express this view, but all other recent authors have accepted that the most parsimonious hypothesis is of a monophyletic group Eucynodontia for all the advanced-grade cynodonts (Sues 1985; Hopson and Barghusen 1986; Rowe, MS), categorized by the following characters:

(1) dentary greatly enlarged, approaching jaw articulation, and with posteroventral angular process; corresponding relative reduction in postdentary bones, including reflected lamina of the angular (further reduced to ear ossicles and ectotympanic in mammals);

(2) secondary articulation between ventrolateral process of the squamosal and surangular (replaced by dentary in mammals);

(3) much more prominent external auditory meatus formed from posterior folding of the squamosal;

(4) quadrate ramus of the pterygoid reduced or absent, and quadrate ramus of the epipterygoid deepened below trigeminal foramen;

(5) paroccipital process with a posteroventral ridge behind the jugular foramen.

(Kemp's (1982) statement that accurate tooth occlusion occurs in all eucynodonts is incorrect, since it is absent from the carnivorous forms such as *Cynognathus* and the chiniquodontids.)

Agreement about the reality of the Eucynodontia notwithstanding, there are still important points of disagreement about the interrelationships amongst its constituent families. The non-mammalian families that are generally recognized are:

(1) Cynognathidae: the monogeneric family for *Cynognathus*, an Early Triassic carnivorous form characterized by large sectorial post-canine teeth lacking cingulum cusps, and a greatly expanded postorbital bone.

(2) Diademodontidae: Early and Middle Triassic herbivorous forms, notably *Diademodon* and *Trirachodon*, with transversely expanded, occluding postcanines.

(3) Traversodontidae: Early to Late Triassic herbivores, again with laterally expanded, but more complex occluding postcanines.

(4) Chiniquodontidae: Middle Triassic carnivores, particularly *Probelesodon* and *Chiniquodon*, with elongated secondary palate.

(5) Probainognathidae: a monogeneric family for the relatively small Middle Triassic carnivore *Probainognathus*.

(6) Tritylodontidae: a family of very advanced, mammal-like herbivores from the Late Triassic to Late(?) Jurassic (Clark and Hopson 1985), with highly specialized dentition.

(7) Tritheledontidae (ictidosaurs): a second very mammal-like family of small, still rather poorly known Late Triassic and Early Jurassic carnivorous forms.

A number of Late Triassic forms known only from isolated teeth and jaw fragments have been included in a family Dromatheriidae (Hahn *et al.* 1984; Oakes and Lucas 1986). Their relationships are unknown, but they may be close to the Tritheledontidae.

To these must be added the Class Mammalia. There is, of course, an arbitrariness about exactly which monophyletic set of taxa should be designated Mammalia. Most current authors take the mammals to be the group that includes the Rhaetic–Liassic Sinoconodontidae, Morganucodontidae, Kuehneotheriidae, and their subsequent relatives (e.g. Crompton and Sun 1985). As such, the Mammalia are now almost universally regarded as monophyletic. The hypothesis that mammals are a polyphyletic assemblage, with the non-therian mammals having evolved mammalian characteristics independently of the therian lineage (Kermack and Kielan-Jaworowska 1971; Kermack and Kermack 1985), is strongly contradicted by the very large number of apparent synapomorphies shared by the two (Hopson and Barghusen

1986), and by the absence of synapomorphies of the non-therian mammals (Kemp 1983).

Rowe (MS) has recently elected to restrict the taxon Mammalia to include only the latest common ancestor of the living mammals, namely the crown group. Since he accepts a close relationship between monotremes and the modern therians (Kemp 1983), this means excluding the Rhaetic–Liassic forms. The more inclusive taxon, which does include these, is termed the Mammaliformes, which is thus synonymous with Mammalia as used here.

Comparison of the four recent cladograms attempting to show the relationships of the main cynodont families and the mammals (Kemp 1982, 1983; Sues 1985; Hopson and Barghusen 1986; Rowe, MS) indicates that there are two major controversies. The first is the position of the Cynognathidae, and the second the relationships of the Tritylodontidae. There is also a question of how well established is the hypothesis that the Tritheledontidae (ictidosaurs) are the sister-group of the Mammalia.

The relationships of the Cynognathidae Kemp (1982) and Rowe (MS) interpreted the Lower Triassic carnivorous genus *Cynognathus* as the sister-group of all the other eucynodontians, based on the presumed synapomorphic status of the following characters of the latter:

(1) the posterior part of the squamosal (lambdoidal crest) deeply incised to give a strongly W-shaped occiput;

(2) emarginated posterior edge of the coronoid process of the dentary;

(3) snout constricted in front of the orbits, so the orbits tend to face more anteriorly;

(4) ventrolateral edges of the parasphenoid slightly flared to produce parasphenoid alae.

Hopson and Barghusen (1986) and Sues (1985), in contrast, interpreted the Cynognathidae as the sister-group only of the herbivorous cynodonts Diademodontidae, Traversodontidae, and Tritylodontidae, on the basis of the common possession of a descending flange on the anterior part of the jugal, a character very variably developed in the traversodontids, and difficult to distinguish from a marked thickening of this region in at least *Probelesodon* amongst the chiniquodontids (*pers. obs.*). Therefore, it must be concluded that the former hypothesis is the more strongly supported and that cynognathids are best interpreted as the sister-group of all the other eucynodonts (Fig. 1.4).

The relationships of the Tritylodontidae Although a number of otherwise uniquely mammalian characters of tritylodontids have always been

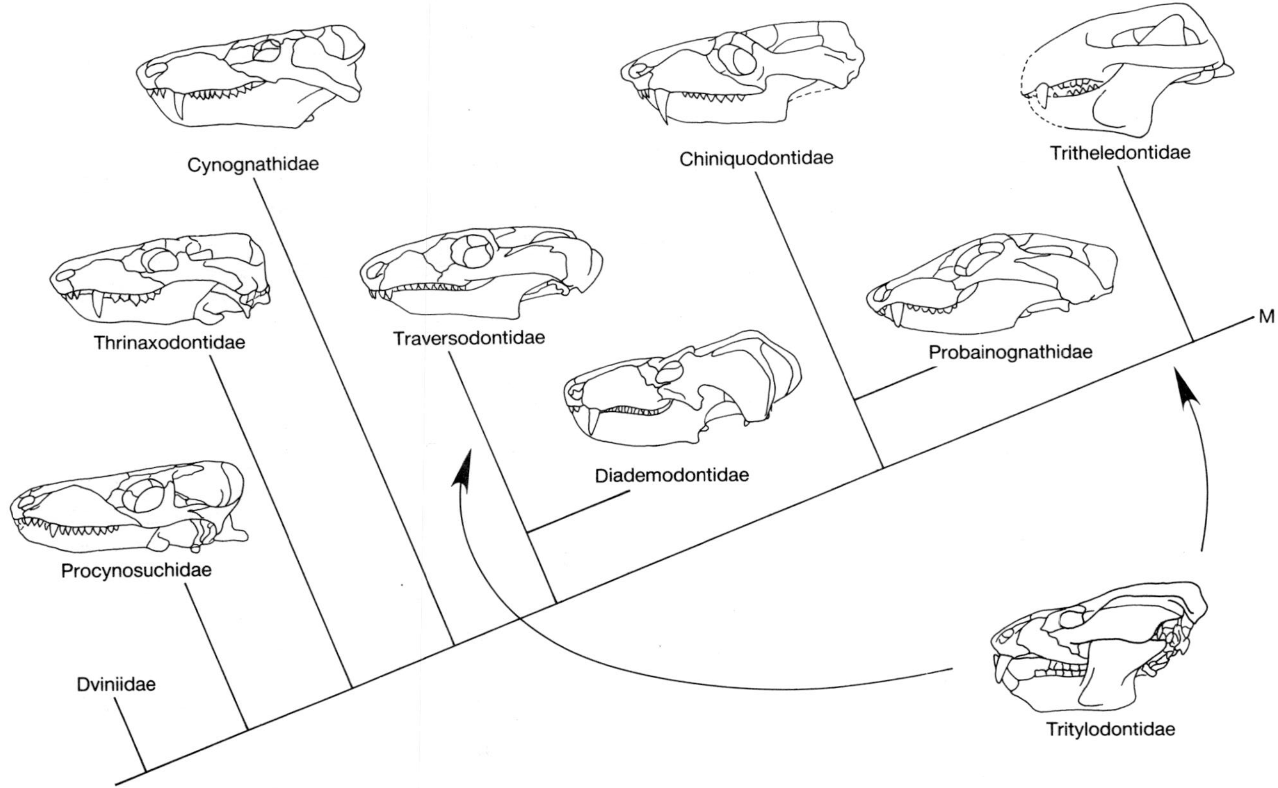

Fig. 1.4. The relationships of the cynodont families. The two hypotheses of the relationships of the Tritylodontidae are shown, of which that with Tritheledontidae plus mammals (M) is preferred.

recognized, such as the loss of the postorbital and prefrontal bones, the group has long been considered to be relateu to the other herbivorous cynodonts, the diademodontids and traversodontids, mainly on the basis of the form of the postcanine teeth (Crompton and Ellenberger 1957; Crompton 1972; Hopson and Kitching 1972). The three families formed a presumed monophyletic group Tritylodontoidea. Kemp (1982, 1983) challenged this view and argued that tritylodontids share many more characters with mammals than with the other herbivorous cynodonts. Therefore, they are more parsimoniously interpreted as the sister-group of mammals (plus, possibly, tritheledontids; see below), and the resemblances between tritylodontids and the diademodontids plus traversodontids, notably in the postcanine teeth, are convergences. Since then, Sues (1985) has reiterated support for the Tritylodontoidea, disputing all but about eight of Kemp's proposed synapomorphies of tritylodontids and mammals, and claiming eleven characters shared by tritylodontids and at least some traversodontids. Hopson and Barghusen (1986) also retain the Tritylodontoidea, although more tentatively. On the other hand, Rowe (MS) has reviewed the evidence and disputed virtually all of Sues's characters. He accepts the majority of Kemp's characters, adds several additional ones, and claims no less than 47 synapomorphies between mammals and tritylodontids.

The situation is actually complicated by the question of the status of the traversodontid cynodonts. Hopson and Barghusen (1986) were unable to state any characters that uniquely defined the group; and several characters quoted by Sues (1985) in support of a relationship between tritylodontids and traversodontids only occur in certain traversodontids, particularly the Late Triassic *Exaeretodon*. Until the interrelationships of the non-tritylodontid herbivorous forms are understood, any arguments in support of the association of tritylodontids with them will be ambiguous. Rowe (MS) actually resolves a number of incongruities by proposing that the supposed traversodontid *Exaeretodon per se* is the sister-group of tritylodontids plus mammals.

Further evaluation must await the publication of Rowe's work, but at the moment a much stronger case exists for relating tritylodontids to mammals than for relating them to the other herbivorous cynodonts.

Tritheledontids Unlike the tritylodontids, the small, carnivorous and, at least superficially, extremely mammal-like tritheledontids (ictidosaurs) are very poorly known indeed. Hopson and Barghusen (1986) list nine characters shared by tritheledontids and mammals. Of these, four are present in tritylodontids (loss of postorbital bar, loss of postorbital bone, loss of prefrontal bone, and form of quadrate), and a fifth, prismatic enamel, occurs in at least *Oligokyphus* amongst tritylodontids (J.-J. Jaeger, *pers. comm.*). Another character, posterior extension of the secondary palate, is found in the chiniquodontids and probainognathids. This leaves only:

(1) basicranium shortened anteroposteriorly;

(2) upper postcanine teeth with external cingulum;

(3) zygomatic arch slender along entire length.

As the known specimens are all small, possibly juvenile, the standing of these characters remains questionable. Until material of the braincase and the postcranial skeleton is described, the position of the tritheledontids as the sister-group of the mammals must be tentative, as Rowe (MS) has stressed.

Conclusions

It is instructive that none of the authors whose recent cladistic analyses have been discussed has felt able to present a complete, rigorously cladistic classification. Although all have insisted that named taxa be strictly monophyletic, several monophyletic groupings are also left unnamed. This is wise in practice because it avoids the extensive new nomenclature that would be necessary for a group as taxonomically complex as this, much of which would concern monotypic taxa. It is also wise in principle, because the last few years have illustrated in the synapsids a phenomenon widespread in the taxonomic literature of many groups. This is the continual revision of cladistic hypotheses, based partly on new information but largely on differing opinions about which is the most parsimonious cladogram, on the basis of the same morphological data. There is still much to be learnt about how to weight morphological characters according to their value in assessing the goodness of cladistic hypotheses. It is, for instance, striking that from Kemp's (1983) description of the similarities between tritylodontids and mammals, Sues (1985) formalized a list of 21 discrete characters; Rowe (MS), in contrast, extracted 35 discrete characters *from exactly the same text*. One feels that their respective prejudices were already showing.

The frail empirical basis of *a priori* weighting by assumed adaptive or other evolutionary significance has been recognized and rightly discarded (e.g. Kemp 1985). But its place has tended to be taken by a simplistic (even when computerized) numbers game of listing and counting characters with no regard to the relative merits of the respective characters. Intuitively, everyone knows that some characters are 'better' than others, and attempts to formalize this have recently begun (Neff 1986; Kemp 1988). Each putative synapomorphy is itself a hypothesis, which in turn is used to test a higher-order hypothesis, i.e. a cladogram. It is possible in principle to make an *a priori* assessment of how good a hypothesis one 'synapomorphy' is, compared to another 'synapomorphy', based entirely on the structure of the characters in question and not on any assumed evolutionary properties of different kinds of characters. Such discrimination need not be tautological. Developments along these lines ought to help resolve outstanding differences of opinion about which of the competing cladograms are best supported by the evidence.

Acknowledgements

I am particularly grateful to Dr Jim Hopson, Dr Gillian King, and Dr Timothy Rowe, who each allowed me the freedom to make use of their then unpublished manuscripts. I hope that I have indicated adequately the parts of this paper that are due entirely to them.

I thank Miss Denise Blagden for preparation of the figures, and Dr King for reading the manuscript and offering a number of helpful suggestions. The referees of this paper, Dr Jim Hopson and Dr Colin Patterson, offered numerous useful suggestions, not all of which I have adopted.

References

Battail, B. (1982). Essai de phylogenie des cynodontes (Reptilia: Therapsida). *Géobios, Mém. Spéc.* **6**, 157–67.

Benton, M. J. (1985). Classification and phylogeny of the diapsid reptiles. *Zool. J. Linn. Soc.* **84**, 97–164.

Brinkman, D. and Eberth, D. A. (1983). The interrelationships of pelycosaurs. *Breviora* **473**, 1–35.

Carroll, R. L. (1986). The skeletal anatomy and some aspects of the physiology of primitive reptiles. In *The Ecology and Biology of Mammal-like Reptiles* (eds. N. Hotton III, P. D. MacLean, J. J. Roth, and E. C. Roth), pp. 25–45. Smithsonian Institution Press, Washington and London.

Clark, J. M. and Hopson, J. A. (1985). Distinctive mammal-like reptile from Mexico and its bearing on the phylogeny of the Tritylodontidae. *Nature (London)* **315**, 398–400.

Cluver, M. A. and King, G. M. (1983). A reassessment of the relationships of Permian Dicynodontia (Reptilia, Therapsida) and a new classification of dicynodonts. *Ann. S. Afr. Mus.* **91**, 195–273.

Crompton, A. W. (1972). Postcanine occlusion in cynodonts and tritylodontids. *Bull. Br. Mus. Nat. Hist. (Geol).* **21**, 29–71.

Crompton, A. W. and Ellenberger, F. (1957). On a new cynodont from the Molteno Beds and the origin of the tritylodontids. *Ann. S. Afr. Mus.* **44**, 1–14.

Crompton, A. W. and Sun, A.-L. (1985). Cranial structure and relationships of the Liassic mammal *Sinoconodon*. *Zool. J. Linn. Soc.* **85**, 99–119.

Currie, P. J. (1979). The osteology of haptodontine sphenacodonts (Reptilia: Pelycosauria). *Palaeontographica, A* **163**, 130–88.

Gaffney, E. S. (1980). Phylogenetic relationships of the major groups of amniotes. In *The Terrestrial Environment and the Origin of Land Vertebrates*, Systematics Association Special Symposium, Vol. 15, (ed. A. L. Panchen), pp. 593–610. Academic Press, London.

Gardiner, B. (1982*a*). Tetrapod classification. *Zool. J. Linn. Soc.* **74**, 207–32.

Gardiner, B. (1982*b*). Mammals, birds and mammal-like reptiles. In *Evolution and the Genetics of Populations, Suppl. Atti Assoc. Genet. Ital.* **29**, 11–17.

Goodrich, E. S. (1916). On the classification of the Reptilia. *Proc. R. Soc. London* **B89**, 261–76.

Hahn, G., Lepage, J. C. and Wouters, G. (1984). Cynodontier-Zähne aus der Ober-Trias von Medernach, Grossherzogtum. *Bull. Soc. Belg. Géol.* **93**, 357–73.

Hopson, J. A. and Barghusen, H. R. (1986). An analysis of therapsid relationships. In *The Ecology and Biology of Mammal-like Reptiles* (eds. N. Hotton III, P. D. MacLean, J. J. Roth, and E. C. Roth), pp. 83–106. Smithsonian Institution Press, Washington and London.

Hopson, J. A. and Kitching, J. W. (1972). A revised classification of cynodonts (Reptilia: Therapsida). *Palaeontol. Afr.* **14**, 71–85.

Kemp, T. S. (1979). The primitive cynodont *Procynosuchus*: functional anatomy of the skull and relationships. *Phil. Trans. R. Soc. London* **B288**, 217–58.

Kemp, T. S. (1982). *Mammal-like Reptiles and the Origin of Mammals*. Academic Press, London and New York.

Kemp, T. S. (1983). The relationships of mammals. *Zool. J. Linn. Soc.* **77**, 353–84.

Kemp, T. S. (1985). Models of diversity and phylogenetic reconstruction. In *Oxford Surveys in Evolutionary Biology*, Vol. 2 (eds. R. Dawkins and M. Ridley), pp. 135–58. Oxford University Press, Oxford.

Kemp, T. S. (1986). The skeleton of a baurioid therocephalian therapsid from the Lower Triassic (*Lystrosaurus* zone) of South Africa. *J. Vertebr. Paleontol.* **6**, 215–32.

Kemp, T. S. (1988). Haemothermia or Archosauria?: the interrelationships of mammals, birds and crocodiles. *Zool. J. Linn. Soc.* **92**, 67–104.

Kermack, D. M. and Kermack, K. A. (1985). *The Evolution of Mammalian Characters*. Croom Helm, London and Sydney.

Kermack, K. A. and Kielan-Jaworowska, Z. (1971). Therian and non-therian mammals. In *Early Mammals* (eds. D. M. Kermack and K. A. Kermack), *Zool. J. Linn. Soc.* **50**(Suppl. 1), 103–15.

King, G. M. (1988). *Anomodontia*. In *Handbuch der Paläoherpetologie* (ed. P. Wellnhofer), Vol. 17C. Gustav Fischer, Stuttgart and New York.

Neff, N. (1986). A rational basis for *a priori* character weighting. *Syst. Zool.* **35**, 110–23.

Oakes, W. and Lucas, S. G. (1986). Triassic cynodonts (Reptilia) from New Mexico. *New Mex. Geol.* **8**, 22.

Olson, E. C. (1962). Late Permian terrestrial vertebrates, U.S.A. and U.S.S.R. *Trans. Am. Phil. Soc.* **52**, 3–224.

Olson, E. C. (1986). Relationships and ecology of the early therapsids and their predecessors. In *The Ecology and Biology of Mammal-like Reptiles* (eds. N. Hotton III, P. D. Maclean, J. J. Roth, and E. C. Roth), pp. 47–60. Smithsonian Institution Press, Washington and London.

Reig, O. A. (1967). Archosaurian reptiles: a new hypothesis of their origins. *Science, N.Y.* **157**, 565–68.

Reisz, R. R. (1980). The Pelycosauria: a review of phylogenetic relationships. In *The Terrestrial Environment and the Origin of Land Vertebrates*, Systematics Association Special Symposium, Vol. 15, (ed. A. L. Panchen), pp. 553–92. Academic Press, London.

Reisz, R. R. (1986). *Pelycosauria.* In *Handbuch der Paläoherpetologie* (ed. P. Wellnhofer), Vol. 17A. Gustav Fischer, Stuttgart and New York.

Romer, A. S. (1966). *Vertebrate paleontology*, 3rd edn. University of Chicago Press, Chicago and London.

Romer, A. S. (1971). Unorthodoxies in reptilian phylogeny. *Evolution* **25**, 103–12.

Romer, A. S. and Price, L. W. (1940). Review of the Pelycosauria. *Spec. Pap. Geol. Soc. Am.* **28**, 1–538.

Rowe, T. (MS). *Osteological diagnosis of Mammalia, L. 1758, and its relationship to extinct Synapsida*, Ph.D. Dissertation. University of California, Berkeley.

Rowe, T. and Van den Heever, J. A. (1986). The hand of *Anteosaurus magnificus* (Dinocephalia: Therapsida) and its bearing on the origin of the mammalian manual phalangeal formula. *S. Afr. J. Sci.* **82**, 641–5.

Sigogneau, D. (1970). *Revisions systématiques des gorgonopsiens.* Éditions du Centre National de la Recherche Scientifique, Paris.

Sues, H.-D. (1985). The relationships of the Tritylodontidae (Synapsida). *Zool. J. Linn. Soc.* **85**, 205–17.

Watson, D. M. S. and Romer, A. S. (1956). A classification of therapsid reptiles. *Bull. Mus. Comp. Zool., Harv. Univ.* **111**, 37–89.

2. A note on the Mesozoic mammals, and the origin of therians

T. S. KEMP
University Museum and Department of Zoology, Parks Road, Oxford, UK

Abstract

It is no longer possible to accept that there was a simple dichotomy of early mammals into therian and non-therian lineages, nor a later dichotomy of advanced therians into marsupials and placentals. In both cases, the discovery of new fossil forms indicates that the pattern of mammalian evolution during the Mesozoic was much more complex, with several short-lived branches. At present, the exact relationships of many of these groups are obscure.

Introduction

There has never been serious doubt that the three groups of living mammals, Monotremata (Prototheria), Marsupialia (Metatheria), and Placentalia (Eutheria), are respectively monophyletic groups; nor that together they constitute a monophyletic group Mammalia with respect to all other living tetrapods. The interrelationships between the three are also uncontroversial at the present time, for the marsupials and the placentals share numerous apparent synapomorphies absent from the monotremes, such as the detailed structure of the tribosphenic molar teeth, the nature of the jaw opening musculature, the loss of the interclavicle, and viviparity (e.g. Marshall 1979). In recent years, only Kühne (1973), following Gregory's (1947) 'palimpsest' theory, has dissented and argued for a sister-group relationship for monotremes and marsupials. In fact, Kühne (1973) only used a single character, the

The Phylogeny and Classification of the Tetrapods, Volume 2: Mammals (ed. M. J. Benton), Systematics Association Special Volume No. 35B, pp. 23–9. Clarendon Press, Oxford, 1988.

supposed mode of tooth replacement in marsupials and *Ornithorhynchus* to support his hypothesis, while the majority of Gregory's characters are better interpreted as plesiomorphic.

There is considerably less agreement about the relationships of the various major Mesozoic fossil groups to the extant mammals (see Lillegraven *et al.* 1979, for a detailed review to that date; and Kemp 1982, for a briefer summary). Often this is for want of adequate information about the fossil taxa, which are still represented in many critical cases by teeth and jaws. But as more finds are reported, it is also becoming clear that mammalian radiation during the Mesozoic was very much more complex than had been supposed. Following the previous article in this volume, by the present author, the taxon Mammalia is extended to apply to the monophyletic group that includes the Rhaeto-Liassic forms *Sinoconodon* (as the most primitive member), morganucodontids, kuehneotheriids, and the obvious relatives of these (Crompton and Sun 1985). As such, the synapomorphies categorizing the Mammalia are listed by Crompton and Sun (1985) as:

(1) dentary condyle articulating with a squamosal glenoid;

(2) anterior lamina forming the side wall to the cavum epiptericum;

(3) floor to cavum epiptericum below primary exit of seventh cranial nerve;

(4) prootic canal;

(5) double-rooted molars aligned longitudinally;

(6) loss of alternate tooth replacement of postcanine teeth;

(7) prominent medial ridge and groove on the dentary for the support of the postdentary bones.

Therian and non-therian mammals

Study of the relationships of the early mammals has been dominated by the hypothesis that a major dichotomy occurred into therian and non-therian (prototherian, or atherian) lineages (e.g. Hopson and Crompton 1969; Kermack and Kielan-Jaworowska 1971; Crompton and Jenkins 1979; Kermack and Kermack 1984). The Theria were categorized by triangulation of the three main cusps of the molars and a presumption that the alisphenoid forms the side wall of the braincase (although the braincase is not known in primitive members). As such, the Theria embraces the Liassic kuehneotheriids, the symmetrodonts, and the eupantotheres of the later Mesozoic, and the forms with tribosphenic molars including the marsupials and the placentals. The non-therians were categorized by a linear arrangement of the main molar cusps, and by the periotic forming the side wall of the braincase. This group contains the morganucodontids, docodonts, triconodonts, multituberculates, haramiyids, and also the monotremes. Kemp (1983) challenged this hypothesis on the grounds that the two characters of the non-therian mammals are plesiomorphic for mammals and cannot therefore categorize a monophyletic group. Furthermore, the structure of monotreme teeth, which in practice means the highly aberrant juvenile dentition of *Ornithorhynchus* and the teeth of its presumed Miocene relative *Obdurodon*, cannot be compared at all closely with the teeth of any other non-therians; indeed, they are no less comparable to the tribosphenic teeth of therians. On the other hand, monotremes and modern therians share a considerable number of apparently synapomorphic characters absent from the non-therian morganucodontids, such as ear ossicles and fused atlas bones. Kemp's (1983) conclusion was therefore that the non-therian mammals are a paraphyletic assemblage, and that the monotremes are related to the therians.

Rowe (MS) has added several more characters in support of a relationship between monotremes and therians. A recent fossil discovery also lends support to this view. Archer *et al.* (1985) described *Steropodon*, from an early Cretaceous jaw fragment from Australia. Although the teeth are obviously comparable to those of *Ornithorhynchus* and *Obdurodon*, they also have certain characters of tribosphenic molars, and therefore support a close relationship between monotremes and therians. Kielan-Jaworowska *et al.* (1987) reached the same conclusion, and suggested that the two lineages diverged from a eupantothere level.

Several other Mesozoic forms have been discovered in recent years whose characters do not support the idea of a simple dichotomy of all mammals. *Dinnetherium*, from the Rhaeto-Liassic of North America (Jenkins *et al.* 1983; Jenkins 1984) is a triconodont-like form but it also has certain advanced characters of the jaw which make it difficult to relate it unambiguously to either therians or non-therians. *Gobiconodon* is an early Cretaceous amphilestine-like mammal that appears to have replacing 'molar' teeth, despite the loss of replacement of the posterior postcanine teeth in primitive therians and non-therians alike (Jenkins 1984). *Hallautherium* (Clemens 1980; Jenkins 1984) has morganucodontid-like lower molars except for an enlarged posterobuccal basin. Even more odd, *Shuotherium* (Chow and Rich 1982) from the Jurassic of China has primitive therian-like teeth, except for development of a 'talonid' at the front of the lower molars. Chow and Rich (1982) and also Kielan-Jaworowska *et al.* (1987) regard *Shuotherium* as an aberrant therian.

Kermack *et al.* (1987), on the other hand, have described *Simsonodon*, a new form of the non-therian Docodonta, which also possesses a talonid-like development at the front of the lower molars. This and certain other points of resemblance lead these authors to suggest that *Shuotherium* is not a therian at all, but is also a docodont.

Clemens (1986) has reviewed most of these and other recent finds of Triassic and Jurassic mammals; he concludes that 'the interpretation of a basic dichotomy separating mammals into therian and non-therian groups probably should be abandoned.'

Primitive therians

Given the probable need to add the monotremes, and the possible need to include the multituberculates (Kemp 1983), especially now that Desui and Lillegraven (1986) have demonstrated the presence of ear ossicles, the Theria is probably a monophyletic group. It is characterized by the triangulated molar teeth, or derivations therefrom, but whether features of the braincase are also synapomorphic for the Theria as a whole awaits information about the cranial anatomy of primitive forms.

Prothero (1981) attempted a cladistic analysis of the non-tribosphenic therians, i.e. the kuehneotheriids, symmetrodonts, and eupantotheres. There are relatively few characters available, and no completely clear-cut result emerges. However, what is certain from his analysis (see also McKenna 1975) is that the eupantotheres are a paraphyletic group, including the ancestors of the advanced tribosphenic groups (and presumably of the monotremes too).

Advanced therians

The possession of tribosphenic molars, with a protocone and a correspondingly enlarged talonid basin, categorizes a good monophyletic group of advanced therians, the Tribosphenida (McKenna 1975). The most primitive of these, with a relatively small talonid (upper molars are not yet known), are the aegialodontids (Kielan-Jaworowska *et al.* 1979; Dashzeveg and Kielan-Jaworowska 1984), which may be taken as the sister-group of the remainder. Amongst those with fully expressed tribosphenic teeth, a number of Cretaceous forms can be assigned to the marsupials or the placentals. However, others do not fit into either of these categories. Clemens and Lillegraven (1986) have recently reviewed these problematic tribosphenids, and conclude that certain forms referred to as deltatheridians are probably a third monophyletic group of advanced therian. However, they are unable to resolve the relationships between Deltatheridia, Marsupialia, and Placentalia.

There are also certain other genera of tribosphenids which do not fit into any of these three groups. Clearly, from an aegialodontid-like ancestor, the pattern of evolution of the advanced therian mammals during the Cretaceous was more complex than supposed. As in the case of the earlier mammalian radiation of the Rhaeto-Liassic, there was no simple dichotomy, but a rapid diversification of several lineages, of which the marsupials and the placentals were but two.

Conclusion

Discoveries of new material during the last decade or so have made it abundantly clear that the pattern of mammalian evolution, at least as represented by dental morphology, is complex. Both during the early phase of diversification from the Rhaeto-Liassic, when the 'non-therian' groups were dominant, and later during the Cretaceous, which saw the onset of the advanced therians, numerous lineages seem to have diverged. Although the interrelationships of many of the better-known groups are reasonably well established, there remains at present a wide range of forms whose relations cannot yet be resolved satisfactorily.

References

Archer, M., Flannery, T. F., Ritchie, A. and Molnar, R. E. (1985). First Mesozoic mammal from Australia–an early Cretaceous monotreme. *Nature (London)* **318**, 363–6.

Chow, M. and Rich, T. H. V. (1982). *Shuotherium dongi*, n. gen. and n. sp., a therian with pseudo-tribosphenic molars from the Jurassic of Sichuan, China. *Aust. Mammal* **5**, 127–42.

Clemens, W. A. (1980). Rhaeto-Liassic mammals from Switzerland and West Germany. *Zitteliana* **5**, 51–92.

Clemens, W. A. (1986). On Triassic and Jurassic mammals. In *The Beginning of the Age of Dinosaurs* (ed. K. Padian), pp. 237–46. Cambridge University Press, Cambridge.

Clemens, W. A. and Lillegraven, J. A. (1986). New Late Cretaceous North American advanced therian mammals that fit neither the marsupial nor eutherian molds. *Contr. Geol. Univ. Wyoming, Spec. Pap.* **3**, 55–85.

Crompton, A. W. and Jenkins, F. A., Jr. (1979). Origin of mammals. In *Mesozoic Mammals: the First Two-thirds of Mammalian History* (eds. J. A. Lillegraven, Z. Kielan-Jaworowska and W. A. Clemens), pp. 59–73. University of California Press, Berkeley and Los Angeles.

Crompton, A. W. and Sun, A.-L. (1985). Cranial structure and relationships of the Liassic mammal *Sinoconodon*. *Zool. J. Linn. Soc.* **85**, 99–119.

Dashzeveg, D. and Kielan-Jaworowska, Z. (1984). The lower jaw of an aegialodontid mammal from the Early Cretaceous of Mongolia. *Zool. J. Linn. Soc.* **82**, 217–27.

Desui, M. and Lillegraven, J. A. (1986). Discovery of three ear ossicles in a multituberculate mammal. *Nat. Geogr. Res.* **2**, 500–507.

Gregory, W. K. (1947). The monotremes and the palimpsest theory. *Bull. Am. Mus. Nat. Hist.* **88**, 1–52.

Hopson, J. A. and Crompton, A. W. (1969). Origin of mammals. *Evol. Biol.* **3**, 15–72.

Jenkins, F. A., Jr. (1984). A survey of mammalian origins. In *Mammals: Notes for a Short Course*, Studies in Geology, Vol. 8 (ed. T. W. Broadhead), pp. 32–47. University of Tennessee, Dept. of Geological Sciences, Knoxville, Tennessee.

Jenkins, F. A., Jr., Crompton, A. W. and Downs, W. R. (1983). Mesozoic mammals from Arizona: new evidence on mammalian evolution. *Science. N.Y.* **222**, 1233–5.

Kemp, T. S. (1982). *Mammal-like Reptiles and the Origin of Mammals*. Academic Press, London and New York.

Kemp, T. S. (1983). The relationships of mammals. *Zool. J. Linn. Soc.* **77**, 353–84.

Kermack, D. E. and Kermack, K. A. (1984). *The Evolution of Mammalian Characters*. Croom Helm, London.

Kermack, K. A. and Kielan-Jaworowska, Z. (1971). Therian and non-therian mammals. In *Early Mammals* (eds. D. M. Kermack and K. A. Kermack), *Zool. J. Linn. Soc.* **50**(Suppl. 1), 103–15.

Kermack, K. A., Lee, A. J., Lees, P. M. and Mussett, F. (1987). A new docodont from the Forest Marble. *Zool. J. Linn. Soc.* **89**, 1–39.

Kielan-Jaworowska, Z., Eaton, J. G. and Bown, T. M. (1979). Theria of metatherian–eutherian grade. In *Mesozoic Mammals: the First Two-thirds of Mammalian History* (eds. J. A. Lillegraven, Z. Kielan-Jaworowska and W. A. Clemens), pp. 182–91. University of California Press, Berkeley and Los Angeles.

Kielan-Jaworowska, Z., Crompton, A. W. and Jenkins, F. A., Jr. (1987). The origin of egg-laying mammals. *Nature, Lond.* **326**, 871–3.

Kühne, W. G. (1973). The systematic position of monotremes reconsidered (Mammalia). *Z. Morphol. Tiere* **75**, 59–64.

Lillegraven, J. A., Kielan-Jaworowska, Z. and Clemens, W. A. (eds.) (1979). *Mesozoic Mammals: the First Two-thirds of Mammalian History*. University of California Press, Berkeley and Los Angeles.

McKenna, M. C. (1975). Towards a phylogenetic classification of the Mammalia. In *Phylogeny of the Primates* (eds. W. P. Luckett and F. S. Szalay), pp. 21–46. Plenum Press, New York.

Marshall, L. G. (1979). Evolution of metatherian and eutherian (mammalian) characters: a review based on cladistic methodology. *Zool. J. Linn. Soc.* **66**, 369–410.

Prothero, D. R. (1981). New Jurassic mammals from Como Bluff, Wyoming, and the interrelationships of non-tribosphenic Theria. *Bull. Am. Mus. Nat. Hist.* **167**, 277–326.

Rowe, T. (MS). Osteological diagnosis of Mammalia, L. 1758, and its relationship to extinct Synapsida, Ph. D. Dissertation. University of California, Berkeley.

3. The major groups of eutherian mammals

MICHAEL J. NOVACEK
Department of Vertebrate Paleontology, American Museum of Natural History, New York, USA

ANDRE R. WYSS
Department of Vertebrate Paleontology, American Museum of Natural History, and Department of Geological Sciences, Columbia University, New York, USA

MALCOLM C. McKENNA
Department of Vertebrate Paleontology, American Museum of Natural History, and Department of Geological Sciences, Columbia University, New York, USA

Abstract

Eutherian mammals are popularly thought to be the product of a great polytomous radiation at the beginning of the Cenozoic. Despite numerous studies, many aspects of this radiation have not been clearly resolved, and the interrelationships of several modern eutherian orders are still poorly understood. Several higher-level relationships for these lineages do, however, seem compelling. There is morphological evidence for the close relationship of rodents and lagomorphs within Glires, proboscideans and sirenians within Tethytheria, hyracoids and tethytheres within Paenungulata, and chiropterans and dermopterans within a superordinal clade. A close affinity between pholidotans and edentates is also suggested by some morphological evidence. Less persuasive cases can be made for a grouping of Glires and macroscelideans, a superclade Archonta that comprises tree-shrews, primates, chiropterans, and dermopterans, and all Eutheria excluding edentates and pholidotans (Epitheria). Although many traditional

The Phylogeny and Classification of the Tetrapods, Volume 2: Mammals (ed. M. J. Benton), Systematics Association Special Volume No. 35B, pp. 31–71. Clarendon Press, Oxford, 1988.

studies support the monophyly of the ungulate clades, we find little morphological evidence consistent with this view. A problem with these proposals for higher-level groups is the general lack of corroboration from molecular studies. However, a comparative analysis shows that molecular results are either based on data sets with significant sampling gaps or are strongly discrepant with each other. Hence, the relevance of many molecular-based trees to morphological results is, at present, ambiguous. There are a few areas of general agreement, including the recognition of Tethytheria, Paenungulata, and the remote position of edentates. Rather than providing a bush with essentially no resolution, comparative morphology and palaeontology offer some important hypotheses for higher-level eutherian relationships that constitute a framework for molecular systematics.

Introduction

Eutherians represent the great majority of mammals living today, and an interest in their higher-level systematics is understandable; indeed, the problem has recently attracted much attention in molecular systematics. That is not to say that the problem is a new one. During the 19th century, the higher classification of mammals was a major issue in systematics, one which absorbed, among others, Cuvier, de Blainville, and T. H. Huxley. Yet for much of the 20th century, this interest in higher-level problems gained little momentum. At least since the publication of Simpson's (1945) classification of mammals, systematic revisions have primarily concerned familial, generic, or species-level taxa. Broader patterns were of interest mainly to those who dealt with perplexing fossil members of so-called 'stem' eutherian groups, such as insectivorans and condylarths (e.g. Butler 1956; McDowell 1958; McKenna 1960; Van Valen 1966). Reflecting these interests, one of us (McKenna 1975) proposed a higher-level phylogenetic classification of mammals that stimulated subsequent activity and debate. What progress has been made in the intervening years is the subject of this report.

At this point, a general assessment of such progress would be helpful. Unfortunately, our review is mixed; in some areas we see illumination, in others the picture continues to be clouded. Those blessed with the opportunity of working with more tractable systematic problems may be disappointed to learn that we cannot confidently identify—from both morphological and molecular evidence—a highly dichotomous tree for the Eutheria. (Shoshani (1986) and Miyamoto and Goodman (1986) give more optimistic conclusions.) Nevertheless, we maintain that recent study has been fruitful in identifying evidence for a few cases of interordinal relationships. Here, we shall attempt to assess the relative strengths of such hypotheses, accounting for the range of evidence—both morphological and molecular—that bears upon them.

Fossils

The range of taxa considered here includes both living and fossil groups. Most reviews are naturally focused on Recent eutherians, where diverse data, including soft anatomy, embryology, and molecular comparisons come into play. Fossil evidence relevant to the problem is basically of two kinds. In the first instance, there are groups whose affinity with a modern order is well established. If well represented by skeletal data, these fossils can have a strong impact on generalizations about the groups in question, often providing insight on the primitive condition of a trait markedly transformed in all Recent members. Likewise, fossils can contradict synapomorphy schemes for two groups based only on living members if the fossil is demonstrably a member of one group and at the same time is known to have more general (primitive) characters than living members of either group. For example, the clavicle is reduced or lost in modern members of the Order Carnivora and many ungulate lineages. However, this derived condition is not attained in certain fossil carnivorans (e.g. 'miacids'), where instead the clavicle is well-developed. The reduction or loss of the clavicle does not then seem an appropriate synapomorphy for relating carnivorans to ungulates.

A second category of fossils considered here comprises those lineages that cannot be firmly associated with one or another of the modern groups. Although such fossils may have little effect on generalizations concerning Recent taxa, they are of intrinsic importance in permitting a classification that accounts for the greatest diversity of taxa. Indeed, one of the obvious shortcomings of molecular or related data is that they cannot be extracted effectively from most fossils, which is a severe limitation in the case of mammals, where the fossil history is particularly rich.

Below, we consider several extinct eutherian lineages whose relationships to other clades have, through the years, engendered much debate. Our treatment of such taxa cannot be very even-handed. The character evidence for some of these groups is poorly known. Although computer programs for cladistic algorithms can accommodate missing information, results are not very illuminating when so many blanks remain in the data matrix. Nevertheless, a few of these problematic fossil groups have enough features to allow assessment in the broader context of eutherian relationships.

Characters

Our approach to the study of relationships emphasizes the identity of homology and the characterization of groups by special traits (cladistics). During the last decade, most workers interested in higher-level mammalian relationships have adopted a similar approach. Exceptions are the distance analyses of immunological data (Sarich 1985; Shoshani 1986, fig. 2), where it is difficult to recognize, objectively, groups by specific character differences. We do not ignore this approach; if there is a strong signal for a particular relationship, it is possible that both distance trees and trait-based trees will identify it. We do note, however, that such congruence is not yet particularly evident (see Shoshani 1986, and remarks below). We are primarily interested in hypotheses based on cladistic analysis of discrete traits, whether morphological innovations or amino acid sequence changes.

Among practitioners of cladistics one finds a diversity of attitudes, and the higher-level classification of mammals has served as an arena for various combatants. Szalay (1977, 1985), for example, emphasized the recognition of groups through the analysis of form-function complexes, rather than through what he characterizes as a superficial summary of trait distributions. On the other hand, there are arguments (Novacek 1980) that a classification based on a single system, no matter how richly described, is highly vulnerable to refutation when a broader range of characters is considered. It is hard to find a real issue here. Both opinions seem to converge on the platitude that the best classifications will be based on the most characters for which we possess the most information. Here, we consider a diversity of traits, some of which are 'atomized', or extracted from a complex of features that might be found to be integrally related. We do not, however, deny the value of detailed theories of transformation involving a complex of characters. Certain features have been investigated in some detail, while others, though constituting evidence for relationships, have received comparatively little study. It seems better to recognize this inequity than to ignore the less 'functionally meaningful' evidence. The latter solution would not be kind to molecular analyses, where the mapping of sequence changes can hardly be characterized as a detailed 'form-function' study of transformation.

Critical to the strategy of character analysis is the question of sampling. In the case of molecular studies, this problem is rather straightforward—sampling is incipient, so each new study generally represents information on more molecules in more taxa (although, as discussed below, critical gaps remain). By comparison, morphologists are confronted with an embarrassment of riches. The problem becomes one of extracting a meaningful data set from a vast array of potential evidence, both described in three centuries of literature, and resident in museum collections.

The strategy for this culling process can vary. One of us (Novacek 1986) recently described skull features of the fossil leptictids (an archaic mammal group of problematic affinities), summarized these in the form of 73 conditions, and compared these with a suite of mammals, including representatives of all the living orders and selected fossil taxa. From this, more than 130 derived traits were described, and various samples of these traits were analysed for the construction of cladograms. This approach was guided simply by a point-by-point comparison with skull features in leptictids; no attempt was made *a priori* to identify 'important', derived traits for particular superordinal groupings. The cladograms based on larger samples of these characters (between 46 and 104 traits) had an expectedly high degree of homoplasy. Nonetheless, aspects of these cladograms are consistent with those based on a more conscious sampling for various morphological evidence (both skeletal and soft anatomical) for specific superordinal groupings (e.g. Novacek and Wyss 1986). This agreement in certain topologies persists even when redundancies in analysed characters are removed (Table 3.3; Fig. 3.5, and discussion below). There is also considerable agreement between these results and those of Shoshani (1986, fig. 3), despite the very different range of skeletal characters analysed by the latter author. In short, there is no foolproof prescription for character sampling, and comparisons of results derived from different approaches can be illuminating.

Once characters have been sampled and a data matrix constructed, it remains to identify the branching patterns most consistent with those data. The premise for finding such trees that is often applied is the parsimony principle, which can be applied with relative ease in the case of smaller data sets. In fact, available software has exhaustive search options for finding all possible tree topologies for nine or fewer taxa. Invariably, however, data for this and other systematic enquiries are more complicated. The solution either requires algorithms for smaller samples or heuristic branch-swapping techniques. The latter methods fail to guarantee a most-parsimonious solution. Nonetheless, it has been demonstrated that heuristic procedures, such as global branch-swapping, usually yield the same solution as 'exact' techniques (Swofford 1985), and they have been applied in this study.

Computer programs used here include PHYSYS (Mickevich and Farris 1982; Phylogenetic Analysis System PHYSYS, Fortran V software system of cladistic and phenetic algorithms) and PAUP (Phylogenetic Analysis Using Parsimony; Swofford 1985). The measurement of homoplasy used here is the consistency index, or CI (*sensu* Kluge and

Farris 1969), where CI is the number of character states minus one, summed over all characters in the data set, divided by the total number of characters displayed by the cladogram in question. A high CI (approaching 1.0) represents a very 'clean' solution with little homoplasy; a low CI indicates rampant homoplasy. In cases where several (or many) 'equally most-parsimonious' trees were found, these were resolved with a program for computing a strict consensus tree (CONTREE) available in the PAUP package.

Some of the morphological results reviewed herein were achieved through manual procedures, i.e. without the aid of computer programs for branch-swapping algorithms (e.g. McKenna 1975; Novacek 1982; Novacek and Wyss 1986). Others relied on either PAUP (with CONTREE) or PHYSYS programs for finding 'most-parsimonious' trees. Molecular studies employ computer programs (PAUP, CONTREE, and Moore's maximum-parsimony algorithm), with the exception of McKenna's (1987) 'manual' treatments of myoglobin and alpha-lens data. Nevertheless, McKenna's results converge on computer-generated schemes (e.g. de Jong 1982; Wyss *et al.* 1987), partly because his analysis assumes a plausible range of alternative relationships before the initiation of a branch-swapping procedure. These 'input trees' comprise a particular sequence of taxa, usually based on the consensus results or previously published studies. The alternative, namely the randomized 'deck-shuffling' of input taxa, is a daunting prospect for either manual operations or computers. Hence, manual procedures are not necessarily inferior to computer-assisted ones. Computer programs are, however, likely to yield shorter trees than manual procedures, and many complicated analyses (consensus trees, optimization, etc.) are greatly facilitated by the use of computers.

Interordinal relationships

The history of higher-level mammalian classification provides many insights into the development of systematics as a whole. This was clearly recognized by Gregory (1910) in his comprehensive review of mammalian classifications before the 20th century. Earlier 20th century investigations were covered in some detail in Simpson's (1945) treatise. More modern works are considered, albeit briefly, in various papers, including McDowell (1958), McKenna (1969, 1975), Van Valen (1966, 1967), Szalay (1977), Simpson (1978), and Novacek (1982, 1986). The subject is in need of a major update, which exceeds our intentions here. Four alternative schemes (Figs. 3.1 and 3.2) illustrate some of the trains of thought in this area. Gregory's (1910) hypothesis, representing the culmination of a variety of 19th century

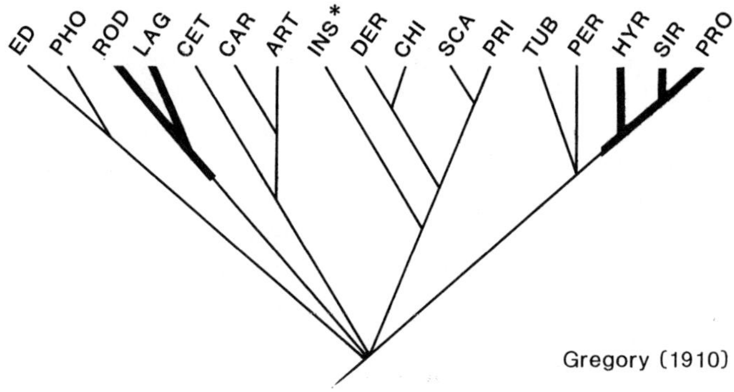

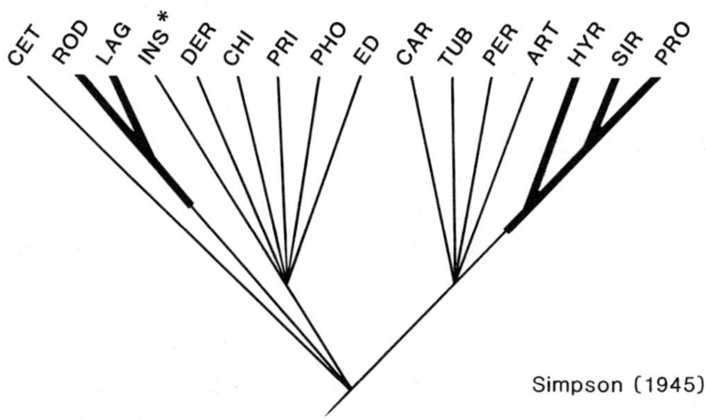

Fig. 3.1. Higher-level hypotheses for eutherian groups as proposed by Gregory (1910) and Simpson (1945). Bold lines indicate segments of agreement between the two trees. Acronyms for groups are ART, Artiodactyla; CAR, Carnivora; CET, Cetacea; CHI, Chiroptera; DER, Dermoptera; ED, Edentata; HYR, Hyracoidea; INS, Insectivora (asterisks indicate that Insectivora includes Macroscelidea in these papers); LAG, Lagomorpha; PER, Perissodactyla; PHO, Pholidota; PRI, Primates; PRO, Proboscidea; ROD, Rodentia; SCA, Scandentia; SIR, Sirenia; TUB, Tubulidentata.

attempts at mammalian classification, is, even from a modern perspective, rather explicit (Fig. 3.1), since it shows comparatively high resolution.

Refinements of Gregory's hypothesis involved prodigious investigations of Early Tertiary insectivorans, carnivorans, and ungulate groups (e.g. Matthew 1937), which revealed the complexity of possible relationships among higher eutherian clades. Subsequently, a new realism—or perhaps pessimism—concerning the whole endeavour seems to have set in, a situation well-represented in Simpson's (1945) rather non-committal arrangement of taxa (Fig. 3.1), as well as his regard for classification as 'art and practice' (*ibid*; p. 22).

Simpson's review did, however, include two bold departures from earlier classifications. He grouped insectivores, dermopterans, chiropterans, primates, pholidotes, and edentates within a single cohort Unguiculata, and he united the Carnivora, primitive condylarths, hyracoids, sirenians, proboscideans, perissodactyls, and artiodactyls within the cohort Ferungulata. Simpson (1945) claimed that the Ferungulata merely codified a concept fostered by Gregory (1910), but the latter author had a more complicated vision—he allied artiodactyls (and, to a lesser degree, cetaceans) closely with carnivorans, but he distinguished all other ungulates as a distinct clade (Fig. 3.1). The interrelationships of these lineages are ambiguously represented in Gregory's (1910) review. In his fig. 32, the cetacean–carnivore–artiodactyl branch shares a common stem with insectivore–anchontans, before it joins the clade for the remaining ungulates. In his fig. 31, the lines of these different clades converge very closely, a pattern more akin with his formal classification (pp. 464–6), and reproduced in Fig. 3.1 herein. At any rate, the proposed refashioning of these patterns to form the groups Ferungulata and Unguiculata has not been broadly accepted—Simpson's own subsequent regard for these groupings is, in fact, moot. However, a cladistic study of brain features (Kirsch *et al.* 1983) in selected eutherian orders does offer some support for a ferungulate clade.

Two more modern attempts at resolving higher eutherians are shown in Fig. 3.2. Historically, these might be viewed as attempts to recapture, through a cladistic approach, some of the more explicit qualities of Gregory's (1910) and earlier works. Despite this intention, these more modern revisions are not impressively resolved, even in comparison to Gregory's original scheme.

Fully resolved trees are not, however, necessarily good ones; they can be built on very flimsy evidence. Given the information available, we do not at present regard a fully resolved tree for the higher Eutheria as a realistic objective. We simply argue that certain interordinal relationships are fairly well substantiated (Fig. 3.3), while ambiguity remains in others. What follows is a rather telegraphic survey of these cases, with citations of papers that give more thorough treatments. Relevant characters are listed (Table 3.1) rather than discussed in any detail. For considerations of homology, distribution (Table 3.2), transformation, and synapomorphy of these characters, see McKenna (1975), Novacek (1982, 1985, 1986), Novacek and Wyss (1986, especially the Appendix), and other papers cited.

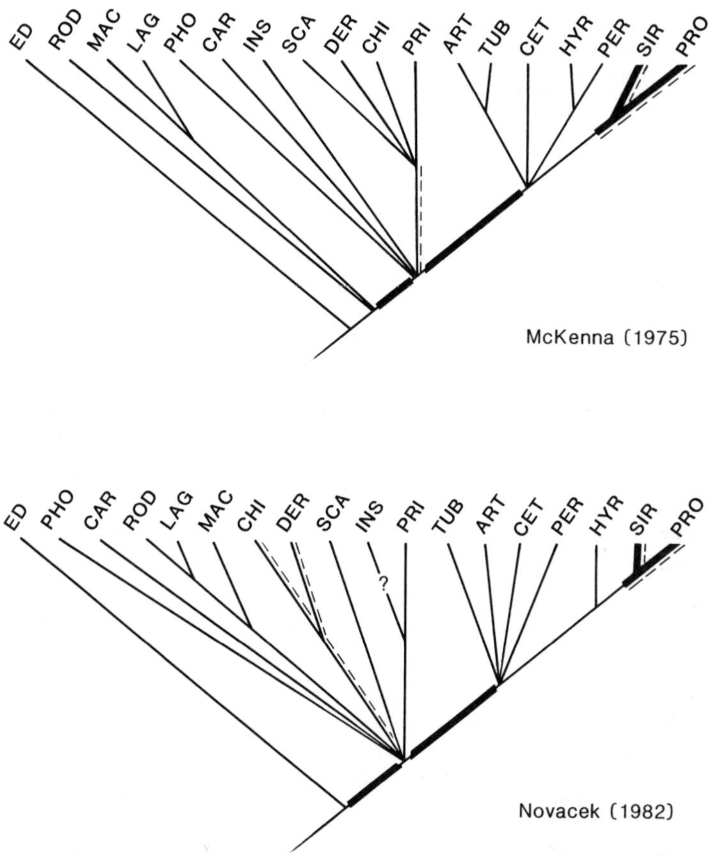

Fig. 3.2. Higher-level hypotheses for eutherian groups as proposed by McKenna (1975) and Novacek (1982). Bold lines indicate agreement between the two trees. Dashed lines indicate agreement with Gregory (1910). Acronyms for groups are MAC, Macroscelidea; for others, see Fig. 3.1.

Eutheria

The Eutheria include all living higher groups of mammals, except the Marsupialia and the Monotremata (see Chapter 2), and also the following fossil taxa (see McKenna 1975): Asioryctidae, Kennalestidae, Anagalidae, Leptictidae, Cimolesta, Creodonta, Arctocyonia, Tillodontia, Dinocerata, Embrithopoda, Litopterna, Notoungulata, Astrapotheria, Xenungulata, Pyrotheria, Condylartha, and Desmostylia. (Note: no claim to monophyly of these groups is implied by our usage of some of these terms.)

Synapomorphies for the Eutheria are many, and much discussed (Fig. 3.3, Table 3.1). However, few of these characters are found in

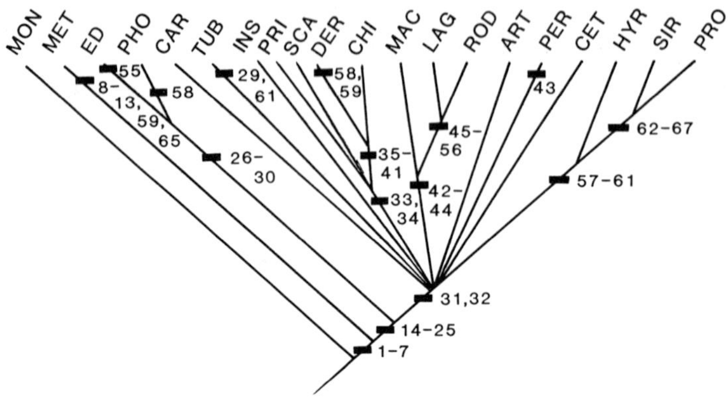

Fig. 3.3. Higher-level hypothesis for groups proposed in this paper. Numbers refer to derived traits shown in Tables 3.1 and 3.2. Acronyms for groups are MON, Monotremata; MET, Metatheria; for others, see Figs. 3.1 and 3.2.

Table 3.1. Derived morphological characters for higher-level groupings shown in Fig. 3 (for explanation of characters, see Novacek and Wyss 1986; and text)

1. Anal and urogenital openings separate in adults
2. Eggshell absent
3. Unilaminar blastocyst stage present
4. Microlecithal egg with holoblastic cleavage
5. Cochlea spiral, with at least one turn
6. Supraspinous fossa of scapula present
7. Mammary glands with teats
8. Development of chorioallantoic membrane suppressed
9. Pseudovaginal canal present
10. Ossified contribution to auditory bulla by alisphenoid
11. Last premolar only cheek tooth replaced
12. Dental enamel with distinct 'tubercles'.
13. Sphenorbital fissures confluent below dorsoventrally compressed orbitosphenoid
14. Modified trophoblast and inner cells mass
15. Definitive chorioallantoic placenta present
16. Prolonged intrauterine gestation
17. Ureters pass lateral to derivatives of Müllerian ducts
18. Corpus callosum connects cerebral hemispheres
19. Epipubic bones and 'marsupium' absent
20. Upper molars with narrow stylar shelves
21. Fusion of Müllerian ducts into median vagina
22. Shell membrane absent
23. Seminal vesicle present
24. Retinal cones of eye simple, lacking oil droplets
25. Optic foramen distinctly separated from sphenorbital fissure.
26. Frontal with pronounced ventral expansion in orbital wall, contacts ventrally confined palatine

Table 3.1. (continued)

27. Alisphenoid ventrally confined in orbital wall, not in broad contact dorsally with parietal
28. Extensive sacral-innominate fusion and complete or nearly complete closure of ischiatic notch to form sacro-ischiatic foramen.
29. Tooth development suppressed, anterior teeth lost, and enamel poorly developed or absent
30. Rectus thoracicus muscle present
31. Stapes bicrurate, stirrup-shaped with large stapedial foramen (or stapes derived from this condition)
32. Foramen ovale completely enclosed within alisphenoid, not notched into posterior margin of this element
33. Pendulous penis suspended by reduced sheath between genital pouch and abdomen
34. Sustentacular facet of the astragalus in distinct medial contact with distal astragalar facets
35. Ribs flattened, especially near vertebral ends
36. Marked elongation of the forelimbs
37. Patagium (flight membrane) continuously attached between digits of the manus
38. Occipitopollicalis and humeropatagalis muscles present in flight membranes
39. Concomitant proximal and distal reduction of ulna
40. Fenestra rotundum of cochlea faces directly posteriorly
41. Subarcuate fossa greatly expanded and dorsal semicircular canal clearly separated from endocranial wall of squamosal
42. Embryonic disc oriented toward mesometrial pole of uterus at time of implantation
43. Canals in external alisphenoid for transmission of masseteric and buccinator branches of maxillary nerve
44. Inferior ramus of the stapedial artery and inferior petrosal nerve run dorsal to tegmen tympani
45. Paired incisive foramina well developed, elongated, open posteriorly on palate, but nasals not recessed
46. Posterodorsal process of the premaxilla well developed, contacts frontal
47. Contact between maxilla and frontal confined by premaxilla above and lacrimal below
48. Infraorbital canal reduced to a foramen high within anterior root of zygoma
49. Premaxilla and maxilla of roughly equal exposure on palate
50. Glenoid fossa set well dorsally of basicranium, concave, anteroposteriorly oriented trough lacking a postglenoid process
51. Postglenoid foramen shifted, opens in lateral eminence of squamosal
52. Two or three closely spaced hypoglossal foramina present
53. Upper and lower I1 lost, large upper and lower, ever-growing deciduous I2 and lower I3 lost (see text)
54. Foetal development involves haemochorial placenta differentiating opposite attachment of blastocyst at abembryonic pole
55. Allantoic diverticulum small, rudimentary
56. Pancreas diffuse
57. M. styloglossus bifurcate

Table 3.1. (continued)

58. Amastoidy: mastoid process concealed in ventral-lateral cranium by expansion and posterior overlap of squamosal on the occiput
59. Jugal extends posteriorly as a prominent ventral crest to anterolateral border of the glenoid fossa
60. Carpals dorsoventrally compressed and serially arranged; lunar-unciform contact very weak or absent
61. Foetal membrane development comprises zonary placenta and free reduced yolk sac; allantoic vesicle broadly developed into four sacculations
62. Bilophodont cheek teeth with tendency to form additional lobe on posterior part of cingulum
63. Forward displacement of orbits
64. Infraorbital canal very short, ventrally situated, floored by a narrow bridge of the anterior zygoma
65. Zygomatic process of the squamosal robust, strongly produced dorsally and laterally
66. Premaxilla with very strong posterodorsal process extending around retracted nasals, nearly contacts frontals
67. Cochlear canaliculis absent.

fossil mammals, and their allocation is based on special affinities with certain Recent eutherian groups, or on possession of derived traits that, while ambiguous for pinpointing relationships, are not found outside the Eutheria (e.g. entotympanic bulla). This leaves several Cretaceous taxa, including *Kennalestes* and *Asioryctes*, whose allocation to the Eutheria (Kielan-Jaworowska *et al.* 1979) is open to question (Novacek 1982).

Edentata, Pholidota, †Palaeanodonta

The Edentata are often equated with the Xenarthra in the older literature (Gregory 1910), namely the group including the New World anteaters (Myrmecophagidae), tree and ground sloths (Bradypodoidea, Megatherioidea, Megalonychoidea), and armadillos (Dasypodoidea). Gregory (1910), however, grouped the Xenarthra with the Pholidota under Edentata, an arrangement that two of us (Novacek and Wyss 1986, Novacek 1986) have endorsed, although we do not accept Gregory's placement of tubulidentates and taeniodonts under Edentata.

In Novacek and Wyss (1986), several characters were suggested as supporting a relationship between edentates (xenarthrans) and pholidotans. Because of errors in accounting for distribution, one of these, the loss of the interparietal, is withdrawn as a synapomorphy (see also *ibid*, note added in proof). Although Gregory (1910), de Beer (1937), and many others claimed that the interparietal is absent in both

Table 3.2. Distribution of derived traits (1) listed in Table 3.1

	1-7	8-13	14-25	26	27	28	29	30	31	32	33	34	35	36	37	38	39	40	41	42	43	44	45	46	47	48	49	50	51	52	53	54	55	56	57	58	59	60	61	62	63	64	65	66	67
MONOTREMATA																																													
METATHERIA	1	1																																			1						1		
EDENTATA	1		1	1	1	1	1	1																									1												
PHOLIDOTA	1		1	1	1	1	1	1																												1									
CARNIVORA	1		1						1	1																																			
TUBULIDENTATA	1		1				1		1	1																													1						
INSECTIVORA	1		1						1	1																																			
PRIMATES	1		1						1	1	1	1																																	
SCANDENTIA	1		1						1	1	1	1																																	
DERMOPTERA	1		1						1	1	1	1	1	1	1	1	1	1	1																	1	1								
CHIROPTERA	1		1						1	1	1	1	1	1	1	1	1	1	1																										
MACROSCELIDEA	1		1						1	1										1	1	1																							
LAGOMORPHA	1		1						1	1										1	1	1	1	1	1	1	1	1	1	1	1	1	1	1											
RODENTIA	1		1						1	1										1	1	1	1	1	1	1	1	1	1	1	1	1	1	1											
ARTIODACTYLA	1		1						1	1																																			
CETACEA	1		1						1	1																																			
PERISSODACTYLA	1		1						1	1											1																								
HYRACOIDEA	1		1						1	1																									1	1	1	1	1						
SIRENIA	1		1						1	1																									1	1	1	1	1	1	1	1	1	1	1
PROBOSCIDEA	1		1						1	1																									1	1	1	1	1	1	1	1	1	1	1

pholidotans and edentates, it is in fact present in fossil pangolins (Emry 1970) and in neonatal skulls of *Manis* (Jollie 1968). The loss of the interparietal thus seems only characteristic of edentates, proboscideans, and monotremes.

To the list of potential synapomorphies for edentates and pholidotans suggested by Novacek and Wyss (1986), we add here a feature of the pelvic girdle (Fig. 3.3; Table 3.1, character 28). The mammalian innominate usually contacts the vertebral column by way of the sacro-iliac joint, a narrow articulation between the ilium and the anterior sacral elements. In lateral view, the dorsal edge of the innominate normally appears to cross the main axis of the vertebral column at an acute angle with the greatly curved posterior portion of the dorsal ilial edge forming the greater ischiatic notch (incisura ischiadica major), which slopes and opens posteriorly. In pangolins and edentates, this angle is greatly reduced and extensive fusion between the pelvis and sacrum occurs. In xenarthrans, this broad ankylosis between the sacral transverse processes and the dorsal edge of the innominate extends far posteriorly and includes fusion between the ischium and the anterior caudal elements. This secondary connection forms a posterior border of the former ischiatic notch, thereby transforming it into what has been referred to as the sacro-ischiatic foramen. This large circular opening is bordered anteriorly and posteriorly by the fused sacrum and innominate, dorsally by emargination into the fused transverse processes, and ventrally by emargination into the innominate itself. This foramen, situated dorsal to the acetabulum, is usually quite large, approaching and sometimes exceeding the size of the obturator foramen.

The arrangement of the pangolin pelvic-sacral region closely resembles that in the xenarthrans: both groups have extensive regional fusion in the pelvic-sacral area, and the large sacro-ischiadic foramen. The only significant difference between the two groups is that pholidotes have an ischial-sacral connection that is not completely fused, but is ligamentous.

There are also a fair number of derived traits shared by pholidotans and the Pilosa (sloths and anteaters) that exclude dasypodids (discussed in Novacek and Wyss 1986), a situation that prompted Engelmann (1977) to argue for a closer relationship between the former groups. This hypothesis is contradicted by the presence of xenarthry (a complicated system of accessory articulations between some of the vertebrae), which constitutes the premier attribute of South American edentates. For some time, however, xenarthry has been suspected as a character whose importance has been inflated (Windle and Parsons 1899); in bradypodid sloths, for example, it is confined only to the last lumbar-first sacral articulation. Hence, a more intimate connection between pholidotans and pilosan edentates is an intriguing, but insufficiently investigated, possibility. For the present, we find support for pangolin–edentate affinities at a higher level. We also provisionally regard the fossil palaeanodonts as close relatives of both edentates and pholidotans. Matthew's (1918) argument for this alliance is certainly not contradicted by more recent reviews (Rose 1978, p. 94), and has received emphatic support from some palaeontologists (e.g. Gingerich 1985). We caution, however, that many of the characters listed in Table 3.1 in support of an edentate–pholidote grouping have not been identified—or are not yet retrievable—in known specimens of archaic palaeanodonts.

Epitheria

In its original formulation (McKenna 1975) the Epitheria are meant to include all living and fossil eutherians except the New World edentates (Xenarthra). Two of us (Novacek and Wyss 1986; Novacek 1986) have also excluded pholidotans from the Epitheria, by virtue of their possible close relationship with xenarthrans. This view was also endorsed by Shoshani (1986), although he simply regarded pangolins and edentates as outgroups to epitheres, without explicitly recognizing them as members of a monophyletic unit.

Despite this agreement, the morphological basis for the Epitheria is somewhat weaker than we originally envisaged. The septomaxilla, an allegedly primitive feature in edentates (McKenna 1975; Novacek 1982), is of ambiguous significance, whether or not pangolins are made a sister-group to edentates (Novacek and Wyss 1986). Likewise, it is not clear that epitheres differ from all edentates in developing a sharply differentiated uterus and vagina (McKenna 1975; Novacek 1982). The latter condition also occurs in juvenile dasypodids (Fernandez 1915; Grassé 1955) and bradypodids (Wislocki 1928). Even marsupials have a distinct cervix in each uterovaginal tract (McGrady 1938). This would suggest that either the cervix was gained independently at least *three* times (marsupials, bradypodid sloths, and all other eutherians, with dasypodids representing a variation of this theme), or that the loss of the cervix is a derived condition within the Theria, occurring only in some myrmecophagids (Van der Horst 1942, p. 427).

Two potential synapomorphies for the Epitheria remain (Novacek and Wyss 1986). The first, the bicrurate stapes, was judged to have evolved independently *within* some edentates (dasypodids and two of the three myrmecophagids) and epitheres. There is also a large stapedial foramen in the stapes of some marsupials (dasyurids, didelphids), but true resemblance of this stapes to the eutherian bicrurate type is uncertain. Some ambiguity remains regarding the primitive stapedial condition for both Eutheria and Theria, although the idea that the bicrurate stapes was derived *within* the Eutheria is still slightly favoured. A second epitherian synapomorphy involves the relationship between the foramen ovale and the alisphenoid (Table 3.1, character 32), discussed and illustrated by Novacek and Wyss (1986).

Shoshani's (1986, table 3) recognition of a third synapomorphy, the loss of a clavicle, is unconvincing. His scheme allows the reappearance of a clavicle only once, under the assumption that carnivorans, primates, tupaiids, dermopterans, chiropterans, insectivorans, macroscelideans, lagomorphs, and rodents are united in a superclade that excludes all other Epitheria (*ibid*, fig. 3). We find his evidence for this superclade very problematic, as several of his proposed synapomorphies (e.g. extensively fused tibia–fibula, unfused mandibular symphysis, digit III not the longest) are extremely labile (*ibid*, table 3) or are more variable among major eutherian lineages than his tabulations would indicate. Hence, the dissolution of this superclade allows for the multiple reacquisition of a clavicle in various lineages, a pattern that seems to us highly unlikely.

Carnivora, †Creodonta, Tubulidentata, Insectivora

Three recent orders and a variety of extinct major clades are particularly resistant to classification. The Carnivora can be associated with the fossil Creodonta by features of the dentition and, perhaps, the astragalo-calcaneal complex (Szalay 1977; Novacek 1982). Living carnivorans, and creodonts, share with pholidotans a thin, obliquely oriented lamina that divides the cerebral and cerebellar fossa, a feature found in few other mammals. Shoshani (1986, fig. 3), however, identified a sister-group relationship between the Carnivora and the Primates based on his survey of skeletal features. Again, the evidence that he recognized for this alliance is dubious; four of his five synapomorphies are conditions that reverse or evolve independently several times on his cladogram. The remaining trait, the absence of fibula–calcaneum contact, is more widespread among eutherians than his tabulations indicate; moreover, this loss does not characterize several carnivoran groups (Flynn and Galiano 1982). Clues to the higher relationships of the Carnivora (and their presumed fossil relatives, the creodonts) are not readily apparent (see Chapter 4).

The Insectivora are classically regarded as 'stem' eutherians, although the Recent insectivorans (the lipotyphlans) comprise an assemblage with several distinct specializations (Butler 1956, 1972, and Chapter 5 of this volume; Novacek 1986). Arguments that lipotyphlans have a special relationship with primates (Szalay 1975; Novacek 1982) relied on basicranial features not common to certain fossil hedgehog-like (erinaceomorph) insectivorans (Novacek *et al.* 1983). Alternative hypotheses are equally unconvincing. Associations between insectivorans, tupaiids, and macroscelideans (Gregory 1910) seem largely a function of their primitive resemblance (Butler 1956; McKenna 1960; Van Valen 1967; Novacek 1982, 1986).

In another scheme, Shoshani (1986, fig. 3) presented the following evidence for associating (lipotyphlan) insectivorans with a macroscelidean–lagomorph–rodent clade: foramen rotundum confluent with orbital fissure, alisphenoid canal present, mandibular condyle oriented posteriorly or almost roundish, tibia–fibula fused proximally and/or distally, astragalar groove deep. However, the alisphenoid canal is variably present in macroscelidids and lipotyphlan insectivorans, and the derived status of this feature is very uncertain. The confluence of the foramen rotundum and (sphen-)orbital fissure is widespread in the Theria, and is probably a primitive condition (Novacek 1986). The mandibular condyle feature is not presented with any explanation, and the postcranial traits seem highly labile. Novacek (1986) resurrected the notion that lipotyphlan insectivores are closely related to the extinct Early Tertiary leptictids, but did not extend this to include any other major groups of Eutheria.

Tubulidentates (aardvarks) pose a particularly interesting problem because of their alleged affinities with all or some of the ungulate clades. Shoshani (1986), for example, allied aardvarks with artiodactyls, perissodactyls, fossil condylarths, and notoungulates. However, we are unable to identify many morphological traits that would suggest this affinity. The shallow subarcuate fossa found in tubulidentates and some ungulate groups (Novacek and Wyss 1986) shows marked homoplasy with respect to the branching patterns preferred here, and Thewissen (1985) strongly questions tubulidentate–'ungulate' relationships. One prohibitive aspect to this enquiry is the scanty comparative information on condylarths and other fossil ungulate lineages, a problem that also plagues investigations of ungulate monophyly and relationships (see below). On the other hand, Thewissen (1985) noted basicranial resemblances between tubulidentates and lipotyphlan insectivorans (see also Novacek 1986), without committing these to a firm proposal for relationship. Aardvarks also display a peculiar quadrilobular allantois (Table 3.1, character 61) characteristic of hyracoids, sirenians, and proboscideans (Mossmann 1957), thus suggesting a tubulidentate–paenungulate relationship, independently promoted in some molecular studies (de Jong 1982; Miyamoto and Goodman 1986). These possibilities are, however, in need of more thorough analysis (McKenna, 1987).

Archonta

Gregory (1910) united primates, scandentians (tree-shrews), dermopterans (flying lemurs), and chiropterans (bats) within a single superclade, the Archonta. This grouping was not reflected in Simpson's (1945) polytomy of unguiculate lineages (see Fig. 3.1), but was resurrected in more recent reviews (McKenna 1975; Szalay 1977). Novacek and Wyss (1986) judged criticisms of the Archonta (Cartmill and MacPhee, 1980; Luckett, 1980; Novacek 1980) to be overstated. They salvaged this clade on the basis of derived traits in penial and tarsal morphology (Fig. 3.1; Table 3.1, characters 33 and 34), also cited by other workers (Szalay 1977, Smith and Madkour 1980). This does not constitute a very strong case for the Archonta, but currently it is the only plausible higher-level hypothesis for relating these lineages (see also Wible and Covert 1987). Shoshani (1986) also recognized a basic clustering of the archontan groups, but linked these with carnivorans for the dubious reasons noted above.

The various subdivisions within the Archonta have elicited much interest, primarily because of their bearing on the question of primate origins. Here, most possibilities remain open, although one of the more rigorous treatments of the character evidence to date (Wible and Covert, 1987) breaks with tradition in distinguishing euprimates (which can be recognized by the unique, wholly petrosal, composition of the auditory bulla) from the paraphyletic grouping of the Early Tertiary plesiadapiformes (including microsyopids). Wible and Covert (1987) further argue that tupaiids may actually be stronger candidates for the near relatives of (eu)primates than the fossil plesiadapiformes, thus resurrecting the famed tree-shrew–primate connection, somewhat maligned in more modern treatments (Cartmill and MacPhee 1980; Luckett 1980; Novacek 1980; but see Chapter 6 of this volume). Such an association is contradicted by the special resemblance in the dentitions of euprimates and plesiadapiformes. However, two of us (Novacek and Wyss) endorse Wible and Covert's recommendation that the Order Primates simply be equated with what is formally known as euprimates, at least until the relationships of (some or all) plesiadapiformes can be more firmly established.

The one archontan subdivision accepted by us is that restricted to dermopterans and chiropterans (Gregory 1910; Novacek 1982, 1986; Wyss 1983; Novacek and Wyss 1986). The synapomorphies for this association include not only features of the patagium (flight membrane), but also special similarities in auditory morphology (Fig. 3.1; Table 3.1, characters 35–41). This view represents a vindication of Leche (1886), whose detailed monograph was criticized (Winge 1917) on the dubious grounds that flying lemurs were not fitting 'direct ancestors' to bats. The patagial structure in other gliding mammals (gliding marsupials and rodents) lacks many of these special similarities (e.g. occipitopollicalis, humeropatagialis muscles, and interdigital flight membrane), weighing against the possibility that such resemblance was acquired independently in bats and flying lemurs (Wyss 1983). Indeed, the intimate connection between these taxa has been pushed to the point of questioning the monophyletic origin of the Chiroptera. Pettigrew (1986) claimed that megachiropterans (fruit bats) share a highly specialized visual (neural) pathway with dermopterans and primates, and strongly advocated a close relationship between these taxa, to the exclusion of microchiropterans. We do not see this or other arguments (Smith and Madkour 1980) as a serious challenge to chiropteran monophyly, given the abundance of synapomorphies for the latter (Novacek 1982; Wible and Novacek, in press). However, the neurological evidence could be regarded from another perspective—the highly specialized visual pathway in megachiropterans, dermopterans, and primates might relate these groups, and the apparently more primitive condition in the Microchiroptera simply might represent a reversal correlated with de-emphasis on the visual system in these echolocating forms. We resist further speculation here, and note that the neuroanatomical systems considered above require more thorough distribution analysis (see also Wible and Covert 1987).

Glires (Lagomorpha and Rodentia), Macroscelidea, †Anagalida, †Pseudictopidae, †Eurymylidae

Gregory's (1910) emphatic support for the special relationship between lagomorphs and rodents was one of the few higher-level groupings maintained (though without strong conviction) in Simpson's (1945) classification (Fig. 3.1). Nonetheless, this concept was criticized and abandoned by many rodent specialists (Wood 1957). A recent revival in support for the Glires (Luckett 1977; Szalay 1977; Novacek 1982; Hartenberger 1985; Chapter 7 of this volume) has been amplified by several papers covering cranial features (Novacek 1985), fetal membrane and tooth development (Luckett 1985), and palaeontological information (Li and Ting 1985). The stronger aspects of this evidence are summarized as a rather lengthy list of synapomorphies in Table 3.1 (characters 45–56). In addition, Shoshani (1986, table 3) listed several skull features corroborating rodent–lagomorph affinities, although some of these were previously published in Novacek (1985).

Despite the comparatively strong morphological evidence for the Glires, this grouping encounters at least two difficulties. First, surprisingly little corroboration is provided by molecular studies (Sarich

1985). Shoshani (1986, fig. 3) grouped rodents and lagomorphs on the basis of skeletal evidence, but associated lagomorphs with edentates and soricomorph insectivores, and rodents with macroscelideans and dermopterans, on the basis of immunodiffusion data (fig. 2). While Miyamoto and Goodman (1986) recognize the Glires in their molecular-based classification, their grouping includes primates—a notion that hardly conforms with the original concept or recent morphological characterization of the Glires (see remarks below). Moreover, analysis of some sequences does not even support the monophyly of leporids and ochotonids within the Lagomorpha, a result that seriously undermines the significance of these studies to date (Wyss *et al.* 1987, and remarks below). In fact, the only case of concordance was found by Penny and Hendy (1985, p. 81), who noted that a close association between lagomorphs and rodents is suggested by their analysis of five trees generated by protein sequence data.

A second difficulty is posed when a variety of fossil candidates for rodent and/or lagomorph relatives are considered. For example, the Pseudictopidae from the Asian Early Cenozoic show lagomorph specializations in foot and dental structure (Bleefeld and McKenna 1985; McKenna 1987) but do not have the enlarged gliriform incisors famously characteristic of rodents and lagomorphs (Table 3.1, character 53). Acceptance of a close pseudictopid–lagomorph relationship would suggest that the gliriform condition and its attendant developmental pattern (Luckett 1985) evolved independently in rodents and lagomorphs. The former grouping is, however, disputed by Lucas (1986), who associates pseudictopids with the extinct uintatheres, arguing that pedal resemblances between pseudictopids and lagomorphs are the result of convergence.

Despite these problems, it is difficult to deny the strength of the case for the Glires. The molecular evidence, though generally incongruent with the Glires, is so internally inconsistent (Table 3.3) that its significance is moot. The fossil evidence may, at most, exclude some of the alleged synapomorphies for the group (see also McKenna 1987), but this still leaves a considerable body of morphological data in support of the Glires clade.

The identity of near relatives of the Glires is less certain, the best candidate among living taxa being the Macroscelidea (McKenna 1975). Originally, macroscelideans were associated with tupaiids as the 'other half' (Suborder Menotyphla) of the broadly defined Insectivora (e.g. Gregory 1910, and earlier studies), a concept upheld in Simpson's (1945) classification. Subsequently, the Menotyphla was dismembered (Patterson 1965; Van Valen 1967; Butler 1972); tupaiids were tentatively linked with the other archontan clades, and macroscelideans were left as a lost branch within the great eutherian radiation.

Subsequently, a number of studies (McKenna 1975, 1987; Novacek 1982, 1986; Wible and Covert 1987) agreed on a close association *at some level* between macroscelideans and the Glires, although the morphological evidence for this argument is not overwhelming. Two supportive features cited by Novacek and Wyss (1986) are the distinctive orientation of the embryonic disc at the time of implantation (Luckett 1985), and the presence of osseous canals for the masseteric and buccinator branches of the maxillary nerve (Fig. 3.3; Table 3.1, characters 42 and 43). In addition, Wible (1984) and Wible and Covert (1987) have described a neurovascular pathway that seems to separate the Glires and macroscelideans from other groups of mammals (Table 3.1, character 44). In this condition, both the inferior petrosal nerve and the ramus inferior of the stapedial artery run dorsal to the tegmen tympani, a rare and presumably derived eutherian trait (although one possibly shared with fossil leptictids; Novacek 1980, 1986). In most eutherians, these structures course through the tympanic cavity ventral to the tegmen tympani.

The relationship between the Recent Glires and macroscelideans masks a more complicated array of alternatives presented by the fossil groups Anagalida, Pseudictopidae, Eurymylidae (including Mimotonidae), and possibly Zalamdalestidae. These taxa, thought to have an origin in the Asian Late Mesozoic or Early Tertiary, were broadly associated by McKenna (1975) in the superclade Ernotheria, a group that also included leptictids, and the Cretaceous genera *Kennalestes* and *Asioryctes*. McKenna's (1975) primary evidence for this clade involved the suggested departure from the primitive eutherian dental replacement pattern. This hypothesis was criticized by some (Szalay 1977; Novacek 1986), but the grouping of the ernotherians, *sans* Tertiary leptictids and Cretaceous forms, and with the addition of the Rodentia, was endorsed by others (Novacek 1982, 1986; Li and Ting 1985).

There is reason to suggest that some of these fossil groups have a more intimate connection with one or another of the modern lineages of the ernotherian clade. This seems certainly true of the eurymylids, where close relationship between some members of this group and rodents has been suspected (Li and Ting 1985: the monophyly of this assemblage and the possibility that some eurymylids are closer to lagomorphs requires further inspection). On the other hand, *Mimotona* has been regarded as a close lagomorph relative (Li and Ting 1985), an association that may also hold for pseudictopids, according to McKenna (1975, 1987) and Bleefeld and McKenna (1985).

Perhaps the most intriguing members of this Asian radiation are the anagalids, whose alleged relationships to tupaiids (Evans 1942) were subsequently rejected (McKenna 1963). More modern studies tend to associate anagalids with macroscelideans (McKenna 1975; Novacek 1982, 1986). However, Novacek's (1985) survey of cranial data suggested that anagalids were closer to the Glires than to the Macroscelidea. This 'manual analysis', however, was not corroborated by a PAUP run of the same data, in which the number of missing data entries for anagalids made the Anagalida an outlier to the Glires, dermopterans(!), macroscelideans, and several other groups. McKenna has recently reprepared the type skull of *Anagale gobiensis*, whose ear region has been misinterpreted in the past. Preliminary indications are that the bulla is entirely ectotympanic in composition, a condition in strong contrast to that in macroscelideans, tupaiids, and primates, but identical to that in the Glires.

Ungulates

The myriad lineages that constitute the ungulates have long been associated as a supposedly monophyletic unit (Gregory 1910; Simpson 1945—though here with carnivorans—see fig. 1). Recognition of this vast assemblage is also found in more modern treatments (McKenna 1975; Novacek 1982; Chapter 8 of this volume). However, the explicit characterization of a monophyletic Ungulata is embarrassingly poor; it hinges on a few rather labile characters of the skull and skeleton and the 'gestalt' design of a cheek-tooth dentition suitable for herbivory (but notably more trenchant in some ungulate clades). This evidence (discussed in Novacek and Wyss 1986) provides us with a rather unenthusiastic position on the argument for ungulate monophyly. Our solution is to recognize the monophyly of some ungulate clades (e.g. the Paenungulata) but not the monophyly of the 'Ungulata' as a whole (Fig. 3.3).

Tethytheria, Paenungulata

While ungulate monophyly cannot be adequately demonstrated, various ungulates can be organized in more compelling, if controversial, groupings. Perhaps the most notable in this regard is the Tethytheria, a group including the sirenians, proboscideans, and the extinct desmostylians (see McKenna 1975, 1987; Tassy 1981; Novacek 1982, 1986; Novacek and Wyss 1986; Shoshani 1986; Domning *et al.* 1986). Several characters supporting this node are shown in Table 3.1 (characters 62–67). A derived trait for the Tethytheria not previously cited is character 67 (Table 3.1), the lack of a cochlear canaliculus typical of other mammals (Fleischer, 1973, pp. 184–5).

Another clade of interest is the Paenungulata, thought to comprise the Tethytheria and the Hyracoidea. The supporting evidence for this group includes both cranioskeletal and foetal membrane features (the latter, of course, cannot be attributed to fossil desmostylians), discussed at some length in Novacek and Wyss (1986). The Paenungulata have also been recognized in some molecular work (e.g. Shoshani 1986, fig. 2; Miyamoto and Goodman 1986; McKenna 1987; Wyss *et al.* 1987). An issue concerning the Paenungulata has, however, come to light with Fischer's (1986) detailed anatomical studies of hyracoids. This author argued that hyracoids and perissodactyls are closely related because they share a complex of locomotor features, a large eustachian sac, and a correlated pathway of the internal carotid and a tuber maxillaris (the inflation of the maxilla around the posterior dental arcade). Fischer (1986) further maintained that essentially all the morphological features supporting the Paenungulata can be disallowed as symplesiomorphies or incorrectly analysed traits.

We find that Fischer's generalizations concerning locomotory similarity between hyracoids and perissodactyls are problematic (see also Novacek and Wyss 1986). Many of the derived characters that he cites are not clearly present in fossil equids (e.g. acromion greatly reduced or lost with concomitant modification of shoulder girdle muscles), and other features are found in many or most 'ungulate' lineages, not just perissodactyls (e.g. reduction of clavicle, confined pronation and supination, elongated fibula, hoofs). Finally, the tridactyl pes in hyracoids and perissodactyls is not found in phenacodonts, an Early Tertiary group which Fischer (1986) acknowledges as a close relative of perissodactyls (Radinsky 1966), although other non-tridactyl condylarths such as hyopsodontids may prove to have closer affinities with perissodactyls (McKenna *et al.* MS). It is possible that hyracoids and perissodactyls share a unique form-function complex of locomotor organs, a complex that subsumes any inconsistencies found in distributions of components of this system. However, we see here no thoroughly-built argument for such a hypothesis, which, at the very least, should account for skeletal morphology in many fossil groups, including the South American notoungulates. Hence, we share Fischer's (1986) own suspicion that these skeletal similarities may represent parallelisms in hyracoids and perissodactyls.

This leaves only the tuber maxillaris and the enlarged eustachian sac (and related carotid pathway) in support of a hyracoid–perissodactyl clade. The former evidence is problematic; a swollen tuber maxillaris that corresponds to Fischer's (1986) description is present in many ungulate groups, including proboscideans and sirenians. Fischer's *ad hoc* argument (based on developmental patterns) that there is an essential difference in the tuber maxillaris condition between tethytheres and ungulates, where a swollen maxilla occurs, is not convincing. The eustachian diverticulum is, in itself, compelling evidence, but it seems effectively offset by the synapomorphies enlisted for the Paenungulata (Table 3.1; Fig. 3.3).

In this context, we reject most of Fischer's (1986) criticisms of paenungulate synapomorphies. He disallows the serial carpal configuration (Table 3.1, character 60) in these forms because of the alleged presence of the more primitive alternating arrangement in extinct desmostylians (Tassy 1981). Novacek and Wyss (1986) have, however, argued that the carpal arrangement in desmostylians is an anomalous condition, effected by the enlargement of the lunar. Thus, the desmostylian carpus is not primitive with respect to other paenungulates; nor does it exclude desmostylians from this group. Fischer (1986) also dismissed the fetal membrane evidence for paenungulates (Table 3.1, character 61—see discussion in Novacek and Wyss 1986) without adequate explanation. His doubts concerning the paenungulate jugal condition (Table 3.1, character 59) seem better founded, but we still maintain that sirenians show a backward extension of the jugal on the zygomatic arch not attained in many groups other than hyracoids and proboscideans (Novacek and Wyss 1986, fig. 10). This argument applies even though, in the latter, the backward extension of the jugal is more pronounced (the morphological condition for desmostylians is difficult to determine from the available fossil evidence). Finally, Fischer (1986) does not consider the 'amastoid' condition (Novacek and Wyss 1986, Appendix) that, to us, represents strong corroborative evidence for paenungulates, and his analysis even discloses a feature of the musculature (bifurcation of the m. styloglossus, see Janis 1983; and Table 3.1, character 57) consistent with this grouping. Most of Fischer's criticisms are, in fact, aimed at ambiguous similarities found in the earlier literature (e.g. enlargement of upper incisors, musculature patterns, testicondy), which we would likewise not recognize. Overall, we find Fischer's argument for a perissodactyl–hyracoid grouping less convincing than the case for the Paenungulata.

Protein sequence results

The impact of molecular systematics has been more apparent in mammals than in any other group. Such work is an extension of the initial forays into the problems of hominid and other primate relationships, involving immunoglobin and polypeptide sequence data (e.g. Sarich and Cronin 1976; Goodman *et al.* 1985). Molecular data are hence appreciably richer for this group than for all others; a recent molecular-based classification of the eutherian orders by Miyamoto and Goodman (1986) incorporates polypeptide sequence data from 21 OTUs, representing most of the major lineages of the primates.

Although a few other orders are represented by a fair sprinkling of taxa (artiodactyls with 14 OTUs, and carnivores with 8 OTUs), there are notable gaps elsewhere: there are only three insectivore OTUs, one talpid, one erinaceid, and one soricid, but no tenrecid or solenodontid. Most problematic is the virtual lack of data on ordinal-level groups, such as the Dermoptera (no information) and the Macroscelidea (a single fragmentary chain of lens α-crystallin A, de Jong 1982), which are pivotal to any higher classification of the Eutheria. This is not meant as a criticism of the exhaustive efforts of molecular systematists to date, but it indicates that molecular results are often difficult to compare with morphological results, where taxonomic sampling has been, for both historical and accessibility reasons, much more effective.

The taxonomic sampling problem in molecular systematics is compounded by the uneven sampling of character (=polypeptide sequence) data from taxon to taxon. In their recent analysis, Miyamoto and Goodman (1986) considered amino acid sequences of seven polypeptides (α- and β-haemoglobins, myoglobin, lens α-crystallin A, fibrinopeptides A and/or B, cytochrome c, and ribonuclease). These were 'tandemly aligned', i.e. combined in a long array of sequences and then analysed as if each OTU were represented by a single gigantic polypeptide. Maximum parsimony analysis of these 'tandemly aligned' sequences was executed by a branch-swapping computer program (Moore 1975) capable of accounting for data gaps. Nonetheless, it is clear that there is a severe imbalance in sequence evidence for various OTUs. While at least some orders are represented by information from all, or nearly all, seven molecules (e.g. Artiodactyla, Carnivora, Cetacea, Lagomorpha, Perissodactyla, Rodentia, and especially Primates) several orders are known from little or no information on cytochrome, fibrinopeptides A and/or B, and ribonuclease (Chiroptera, Edentata, Hyracoidea, Pholidota [known only from lens α-crystallin A], Scandentia [tree-shrews], Sirenia and Tubulidentata). In fact, 35 cells of a 112-cell data matrix for eutherian orders (Miyamoto and Goodman 1986, table 1) are empty. This is actually a rather generous portrayal of the data, as only artiodactyls, perissodactyls, and primates are known from nearly every polypeptide sequence for several (but not all) representative members. By analogy, expectations would be rather low for a morphologically-based classification where more than 30% of the relevant data were missing. Again, this is not meant as a criticism of a molecular approach. We do, however, hesitate in accepting a formal classification of the Recent Eutheria based on a molecular study that fails to account for all the Recent orders of mammals (e.g. Miyamoto and Goodman 1986, p. 235, table 2) and which is established from cladograms (*ibid*, figs 2 and 3, pp. 234–5) that present no explicit character information for their various nodes.

This problem in evaluation has prompted us to analyse the relationships between 'raw' sequence data and the trees constructed from them (Wyss *et al.* 1987; McKenna 1987). Our reassessments demonstrated that the degree of homoplasy (as measured by CI values) was comparable in molecular and morphological data sets. However, also revealed were a number of important discrepancies in trees derived from different protein sequences. In fact, an ordered character analysis (based on outgroup parsimony, see Maddison *et al.* 1984) of sequences for each of four different molecules (α-crystallin A, myoglobin, and α- and β-haemoglobins) produced consensus trees with markedly contrasting topologies (Wyss *et al.* 1987). These trees, derived through the use of Swofford's (1985) PAUP program, are compared in detail in Wyss *et al.* (1987), and their degree of congruence is summarized in Tables 3.3 and 3.4. The notable lack of consistency among molecular sequence results is readily apparent; this problem is discussed in the following section of this paper.

Congruence

One measure of a particular hypothesis of relationships is the degree to which it is congruent with other results. Congruence does not, of course, carve a cladogram in stone, but it is one of the few ways of assessing the predictive strength of a particular hypothesis. In this section we compare the topologies of trees found in several morphological and molecular studies (summarized in Tables 3.3 and 3.4).

Morphologically-based trees come from several sources, spanning a long history of investigations. Methodologically, Simpson (1945) and Gregory (1910) could hardly be called cladistic studies, but Gregory's review contains many explicit characterizations of ordinal and superordinal groupings. Moreover, Gregory (1910, p. 105) expressed the critical importance of distinguishing primitive from derived characters, although he did not always rigorously apply this principle (e.g. his characterization of Archonta, *ibid*, p. 322). Most of the morphologically-based solutions reflect diverse anatomical data, as they incorporate—or at least acknowledge—soft anatomical, fetal, skeletal, cranial, and dental features (e.g. Gregory 1910; Simpson 1945; McKenna 1975; Novacek 1982; Novacek and Wyss 1986; this chapter). The remaining analyses are more restricted in the range of anatomical traits. Novacek's (1986) cladograms are based on features of the mammalian skull (no dental evidence), and Shoshani's (1986) analysis is based on non-dental skeletal features.

Molecular studies of sequence data are either focused on individual proteins (de Jong 1982; McKenna 1987; Wyss *et al.* 1987) or are

Table 3.3. Comparisons of topologies in morphological and molecular-based trees with Fig. 3.4A (see text for explanation)

	Nodes in Fig. 3.4A										CI
Studies	a	b	c	d	e	f	g	h	i	j	0.81
Morphological											
This study (Fig. 3.3)[1,2]	?	+	−	−	+	+	+	+	+	+	0.87
Shoshani 1986 (fig. 3)[1,3–5]	−	+	−	−	+	+	+	+	−	+	0.69
Novacek 1986 (fig. 34)[6]	+	−	+	+	+	+	+	+	−	+	0.48
Novacek 1986 (fig. 35)[7]	+	+	−	+	+	+	+	+	+	+	0.32
CONTREE (Fig. 3.6)[8]	+	−	+	+	+	+	+	+	+	+	0.84
Gregory 1910 (Fig. 3.1)[9]	+	−	−	−	+	+	−	+	+	+	?
Simpson 1945 (Fig. 3.1)[9]	−	−	−	−	+	+	−	+	−	−	?
McKenna 1975 (Fig. 3.2)[9]	−	+	+	−	−	+	+	−	+	−	?
Novacek 1982 (Fig. 3.2)[9]	−	+	+	−	+	+	+	+	−	+	?
Molecular											
McKenna 1987 (Myoglobin)[10,11]	NA	NA	−	−	NA	NA	NA	−	−	NA	0.50
McKenna 1987 (A-lens)	−	−	−	−	−	−	−	−	−	NA	0.70
Wyss *et al.* 1987 (β-haemoglobin	NA	−	−	−	+	NA	NA	−	−	NA	0.60
Wyss *et al.* 1987 (α-haemoglobin)[13]	NA	−	−	−	+	NA	NA	−	NA	NA	0.49
Wyss *et al.* 1987 (Myoglobin)	NA	NA	+	−	NA	NA	NA	NA	−	NA	0.59
Wyss *et al.* 1987 (A-lens)[14–16]	−	−	−	−	−	−	−	−	−	NA	0.63
Shoshani 1986 (fig. 2)[17–18]	−	−	−	−	−	−	−	−	−	−	NA
Goodman *et al.* 1985 (fig. 3)[19,20]	NA	−	−	−	NA	NA	NA	−	−	NA	?
de Jong 1982 (A-lens)[16,21,22]	−	−	−	−	−	−	NA	−	−	NA	0.72
Goodman *et al.* 1982 (fig. 5)[23,24]	NA	NA	−	−	NA	NA	NA	−	−	NA	?

[1]Monophyly of ungulate clade not recognized. [2]Monophyly of PER, CET, and paenungulates not recognized. [3]ED+PHO+all other Eutheria in trichotomy. [4]PER isolated from CET+paenungulate clade. [5]CAR included within archontan clade. [6]PHO+ED included within the archontan clade with CAR. [7]TUB nearest sister-group of all ungulates. [8]PHO+ED nearest sister-group of all ungulates. [9]See Fig. 3.1 and Fig. 3.2. [10]PRO excluded from archontan, glire, and ungulate clades. [11]ROD, LAG, and INS+CHI in trichotomy. [12]ED, MAC, PRO, and all other eutherians in tetrachotomy. [13][CET+ROD+LAG] monophyletic group excluding other eutherians. [14]ED, MAC, and all other eutherians in trichotomy. [15]Paenungulate grouping includes TUB. [16]PRO outgroup to [TUB+HYR+SIR]. [17]SIR outgroup to [HYR+TUB+PRO]. [18]DER associated with [LAG+ROD]. [19]ED+PROB isolated from all other eutherian clades. [20]ROD excluded from [LAG+PRI]. [21]PHO, CAR, and [ART+PER+CET] in trichotomy. [22]ED, paenungulates, and all other Eutheria in trichotomy. [23]PRO excluded from all other Eutheria. [24]ROD excluded from [LAG+SCA+PRI). For acronyms of the above-mentioned groups, see Figs. 3.1 and 3.2.

developed from 'tandem alignments' of several sequences (Goodman *et al.* 1982, 1985; Miyamoto and Goodman 1986). One solution listed here is not a 'character trait' analysis of sequence data, but a distance analysis of immunodiffusion data (IMDFM) for albumin and whole sera (Shoshani 1986, fig. 2).

Comparisons of these several solutions naturally lead to problems. Not all data sets include equal numbers or similar terminal taxa. Such cases are indicated by NA, where a taxon has not been analysed in the study in question. For example, in Table 3.3 the published cladogram of Novacek and Wyss (1986, fig. 1) is used as a reference for comparison (reproduced in Fig. 3.4A here); the postulated sister-group relationship of the Dermoptera and the Chiroptera (Fig. 3.4A, Node J) in this

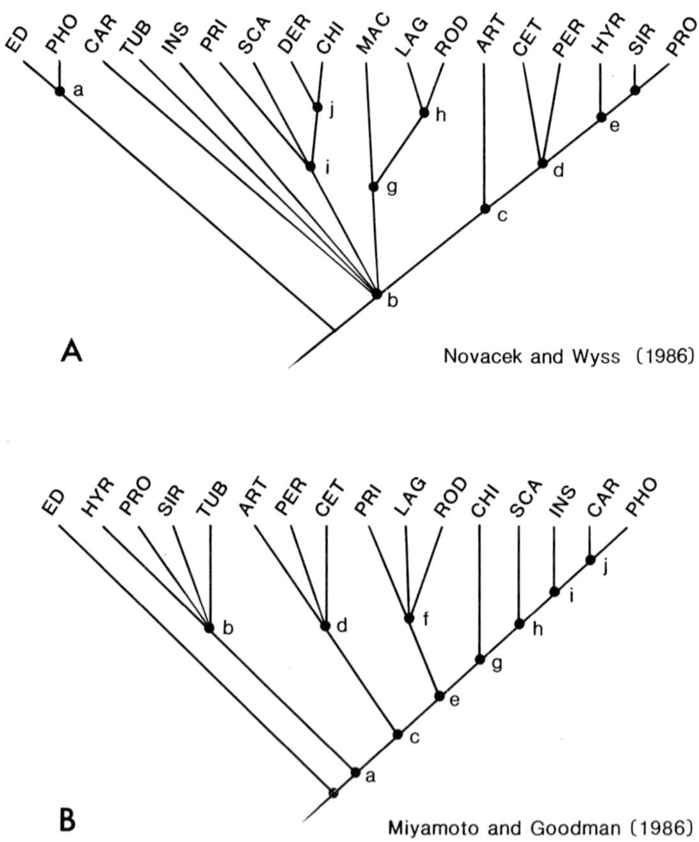

Fig. 3.4. Higher-level hypotheses for eutherian groups proposed by (A) Novacek and Wyss (1986) based on morphological data and (B) Miyamoto and Goodman (1986) based on molecular sequence data. Letters refer to nodes listed in Tables 3.3 and 3.4. For acronyms of groups, see Figs. 3.1 and 3.2.

analysis cannot be compared with any protein sequence study because these have yet to sample the Dermoptera, and NA appears in the column under J for the sequence studies.

Agreement between the two results for a particular grouping is indicated in Tables 3.3 and 3.4 by '+'. Our convention for agreement is rigorous; it must reflect perfect correspondence for the groups in question. This leaves several possibilities for a discrepant result indicated by '−'. A discrepancy is identified if:

Case 1. A taxon that is a member of Group A in the reference cladogram is a member of another group (Group B) in the compared result (with the implication that Group A is diphyletic): $A(x,y)+B(j,h) \neq A(x)+B(y,j,h)$

Case 2. A taxon that is a member of Group A in the reference cladogram is neither a member of Group A nor Group B in the compared result (with the implication that Group A is paraphyletic): $A(x,y)+B(j,h) \neq A(x)+y+B(j,h)$

Hence, these formulations account for strict contradictions (evidence for alternative groupings) as well as cases where topologies show less resolution (polytomies, etc.) than the topology of the reference cladogram.

For clarification, consider Node (or grouping) a in Fig. 3.4A, which suggests a close (sister-group) relationship between edentates and pholidotans. Discrepancies with this grouping take the following forms:

(1) Pholidotan relationships with any other group not identified—Novacek 1982 (Case 2 discrepancy).

(2) Pholidotans and edentates and all other Eutheria comprise a trichotomy—Shoshani 1986, fig. 3 (Case 2).

(3) Pholidotans are members of a clade comprising carnivorans, insectivorans, archontans, and ungulate lineages, that excludes edentates—McKenna 1975 (Case 1).

(4) Pholidotans are in an unresolved polytomy with several eutherian lineages, including edentates—Simpson 1945 (Case 2).

(5) Edentates are a sister-group of lagomorphs; pholidotans are a sister-group of carnivores—Shoshani 1986, fig. 2 (Case 1).

(6) Pholidotans are members of a clade comprising carnivorans, cetaceans, perissodactyls, and artiodactyls, that excludes edentates—de Jong 1982; McKenna 1987; Wyss *et al.* 1987 (Case 1).

(7) Pholidotans are a sister-group of carnivores, and edentates are an

outgroup of all other Eutheria, including pholidotans—Miyamoto and Goodman 1980 (Case 1).

We acknowledge that a contradiction that implies diphyly is different from one that implies paraphyly. However, it seems misleading simply to equate non-statements concerning the position of a taxon (e.g. polytomies) with an explicit hypothesis for a monophyletic group. For example, a solution advocating the group Glires to include primates, lagomorphs, and rodents (Miyamoto and Goodman 1986) cannot be equated with a group Glires that is restricted to lagomorphs and rodents among recent taxa (Gregory 1910; Novacek 1982, 1986; Novacek and Wyss 1986; McKenna 1987). Agreement in this case would only apply if primates were the nearest outgroup of lagomorphs and rodents (at what level, then, the name Glires applies would be arbitrary). In reality, the contradictory results do not clearly identify the primates as the nearest sister-group of rodents and lagomorphs. This holds whether or not primates are associated with an alternative grouping. Explanations for particular discrepancies are given as footnotes to Tables 3.3 and 3.4.

A reference cladogram for a morphologically-based result, such as the one from Novacek and Wyss (1986), finds many instances of agreement from other morphological trees, even though some of these have been published by the authors of this paper. However, we claim the pattern of congruence among results to be of interest for several reasons. Shoshani's (1986) cladogram shows agreement at several nodes with the reference cladogram, even though Shoshani's analysis was independently derived from a different data set.

Agreement at several nodes also holds when morphological data sets are culled *a posteriori* in order to eliminate character redundancy. Seventeen characters that were incorporated in Novacek's (1986, fig. 35) and Novacek and Wyss' (1986, fig. 1) analyses were deleted from the data matrix used in the latter study. The consensus tree derived from the reduced data set in Novacek and Wyss (1986) retains a topology fairly conformable with the tree exclusively based on the cranial data set (Fig. 3.5), which includes the 17 as well as 87 other traits. The discrepancies arise primarily because the consensus tree does not resolve many branches specified in the cranial-data tree; the latter is one of the 'most-parsimonious' solutions on a PHYSYS run, but it has a very low CI score (0.32). The major Case 1 contradiction is the joining in the consensus tree of the edentate–pholidotan clade to the ungulate (*sans* Tubulidentata) clade, a discrepancy that betrays the scanty evidence for the edentate–epithere dichotomy. However, many of the same groups (e.g. Glires, edentates+pholidotans, Archonta, ungulates, paenungulates) are recognized in both trees.

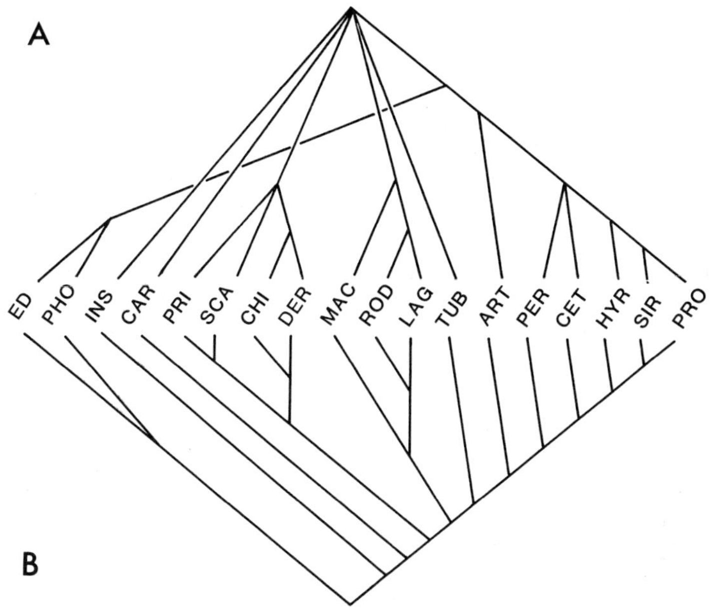

Fig. 3.5. Topology comparison of a CONTREE output (A, above) of data sampled in Novacek and Wyss (1986) with a PHYSYS-generated tree (B, below) for data in Novacek (1986, fig. 35). Cranial traits originally used in both studies were removed *a priori* from the CONTREE input in order to eliminate redundancy. For acronyms of groups, see Figs. 3.1 and 3.2.

It is also noteworthy that the consensus tree in Fig. 3.5 shows exact correspondance—with the exception of the edentate–pholidotan and ungulate grouping—to a CONTREE run of the original data set in Novacek and Wyss (1986). In the CONTREE output (not illustrated), edentates and pholidotans are merely indicated as another lineage of an unresolved polytomy including carnivorans, tubulidentates, insectivorans, archontans, macroscelideans+Glires, and ungulates. This third variation of the placement of the edentate–pholidotan group again confutes the edentate–epitherian dichotomy. Otherwise, the overall symmetry of the trees is apparent, as is also indicated by the agreement between the CONTREE reduced data set (Fig. 3.5, Table 3.3) and the original 'manually'-derived tree of the whole data set (Novacek and Wyss 1986; Fig. 3.4A herein).

This is not to deny the discrepancies among morphological results. McKenna's (1975) earlier study (Fig. 2) offers several contradictions to the reference cladogram (Fig. 3.4A), primarily because he did not attempt to resolve several higher-level sequences of epitherian clades. More marked is the lack of resolution in Simpson's (1945) classification, and the promotion in that study of certain higher clades

(Ferungulata, Unguiculata) that, as noted above, have found little acceptance. Remarkably, Gregory's (1910) results are in rather close agreement with more recent analyses.

There is also a pattern of congruence with respect to groupings. Most popular are the tethythere clade comprising sirenians and proboscideans, the paenungulate clade including hyracoids and tethytheres (but note Fischer 1986), the Glires (lagomorphs+rodents), and the Dermoptera+Chiroptera. Less popular are the Archonta, Epitheria, Pholidota+Edentata, a monophyletic Ungulata, and the macroscelidean link with the Glires. Some of the scepticism is reflected in the revised analysis (Fig. 3.3) of this paper.

Significantly, this general pattern of consistency is in stark contrast to molecular results. The morphologically-based reference cladogram in Fig. 3.4A has many nodes for which molecular studies either provide no information (because the relevant taxon was not sampled) or produce contradictory results (Table 3.3). There are, indeed, few instances of agreement. Paenungulata *sans* Tubulidentata is supported by studies of individual proteins (Wyss *et al.* 1987), but not by 'tandem alignment' studies (e.g. Miyamoto and Goodman 1986) where Tubulidentata are included in this grouping. This relatively minor difference is also difficult to dispense with in morphological studies (as noted above, tubulidentates do share a peculiar fetal trait with the paenungulates, but this relationship is not very convincing from other morphological evidence; see Thewissen 1985). A robust attitude would allow us to claim, then, some support for Paenungulata from molecular studies. Likewise, we would recognize molecular support (Table 3.3) for the Edentata–Epitheria dichotomy (Miyamoto and Goodman 1986) and the ungulate grouping (Wyss *et al.* 1987, myoglobin analysis).

How should such variation in results be judged? It is fallacious to criticize molecular studies simply because they fail to agree with an arbitrarily chosen reference cladogram based on morphological traits. There is, however, a fundamental difference between the morphological and molecular results considered here. Despite the lack of morphological consensus emphasized in some molecular studies (Goodman *et al.* 1985; Miyamoto and Goodman 1986), that problem seems more pertinent to molecular work itself. Apart from the abundance of data gaps, there is very little agreement among different molecular results (Table 3.4). To illustrate this point we use Miyamoto and Goodman (1986, fig. 3) as a molecular reference cladogram (Fig. 3.4B here), since this cladogram summarizes the most sequence information in the most mammalian taxa to date. From Table 3.4 it is evident that few nodes of this cladogram are well-supported by other analyses—the exceptions are (a) Paenungulata; (b) Cetacea, Perissodactyla, Artiodactyla; Table

Table 3.4 Comparisons of topologies in molecular-based trees with Fig. 3.4B (see text for explanation)

	Nodes in Fig. 3.4B										CI
Studies	a	b	c	d	e	f	g	h	i	j	?
Wyss *et al.* 1987 (β-haemoglobin)[1,2,3]	–	NA	+	–	+	–	–	–	+	NA	0.60
Wyss *et al.* 1987 (α-haemoglobin)[4,5]	–	NA	–	–	–	–	–	–	–	NA	0.49
Wyss *et al.* 1987 (Myoglobin)[6]	NA	NA	–	–	–	–	–	–	–	NA	0.59
Wyss *et al.* 1987 (A-lens)[7,8]	–	+	–	–	–	–	–	–	–	NA	0.63
McKenna 1987 (Myoglobin)[2]	NA	NA	+	–	+	–	–	–	–	NA	0.50
McKenna 1987 (A-lens)[9,10]	–	–	+	+	–	–	–	–	–	–	0.70
Shoshani 1986 (fig. 2)	–	+	–	+	–	–	–	–	–	+	NA
Goodman *et al.* 1985 (fig. 3)[2,11–14]	–	–	+	+	+	–	–	–	–	NA	?
de Jong 1982 (A-lens)[15–16]	–	+	+	+	–	–	–	–	–	+	0.72
Goodman *et al.* 1982 (fig. 5)[17–19]	NA	NA	–	+	–	–	–	–	–	NA	?

[1]ED, PRO, and all other eutherians in trichotomy. [2]Some relevant taxa not analysed. [3]PHO not analysed. [4]ED in unresolved polytomy (includes marsupials). [5]CET with [LAG+ROD]. [6][PRO +CET] grouped with ART and PER.[7]ED, MAC, and all other Eutheria in trichotomy. [8]CET, [ART+PER], and [CAR+PHO] in trichotomy. [9]ED, MAC, PRO, and all other Eurtheria in unresolved tetrachotomy. [10]CAR, PHO, and [ART+PER+CET] in unresolved trichotomy. [11]ED sister-group of PRO. [12]But monophyly of ART not recognized! [13]PRI+LAG+ROD also includes SCA. [14]INS sister-group of CHI. [15]See notes 16, 21 and 22 in footnote to Table 3.2. [16]Ambiguity in CAR+PHO grouping, see Wyss *et al.* (1987). [17]PRO outgroup to all other Eutheria but many relevant taxa (including ED) missing. [18]INS and CHI outgroups to all other Eutheria except PRO. [19]ROD excluded from [LAG+SCA+PRI]. For acronyms of the above-mentioned groups, see Figs. 3.1 and 3.2.

3.4 also shows the more remote divergence of edentates and group (a) than the rest of the Eutheria (a result not at all supported by CONTREE analysis of individual proteins, see Wyss *et al.* 1987). Such a marked discrepancy could be mitigated if Case 2 problems (y is neither a member of A nor B) were merely counted as ambiguous. This would, however, notably increase the concordance of the morphological results as well, to the effect that the same contrast between morphological and molecular studies would hold.

In the light of these comparisons, it seems overly optimistic to claim that a highly resolved molecular cladogram provides 'considerable hierarchical structure to evaluate the eutherian bush with its many unresolved problems' (Miyamoto and Goodman 1986, p. 235). More pertinent here is the nature of the evidence supporting this cladogram. Many unresolved problems reside in molecular results as well, if investigators choose to recognize them. In this respect, we find that none of the ten nodes recognized by Miyamoto and Goodman (1986, fig. 3) for major eutherian clades is consistent with more than one of the four protein sequences reanalysed by Wyss *et al.* (1987). Only five nodes are consistent with the results of one of the four sequences. In the case of two nodes, three of the four molecules provide no data at all, and five nodes are inconsistent with the results derived from all four protein sequences (Table 3.4). It is acknowledged that Miyamoto and Goodman (1986) considered the sequences of three additional molecules (cytochrome, fibrinopeptide, and ribonuclease), but the lack of consistency revealed by our analyses of the better sampled molecules is, to say the least, disturbing.

This is not to say that such data are meaningless; in certain cases they have elucidated a basis for groupings and they offer great potential for futher studies in systematics. Moreover, we do not criticize molecular studies because the results derived from one molecule may be discrepant with those derived from another. Such a criticism would disallow study of any kind of data. Although some congruency between cladograms developed from different morphological data sets can be demonstrated (Fig. 3.5, Table 3.3), it would be absurd to claim this approach to be free of contradiction and debate. Morphologists do, however, readily dispute a particular issue by reviewing competing assessments of particular characters. In contrast, we find that molecular studies of higher mammalian systematics are, as a rule, insufficiently frank in their disclosure of shortcomings. We therefore recommend that molecular analysts strive to: present more apomorphy lists (e.g. de Jong 1982) and character data matrices based on explicit procedures for ordering polarity; show exactly what proportion of the input data matrix is missing data (a very useful feature of the recent study by Miyamoto and Goodman 1986); provide a consistency index or some related measure of homoplasy in a tree (e.g. Romero-Herrera *et al.* 1978); disclose the number (or the program's arbitrary maximum number) of most-parsimonious trees for a given data set; make explicit the basic discrepancies between results derived from different proteins; forgo the recruitment of hypothetical gene events to explain away 'peculiar' results (see also the comments of Fitch 1979); and regard molecular studies as another perspective but not a demonstrable cure-all for current systematic problems.

Conclusions

In this study, we have addressed certain claims about the importance, to the higher-level systematics of mammals, of one kind of data over another. We have compared different analyses for their variance in methods, data, and results, but this is not an advocacy paper for a particular study or a particular data set. Hence, we return to the basic issue, namely the identity of the stronger, higher-level hypotheses among the eutherian orders. Several morphological studies coincide in their support for the superordinal clades Tethytheria (Sirenia, Proboscidea, and Desmostylia), Paenungulata (Tethytheria and Hyracoidea—but see Fischer 1986), Glires (Lagomorpha and Rodentia), and a group comprising the Dermoptera and the Chiroptera. We emphasize that this is not merely a question of popularity; the character evidence for these clades is, to us, compelling. Two of us (MJN, APW) also advocate a grouping of the Pholidota and the Edentata, and their connection with the Early Tertiary Palaeanodonta.

Of these groups, only the Tethytheria and the Paenungulata are supported by molecular results (Penny and Hendy's 1985 analysis of sequence data does support, however, a traditional Glires concept, not the 'Glires' of Miyamoto and Goodman 1986, which also includes the Primates). At present, we would judge such a lack of corroboration as ambiguous; relevant molecular evidence is either not available or is markedly inconsistent.

We make a less emphatic claim, based on morphological evidence, for the unity of the Archonta, the association of macroscelideans with the Glires, and the distinction of the Epitheria (Eutheria–[Edentata+Pholidota]). Surely, the least secure is any endorsement for the Ungulata. The vast range of morphology and taxic diversity represented by ostensible members of this clade, both fossil and Recent, is a body of data barely sampled (but see Chapters 8–11 of this volume). Decisive tests of ungulate monophyly, for that reason, seem somewhat remote. Molecular results are again inconclusive in most of these cases. We do, however, note that only one molecular review firmly supports an epithere–edentate (*sans* pholidotan) dichotomy (Miyamoto and Goodman 1986). By contrast, there is some coincidence in molecular support for the grouping of cetaceans, artiodactyls, and perissodactyls as a clade apart from the paenungulates (Fig 3.4B, Tables 3.3 and 3.4). This, one of the few results that approaches consistency in molecular studies, further weakens our confidence in ungulate monophyly.

We have also considered the affinities of several major extinct clades. A case for the close relationship between the Asian anagalids with macroscelideans (McKenna 1975; Novacek 1986) may be overstated; anagalids are perhaps better allied somehow with the Glires (see also Novacek 1985, table 2). The suite of primitive insectivores, often dubbed the 'proteutherians', elude clear placement within 'true' Insectivora (lipotyphlans) or other groups. However, the archaic leptictids do show cranial traits that suggest lipotyphlan affinities (Novacek

1986). The creodonts and carnivorans remain unified, but only on the basis of a few similarities in intracranial and dental morphology. We follow Wible and Covert (1987) in regarding the Plesiadapiformes (including microsyopids) as non-monophyletic. Firm support for the monophyly of the extinct South American ungulates (McKenna's 1975 Meridiungulata) or the various Early Tertiary groups, lumped as condylarths, has yet to be demonstrated; not a surprising conclusion given the problems with the Ungulata noted above. Finally, recent work confirms the association of the Desmostylia with the Tethytheria (McKenna 1975), and further argues for a special relationship between desmostylians and proboscideans (Domning *et al.* 1986). A number of Recent groups remain (Tubulidentata, Carnivora, and Insectivora), for which we have no hypotheses of relationship presently worth promoting.

For the present, an elegant, completely resolved pattern of relationships of the eutherian orders seems a remote solution. Nonetheless, a search for a complete dichotomous tree should not be blind-sided; one ought to weigh continuously the strength of the hypotheses that one promotes. In this spirit, we have re-examined several of the more persuasive cases for higher eutherian relationships. That some of these remain persuasive may, after all, be some indication of progress.

Acknowledgements

We thank Michael Benton for his kind invitation to participate in this symposium; J. J. Flynn and T. S. Kemp for their review and criticism of an earlier draft; D. Prothero, M. Fischer, W. Maier, H. Shoshani, and U. Zeller for stimulating discussion pertinent to items addressed in this paper; and our friends and colleagues at the British Museum of Natural History for their warmth and hospitality during our visit to London.

References

Bleefeld, A. R. and McKenna, M. C. (1985). Skeletal integrity of *Mimolagus rodens* (Lagomorpha, Mammalia). *Am. Mus. Novit.* **2806**, 1–5.

Butler, P. M. (1956). The skull of *Ictops* and the classification of the Insectivora. *Proc. Zool. Soc. London* **126**, 453–81.

Butler, P. M. (1972). The problem of insectivore classification. In *Studies in Vertebrate Evolution* (eds. A. Joysey and T. S. Kemp), pp. 253–65. Winchester Press, New York.

Cartmill, M. and MacPhee, R. D. E. (1980). Tupaiid affinities: the evidence of the carotid arteries and cranial skeleton. In *Comparative Biology and*

Evolutionary Relationships of Tree Shrews (ed. W. P. Luckett), pp. 95–132. Plenum Press, New York.

de Beer, G. R. (1937). *The Development of the Vertebrate Skull.* Clarendon Press, Oxford.

de Jong, W. (1982). Eye lens proteins and vertebrate phylogeny. In *Macromolecular Sequences in Systematic and Evolutionary Biology* (ed. M. Goodman), pp. 75–114. Plenum Press, New York.

de Jong, W., Sweers, A. and Goodman, M. (1981). Relationship of aardvarks to elephants, hyraxes, and sea cows from alpha-crystallin sequences. *Nature (London)* **292**, 538–40.

Domning, D. P., Ray, C. E. and McKenna M. C. (1986). Two new Oligocene desmostylians and a discussion of tethytherian systematics. *Smithson. Contrib. Paleobiol.* **59**, 1–56.

Emry, R. J. (1970). A North American Oligocene pangolin and other additions to the Pholidota. *Bull. Am. Mus. Nat. Hist.* **142**, 457–510.

Engelmann, G. F. (1977). The Logic of Phylogenetic Analysis and the Phylogeny of Xenarthra (Mammalia), Ph. D. Thesis. Columbia University, New York.

Evans, F. G. (1942). The osteology and relationships of the elephant shrews (Macroscelididae). *Bull. Am. Mus. Nat. Hist.* **80**, 85–125.

Fernandez, M. (1915). Die Entwicklung der Mulita. *Rev. Mus. La Plata* **21**, 1–519.

Fischer, M. S. (1986). Die Stellung der Schliefer (Hyracoidea) im phylogenetischen System der Eutheria. *Cour. Forschinst. Senckenberg* **84**, 1–129.

Fitch, W. M. (1979). Cautionary remarks on using gene expression events in parsimony procedures. *Syst. Zool.* **28**, 375–9.

Fleischer, G. V. (1973). Studien am Skelett des Gehörorgans der Säugetiere, einschliesslich des Menschen. *Säugetierkundl. Mitt.* **21**, 131–239.

Flynn, J. J. and Galiano, J. (1982). Phylogeny of Early Tertiary Carnivora, with a description of a new species of *Protictis* from the Middle Eocene of northwestern Wyoming. *Am. Mus. Novit.* **2725**, 1–64.

Gingerich, P. D. (1985). South American mammals in the Paleocene of North America. In *The Great American Biotic Interchange* (eds. F. G. Stehli and S. D. Webb), pp. 123–35. Plenum Press, New York.

Goodman, M., Weiss, M. L. and Czelusniak, J. (1982). Molecular evolution above the species level: branching pattern, rates, and mechanisms. *Syst. Zool.* **31**, 376–99.

Goodman, M., Czelusniak, J. and Beeber, J. E. (1985). Phylogeny of primates and other eutherian orders: a cladistic analysis using amino acid and nucleotide sequence data. *Cladistics* **1**, 171–85.

Grassé, P. E. (ed.) (1955). *Traité de Zoologie*, Vol. 17. Masson et Cie., Paris.

Gregory, W. K. (1910). The orders of mammals. *Bull. Am. Mus. Nat. Hist.* **27**, 1–524.

Hartenberger, J.-L. (1985). The order Rodentia: major questions on their evolutionary origin, relationships and superfamilial systematics. In *Evolutionary Relationships among Rodents* (eds. W. P. Luckett and J.-L. Hartenberger), pp. 1–34. Plenum Press, New York.

Hendy, M. D. and Penny, D. (1982). Branch and bound algorithms to determine minimal evolutionary trees. *Math. Biosci.* **59**, 277–90.

Janis, C. (1983). Muscles in the masticatory apparatus in two genera of hyraxes (*Procavia* and *Heterohyrax*). *J. Morphol.* **176**, 61–87.

Jollie, M. (1968). The head skeleton of a newborn *Manis javanica* with comments on the ontogeny and phylogeny of the mammal head skeleton. *Acta Zool.* **49**, 227–305.

Kielan-Jaworowska, Z., Bown, T. M. and Lillegraven, J. A. (1979). Eutheria. In *Mesozoic Mammals, the First Two-thirds of Mammalian History* (eds. J. A. Lillegraven, Z. Kielan-Jaworowska, and W. A. Clemens), pp. 221–59. University of California Press, Berkeley.

Kirsch, J. A. W., Johnson, J. I. and Switzer, R. C. (1983). Phylogeny through brain traits: the mammalian family tree. *Brain Behav. Evol.* **22**, 70–74.

Kluge, A. G. and Farris, J. S. (1969). Quantitative phyletics and the evolution of anurans. *Syst. Zool.* **18**, 1–32.

Leche, W. (1886). Ueber die Säugetiergattung *Galeopithecus.* Eine morphologische Untersuchung. *K. Svenska Wetenskap. Akad.* **21**, 1–92.

Li, C.-K. and Ting, S.-Y. (1985). Possible phylogenetic relationships: eurymylid–rodent and mimotonid–lagomorph. In *Evolutionary Relationships among Rodents* (eds. W. P. Luckett and J.-L. Hartenberger), pp. 35–58. Plenum Press, New York.

Lucas, S. G. (1986). Pyrothere systematics and a Caribbean route for land mammal dispersal during the Paleocene. *Rev. Geol. Am. Central* **5**, 1–35.

Luckett, W. P. (1977). Ontogeny of amniote fetal membranes and their application to phylogeny. In *Major Patterns in Vertebrate Evolution* (eds. M. K. Hecht, P. C. Goody, and B. M. Hecht), pp. 439–516. Plenum Press, New York.

Luckett, W. P. (1980). The use of reproductive and developmental features in assessing tupaiid affinities. In *Comparative Biology and Evolutionary Relationships of Tree Shrews* (ed. W. P. Luckett), pp. 245–66. Plenum Press, New York.

Luckett, W. P. (1985). Superordinal and intraordinal affinities of rodents: developmental evidence from dentition and placentation. In *Evolutionary Relationships among Rodents* (eds. W. P. Luckett and J.-L. Hartenberger), pp. 227–76. Plenum Press, New York.

Maddison, W. P., Donoghue, M. J. and Maddison, D. R. (1984). Outgroup analysis and parsimony. *Syst. Zool.* **33**, 83–103.

Matthew, W. D. (1918). A revision of the lower Eocene Wasatch and Wind River faunas. Part 5. Insectivora (continued), Glires, Edentata. *Bull. Am. Mus. Nat. Hist.* **38**, 565–657.

Matthew, W. D. (1937). Paleocene faunas of the San Juan Basin, New Mexico. *Trans. Am. Philos. Soc.* **30**, 1–510.

McCrady, E., Jr. (1938). The embryology of the opossum. *Am. Anat. Mem.* **16**, 1–233.

McDowell, S. B. (1958). The Greater Antillean insectivores. *Bull. Am. Mus. Nat. Hist.* **115**, 113–214.

McKenna, M. C. (1960). Fossil Mammalia of the early Wasatchian Four Mile fauna, Eocene of northwest Colorado. *Univ. Calif. Publ. Geol. Sci.* **37**, 1–130.

McKenna, M. C. (1963). New evidence against the tupaioid affinities of the mammalian family Anagalidae. *Am. Mus. Novit.* **2158**, 1–16.

McKenna, M. C. (1969). The origin and early differentiation of therian mammals. *Ann. N. Y. Acad. Sci.* **167**, 217–40.

McKenna, M. C. (1975). Toward a phylogenetic classification of the Mammalia. In *Phylogeny of the Primates: a Multidisciplinary Approach* (eds. W. P. Luckett and F. S. Szalay), pp. 21–46. Plenum Press, New York.

McKenna, M. C. (1987). Molecular and morphological analysis of higher-level mammalian relationships. In *Molecules and Morphology* (ed. C. Patterson), pp. 55–93. Cambridge University Press, Cambridge.

Miyamoto, M. M. and Goodman, M. (1986). Biomolecular systematics of eutherian mammals: phylogenetic patterns and classification. *Syst. Zool.* **35**, 230–40.

Moore, G. W. (1975). Proof of the maximum parsimony ('Red King') algorithm. In *Molecular Anthropology, Genes and Proteins in the Evolutionary Ascent of the Primates* (eds. M. Goodman, R. E. Tashian, and J. H. Tashian), pp. 117–37. Plenum, New York.

Mossmann, H. W. (1957). The foetal membranes of the aardvark. *Mitt. Naturforsch. Ges. Bern* **14**, 119–27.

Novacek, M. J. (1980). Cranioskeletal features in tupaiids and selected eutherians as phylogenetic evidence. In *Comparative Biology and Evolutionary Relationships of Tree Shrews* (ed. W. P. Luckett), pp. 35–93. Plenum Press, New York.

Novacek, M. J. (1982). Information for molecular studies from anatomical and fossil evidence on higher eutherian phylogeny. In *Macromolecular Sequences in Systematic and Evolutionary Biology* (ed. M. Goodman), pp. 3–41. Plenum Press, New York.

Novacek, M. J. (1985). Cranial evidence for rodent affinities. In *Evolutionary relationships among Rodents* (eds. W. P. Luckett and J.-L. Hartenberger), pp. 59–81. Plenum Press, New York.

Novacek, M. J. (1986). The skull of leptictid insectivorans and the higher-level classification of eutherian mammals. *Bull. Am. Mus. Nat. Hist.* **183**, 1–111.

Novacek, M. J. and Wyss, A. R. (1986). Higher-level relationships of the Recent eutherian orders: morphological evidence. *Cladistics* **2**, 257–87.

Novacek, M. J., McKenna, M. C., Neff, N. A. and Cifelli, R. L. (1983). Evidence from earliest known erinaceomorph basicranium that insectivorans and primates are not closely related. *Nature (London)* **306**, 683–4.

Patterson, B. (1965). The fossil elephant shrews (family Macroscelididae). *Bull. Mus. Comp. Zool.* **133**, 295–385.

Penny, D. and Hendy, M. D. (1985). The use of tree comparison metrics. *Syst. Zool.* **34**, 75–82.

Pettigrew, J. D. (1986). Flying primates? Megabats have advanced pathway from eye to midbrain. *Science, N.Y.* **231**, 1304–6.

Radinsky, L. B. (1966). The adaptive radiation of the phenacodontid condylarths and the origin of the Perissodactyla. *Evolution* **20**, 408–17.

Romero-Herrera, A. E., Lehman, H., Joysey, K. A. and Friday, A. E. (1978). On the evolution of myoglobin. *Phil. Trans. R. Soc. London* **283**, 61–163.

Rose, K. D. (1978). A new Paleocene epoicotheriid (Mammalia), with comments on the Palaeanodonta. *J. Paleontol.* **52**, 658–74.

Sarich, V. M. (1985). Rodent macromolecular systematics. In *Evolutionary Relationships among Rodents* (eds. W. P. Luckett and J.-L. Hartenberger), pp. 423–52. Plenum Press, New York.

Sarich, V. M. and Cronin, J. E. (1976). Molecular systematics of the primates. In *Molecular Anthropology* (eds. M. Goodman and R. E. Tashian), pp. 141–70. Plenum Press, New York.

Shoshani, J. (1986). Mammalian phylogeny: comparison of morphological and molecular results. *Mol. Biol. Evol.* **3**, 222–42.

Simpson, G. G. (1945). The principles of classification and a classification of mammals. *Bull. Am. Mus. Nat. Hist.* **85**, 1–350.

Simpson, G. G. (1978). Early mammals in South America: fact, controversy, and mystery. *Proc. Am. Philos. Soc.* **122**, 318–28.

Smith, J. D. and Madkour, G. (1980). Penial morphology and the question of chiropteran phylogeny. In *Proceedings of the Fifth International Bat Research Conference* (eds. D. E. Wilson and A. L. Gardiner), pp. 347–65. Texas Tech Press, Lubbock, Texas.

Swofford, D. L. (1985). *Phylogenetic analysis using Parsimony (PAUP) Version 2.4*, Program manual. Illinois Natural History Survey, Champaign.

Szalay, F. S. (1975). The origin of primate higher categories; an assessment of basicranial evidence. In *Phylogeny of the Primates: a Multidisciplinary Approach* (eds. W. P. Luckett and F. S. Szalay), pp. 91–123. Plenum Press, New York.

Szalay, F. S. (1977). Phylogenetic relationships and a classification of the eutherian Mammalia. In *Major Patterns in Vertebrate Evolution* (eds. M. K. Hecht, P. C. Goody, and B. M. Hecht), pp. 315–74. Plenum Press, New York.

Szalay, F. S. (1985). Rodent and lagomorph morphotype adaptations, origins, and relationships: some postcranial attributes analysed. In *Evolutionary Relationships among Rodents* (eds. W. P. Luckett and J.-L. Hartenberger), pp. 83–132. Plenum Press, New York.

Tassy, P. (1981). Le crâne de *Moeritherium* (Proboscidea, Mammalia) de l'Éocène de Dor el Talha (Libye) et le problème de la classification phylogénétique du genre dans les Tethytheria McKenna, 1975. *Bull. Mus. Natl. Hist. Nat., Ser.* 4 **C3**, 87–147.

Thewissen, J. G. M. (1985). Cephalic evidence for the affinities of Tubulidentata. *Mammalia* **49**, 257–84.

Van der Horst, C. J. (1942). Some observations on the structure of the genital tract of *Elephantulus*. *J. Morphol.* **70**, 403–29.

Van Valen, L. (1966). Deltatheridia, a new order of mammals. *Bull. Am. Mus. Nat. Hist.* **132**, 1–126.

Van Valen, L. (1967). New Paleocene insectivores and insectivore classification. *Bull. Am. Mus. Nat. Hist.* **135**, 217–84.

Wible, J. R. (1984). *The Ontogeny and Phylogeny of the Mammalian Cranial Arterial Pattern*, Ph. D. Dissertation. Duke University, New York.

Wible, J. R. (1987). The eutherian stapedial artery: character analysis and implications for superordinal relationships. *Zool. J. Linn. Soc.* **91**, 107–35.

Wible, J. R. and Covert, H. H. (1987). Primates: cladistic diagnosis and relationships. *J. Human Evol.* **16**, 1–22.

Wible, J. R. and Novacek, M. J. (in press). Cranial evidence for the monophyletic origin of bats. *Am. Mus. Novit.*

Windle, B. C. A. and Parsons, F. G. (1899). On the myology of the Edentata. *Proc. Zool. Soc. London* **1899**, 314–88, 990–1017.

Winge, H. (1917). Udsigt over Insektaedernes indbyrdes Slaegtskab. *Saertryk Vidensk, Meddel. Dan. Naturhist. Foren.* **68**, 83–203.

Wislocki, G. B. (1928). Observations on the gross and microscopic anatomy of the sloths (*Bradypus griseus griseus* Gray and *Choloepus hoffmanni* Peters). *J. Morphol. Physiol.* **46**, 317–97.

Wood, A. E. (1957). What if anything, is a rabbit? *Evolution* **11**, 417–25.

Wyss, A. R. (1983). Systematics and the phylogeny of Dermoptera (Eutheria). MS Thesis, San Diego State University, San Diego.

Wyss, A. R., Novacek, M. J. and McKenna, M. C. (1987). Amino acid sequence versus morphological data and the interordinal relationships of mammals. *Mol. Biol. Evol.* **4**, 99–116.

4. Phylogeny of the Carnivora

JOHN J. FLYNN
*Department of Geological Sciences, Rutgers University, New Brunswick, New Jersey, USA**

NANCY A. NEFF
Department of Ecology and Evolutionary Biology, University of Connecticut, Storrs, Connecticut, USA

RICHARD H. TEDFORD
Department of Vertebrate Paleontology, American Museum of Natural History, New York, USA

Abstract

In this study we survey the distribution of character conditions among carnivorans, and possible sister taxa, and a range of alternative phylogenetic hypotheses. We then identify those hypotheses that are most strongly supported, as well as areas of ambiguity or conflict in carnivoran phylogeny.

The Order Carnivora is a monophyletic group, which appears to be most closely related to the Creodonta (forming the Ferae). Within the Carnivora, two monophyletic groups are clearly recognizable—the Feliformia (including the living families Viverridae, Herpestidae, Hyaenidae, and Felidae) and the Caniformia (including the living families Canidae, Otariidae, Phocidae, Odobenidae, Ursidae, Procyonidae, and Mustelidae). Within the Caniformia, the Canidae (infraorder Cynoidea) are a sister-group to the remaining caniforms, the infraorder Arctoidea. Simpson's Fissipeda is paraphyletic; although controversial, the Pinnipedia may be a monophyletic group

*Present address, Department of Geology, Field Museum of Natural History, Roosevelt Road at Lake Shore Drive, Chicago, Illinois 60605, USA.

The Phylogeny and Classification of the Tetrapods, Volume 2: Mammals (ed. M. J. Benton), Systematics Association Special Volume No. 35B, pp. 73–116. Clarendon Press, Oxford, 1988.

within the Arctoidea. The Caniformia–Feliformia dichotomy may be recognizable in the members of the paraphyletic Early Cenozoic 'Miacoidea'. The phylogenetic relationships of the Nimravidae are equivocal, but they do not appear to be members of the Feliformia.

Introduction

The current diversity among proposed hypotheses for the phylogeny of the Order Carnivora reflects two areas of controversy. First, researchers give different relative weight to competing characters or systems (e.g. biochemical vs. morphological, or basicranial vs. dental), thus generating different phylogenies. Secondly, differences in the systematic treatment of the fossil taxa, especially those taxa that are largely primitive with respect to characters defining modern families or subfamilies, results in differences in the phylogenies incorporating those extinct taxa. Ultimately, many of the disputes reflect disagreement at the level of character definition and analysis.

Because of the present diversity of phylogenetic hypotheses for this order, we have concentrated on surveying the range of hypotheses and the variety of character interpretations rather than arguing for any hypothesis exclusively. The major areas of agreement are thus defined, as are also the outstanding questions.

The principal method that we have employed for deducing a phylogeny from the morphological evidence is cladistic analysis (see Eldredge and Cracraft 1980). Thus, we evaluate the definition and distribution of character conditions among carnivorans in their possible sister taxa in order to suggest character polarity. Contradictions among phylogenies are often directly traceable to differences in the interpretation of the characters themselves (rather than any argument about phylogenetic method), suggesting that one's methods of character analysis and weighting (Neff 1986) are equally or more critical.

Within the text we denote derived character states by a number and taxonomic abbreviation indicating the taxonomic level at which, or within which, the character is derived. Taxonomic abbreviations are: FER=Ferae, CAR=Carnivora, FEL=Feliformia, CAN=Caniformia. Our review is in four parts: first we define the order, secondly consider the relationships of the Carnivora to other taxa, thirdly consider the phylogeny of the extant carnivorans, and finally evaluate the relationships of the fossil forms.

Definition of the order

The present composition of the Order Carnivora, generally agreed upon among both palaeontologists and neontologists, is the result of the reflection of phylogenetic relationships within the classification. Many modern terrestrial carnivorans, as well as the pinnipeds, and non-carnivorans, were assigned to the order Ferae by Linneaus (1758), thus more closely approximating the modern sense of the order than in many other arrangements in the late 1700s or early 1800s. The families of extant terrestrial carnivorans were all established by either Fischer de Waldheim in 1817, or Gray in 1821 or 1825, and indeed the current name of the order, Carnivora, was proposed by Bowdich in 1821. Although the modern names thus were established rather early, there remained a couple of areas of dispute—the pinnipeds, and the fossil taxa, especially the creodonts.

The separation between the terrestrial and aquatic carnivorans was generally more sharply made than in Linneaus (1758), and they were sometimes even assigned to different orders; Simpson (1945) recognized the Fissipeda and Pinnipedia as suborders, arguing that the recognition of two such groups had persisted without much question. Since 1945, the pinnipeds, whether they are believed to be a monophyletic or a diphyletic group, are universally considered to be more closely related to the arctoid carnivorans (see Fig. 4.1) than to the aeluroids, thus rendering Simpson's Fissipeda paraphyletic under modern criteria of classification. Nevertheless, many recent summaries persist in following Simpson's allocation into suborders. The phylogenetic questions of pinniped origins will be discussed further below, under the Caniformia.

The other major rearrangement in the history of carnivoran classification has involved the treatment of the creodonts. Cope (1875) proposed the Creodonta as a suborder of the Order Insectivora, and later (1880*a*) recognized five families within this order. The measure of the subsequent rearrangements can be appreciated when one notes that two of these families are now recognized as primitive ungulates (placed in the Order Condylarthra), two are still kept within the Order Creodonta, and the fifth, the Miacidae, is a basal family within the Order Carnivora. Cope (1880*a*) himself recognized this diversity within his suborder, and suggested that various portions of the Creodonta were ancestral to the modern Insectivora, the modern Carnivora, and the extinct Hyaenodontidae (later included with the Creodonta). The morphological definition of the Order Carnivora is often given merely as those eutherians with the P^4 and M_1 specialized for shearing as a carnassial pair. The creodonts, in contrast, are typically said to have a carnassial pair formed by the M^1 over M_2 (Oxyaenidae) or M^2 over M_3 (Hyaenodontidae). We believe that it is much more significant that in creodonts the entire molar row (and often the P^4 as well) was modified for some degree of shearing, so that the identification of 'a carnassial pair' is merely a matter of degree, or it was variable through ontogeny. It is often not as clear as the simplistic characterization of the two families would suggest. In contrast, the Carnivora have *restricted* the shearing function to the P^4 over M_1 carnassial pair.

This focus of the carnassial function in true Carnivora has permitted a diversity of dental adaptations within the order. Hypercarnivorous taxa (e.g. Felidae, Mustelinae) have tended to lose the postcarnassial molars and emphasize the carnassial blades. Other intermediate forms (e.g. Canidae) have retained both the crushing function of the molars plus the shearing function of the carnassial pair. Hypocarnassial taxa (e.g. Ursidae, Procyonidae) have enlarged the postcarnassial molars, while losing, in extreme cases, most or all of the shearing function. Thus, a great diversity of dental adaptations were permitted by the restriction of the carnassial shear to just one locus in the cheek tooth row (P^4 and M_1), while other teeth exhibited contrasting morphologies. The presence of large canines so typical of carnivorans is surely a primitive trait (cf. the generalized Marsupialia and Primates) but one that few eutherian orders retain.

Carnivorans differ from creodonts and all other eutherians in the following dental features: (1) P^4–M_1 modified as the only carnassial teeth in adults [CAR 1]; (2) P^4 protocone located far anterolingual of the paracone [CAR 2]; (3) elongate, well-developed P^4 metastyle wing and blade with a deep carnassial notch [CAR 3]; (4) elongate P_4 talonid with at least two posterior accessory cusps on the posterior flank of the main cusp [CAR 4]; (5) M_1 much larger than M_{2-3}, progressive decrease in size from M_{1-3} [CAR 5].

Living carnivorans all have an auditory bulla composed only of entotympanic and ectotympanic contributions. This may be a derived feature for all carnivorans, although there is a wide distribution of entotympanic and ectotympanic bullae within the Eutheria (Novacek 1977, 1986; MacPhee 1981). Hunt (1974*a*, 1987) has indicated the presence of at least two, and possibly three, components of entotympanic ossification in living carnivorans (but see Wible 1984). A specialization of the entotympanic ossification that appears to be unique to Carnivora is the presence of a distinct rostral entotympanic element associated with the anteriormost portion of the extracranial path of the internal carotid artery [CAR 6; Hunt 1974*a*]. The presence of this element and its association with the internal carotid artery has been demonstrated for all extant families of Carnivora, and for the

Amphicyonidae. A rostral entotympanic has been identified in a nimravid (Hunt 1987), but it is not clear whether it transmitted the internal carotid artery. A rostral entotympanic is not known for the 'miacoids' because of the lack of a preserved ossified entotympanic bulla.

The bulla is completely ossified in all living taxa, except *Nandinia*. Complete ossification of the bulla [CAR 7] sometimes has been used to unite living carnivorans and their closest fossil relatives (e.g. Neff 1983) to the exclusion of the early Tertiary 'miacoid' carnivorans in which no entotympanic is known. (We place 'miacoids' in quotation marks to emphasize the controversy about their monophyly; see Flynn and Galiano 1982.) However, as Novacek (1986) points out, reduction or incomplete ossification of the entotympanic, weak attachment of an ossified entotympanic to the basicranium, or breakage may all result in lack of preservation of the ossified bulla in fossils. An illustration of this point is the lack of an ectotympanic in any specimen of a 'miacoid', although it is extremely unlikely that these animals lacked an ectotympanic in life. In skulls of some early Tertiary carnivorans (e.g. *Viverravus*) there is a rugosity at the basioccipital–petrosal junction, which may have been a point of attachment of an ossified entotympanic bulla.

Circulatory patterns have been used extensively in reconstructing carnivoran and eutherian phylogeny. Matthew's (1909) eutherian basicranial morphotype relied heavily on inferred circulation patterns in early Tertiary carnivorans. He proposed a primitive eutherian morphotype in which there were two main branches of the internal carotid, a medial internal carotid branch and a lateral branch bifurcating into a promontory artery and a stapedial artery. Matthew's morphotype was widely accepted for many years, even though no living eutherian had been shown to possess all three branches. Subsequent work has shown that eutherians primitively have only two branches of the internal carotid, a stapedial artery and a single internal carotid that may vary in position from lateral to medial (with respect to the middle ear cavity; Presley 1979).

All living carnivorans are derived in the lack of a stapedial artery [CAR 8]. Although a stapedial artery has been hypothesized to be present in some early Tertiary 'miacoids' (Matthew 1909) based on the presence of grooves on the promontorium, these grooves are not present in many specimens and are only weakly developed in others. There is no strong evidence for a stapedial artery in the early Tertiary carnivorans. Therefore, although it is likely that carnivorans lacked a stapedial artery (or possessed only a vestigial remnant) ancestrally, it is difficult to determine this with certainty for most early Tertiary taxa. In creodonts, the relatively deep groove crossing the promontorium towards the fenestra ovale may have carried a stapedial artery.

The course of the internal carotid artery has also been considered very important in carnivoran and eutherian phylogeny, yet the polarity of this character remains quite controversial. Recent workers have recognized that eutherians have only a single branch of this vessel; Presley (1979) and Archibald (1977) suggested that a course medial to the promontorium was primitive, as in monotremes, marsupials, and edentates (except *Myrmecophaga* and *Tamandua*). However, Wible (1986) presents an alternative argument that the more lateral, transpromontorial course, corresponding to the ontogenetically early 'indifferent position', is primitive for Mammalia, and Eutheria; the medial position of the internal carotid artery may have been acquired independently by Monotremata, Marsupialia, and several times within the Eutheria.

The presence of shallow grooves on the promontorium has been used to suggest a transpromontorial course of the internal carotid in some 'miacoids' and a variety of creodonts (Matthew 1909), and in nimravids (Neff 1983). Wible (1986) also cites a transpromontorial course of the artery in some feloid carnivorans. We find that there are at least three independently varying characters to consider, however. The first is the position of the internal carotid artery relative to the middle ear cavity. A second, related character is whether or not the carotid is enclosed in a bullar canal for any part of its length, and a third feature is the position of the anterior carotid foramen. Several combinations of states of these three characters occur within the Carnivora and outgroups, complicating the assignment of a primitive state for the order.

The internal carotid artery may be placed in one of at least five different categories:

(1) entirely medial to the cavity and the bulla (Wible's 'extrabullar'; not observed in carnivorans);

(2) placed medially and (a) in a canal in the auditory bulla or (b) between the bulla and petrosal or basioccipital, without entering the tympanic cavity (Wible's 'perbullar'; condition in the cynoid and arctoid carnivorans);

(3) transpromontorial, close to or medial to the indifferent position, but anteriorly in a short canal in the rostral entotympanic, and posteriorly often in a groove or canal in the caudal entotympanic; still passing through a medially placed anterior carotid foramen (Wible's 'transpromontorial', in part; condition in feloid carnivorans);

(4) a transpromontorial path close to the indifferent position, passing through a medially placed anterior carotid foramen, but not passing through any tube in the bulla (Wible's 'transpromontorial'

in part; inferred to perhaps have been the condition in some 'miacoids'; may be primitive for Eutheria); and

(5) a transpromontorial course across the lateral aspect of the promontorium, entering the cranium through a more laterally placed foramen (the 'promontory artery' of many previous authors, Wible's 'transpromontorial' in part; inferred from grooves possibly to have been the condition in creodonts, some 'miacoids' and nimravids).

Condition 4 may be primitive for the order, but the question still remains open whether the caniform condition (condition 2) arose independently from condition 4 or passed through a state equivalent to condition 3, seen in the feloid carnivorans.

In almost all postcranial features, carnivorans are surprisingly generalized. Except for digitigrady and a concomitant reduction in the number of digits to four in the hind limbs in some forms (others remained or returned to plantigrady), the postcranial skeleton is largely primitive in morphology.

Although most references cite retractile claws as a felid, or an aeluroid feature, retractile claws probably represent a synapomorphy for the entire order [CAR 9]. Fully sheathed, retracted claws are widely distributed within the Carnivora, being present in Felidae, Viverridae s.s. (Gonyea and Ashworth 1975), *Bassariscus astutus* within the Procyonidae (personal observation), and *Martes* within the Mustelidae (retractable but not sheathed; Powell 1982). Gonyea and Ashworth (1975) demonstrated the presence of the ligaments associated with the claw retractile mechanism in all 21 of the species of carnivorans they examined, including members of the Procyonidae, Mustelidae, Canidae, Viverridae s.l., and Felidae; these 21 species included species with a symmetrical distal interphalangeal joint and permanently protracted claws. Fossil taxa broaden the distribution of this feature still further: the asymmetry of the distal articular surface and the concavity of the shaft of the medial phalanx in each digit implies retractable claws in Nimravidae, and the canid *Hesperocyon*.

Another complication is that the claw condition is a character complex rather than a single character (Neff 1982). The features that can vary independently and should therefore be evaluated as separate characters include: presence of elastic ligaments across the distal interphalangeal joint; the degree to which the distal phalanx is actually held retracted, next to the medial phalanx; the degree of development of a fleshy, furred sheath capping the retracted claw; the degree of curvature of the distal phalanx and its keratinous sheath; and the uses to which the claws are typically put and the amount of wear resulting to the sheath (e.g. extended for traction in running versus reserved for hooking prey). Failure to recognize this complexity has led to inaccurate assessment of some character states. The cheetah, *Acinonyx*, for example, is widely cited as having nonretractile claws, whereas Pocock (1916*a*) demonstrated clearly that its distal phalanges are retracted. The typical felid fleshy sheath is lacking, so that the worn claws are visible in the relaxed, retracted position on top of the digits. The wide distribution of retractile claws within the order and absence of the feature outside the Carnivora suggests that retracted claws may have been a feature which evolved early within the order and was subsequently modified or lost within the more terrestrial or cursorial carnivorans.

Relationships of the Carnivora to other taxa

Carnivorans and living eutherians

Determining the relationships of carnivorans to other living eutherians is very difficult. Most of the taxa that have been considered close relatives include fossil groups for which there are no living representatives (such as the 'palaeoryctoids', 'pantolestoids', Cretaceous 'insectivores', creodonts, mesonychids, etc.). However, the 'Ferungulata' (of Simpson 1945) grouped the 'Ferae' (including carnivorans, creodonts, etc.) with many groups of living and fossil ungulates (including Paenungulata, Perissodactyla, Artiodactyla, South American ungulates, etc.). No strong evidence has been presented to support the Ferungulata as a monophyletic group (see Chapter 3). Several recent morphological or biochemical phylogenies that attempt to define monophyletic groups within the Eutheria (McKenna 1975; Goodman *et al.* 1985; Novacek 1986; Shoshani 1986; Novacek and Wyss 1986; Chapter 3 of this volume) show very poor resolution of the relationship of carnivorans to other eutherians. Proposed relationships of the carnivorans include an unresolved tetrachotomy with the Insectivora, Archonta, and Ungulata (McKenna 1975); unresolved hexachotomy with all other eutherians, except edentates plus pangolins (Novacek and Wyss 1986); sister-group to the Insectivora plus Chiroptera (Goodman *et al.* 1985); and sister-group to the Primates, with carnivorans and primates as a sister-group to all other Archonta (Shoshani 1986). Given such widely discrepant placement of the Carnivora within these mammalian phylogenies, we prefer to consider the relationships of carnivorans to other living eutherians to be unresolved.

'Palaeoryctoids' and other primitive eutherians

Many workers have proposed a special relationship between carnivorans and various Cretaceous and Early Cenozoic 'palaeoryctoids'. *Cimolestes* generally is regarded as the closest relative, or 'ancestor', of carnivorans and other Ferae. In fact, McKenna (1975) stated: 'From various species currently placed in a single genus, *Cimolestes*, or at least from close allies, the didelphodontines, Pantodonta (including *Deltatherium*), pantolestids, apatemyoids, taeniodonts, creodonts (oxyaenoids and hyaenodontoids), and Carnivora arose at about the beginning of the Tertiary.' *Cimolestes* is obviously a paraphyletic assemblage of quite different taxa. There are many similarities between carnivorans, as well as other Ferae, and some species of *Cimolestes*, but the species of this genus are known only from incomplete dentitions, which makes determination of its precise phylogenetic relationships difficult. Cranial and postcranial material of *Cimolestes* would be critical for a more definitive analysis of the phylogenetic relationships of the species of this genus to the Ferae.

Neither *Cimolestes* nor any other 'palaeoryctoid' has any uniquely derived feature that clearly would unite them with the Carnivora or the Creodonta. Although some other Cretaceous and Early Cenozoic eutherians have some similarities to members of the Ferae, none appears to be closely related to the Ferae.

Carnivora and Creodonta

Many workers have hypothesized a close relationship between the Carnivora and Creodonta, but there are very few derived characters that support a close relationship between them. In fact, documenting the monophyly of the Creodonta is exceedingly difficult.

All carnivorans and creodonts restrict carnassial shear to a small part of the postcanine dentition (FER 1), although the particular teeth involved in shear are different. Hyaenodontids and oxyaenids have a migrating carnassial locus during individual ontogenetic development (Mellett 1977), but at any point in ontogeny, only a small part of the dentition forms the major locus of shear.

The cranial and basicranial region generally has not been considered important in uniting the two groups, but several interesting features are shared by carnivorans and creodonts (at least some hyaenodontids). Other aspects of the basicranium of creodonts are important in defining the polarity of some basicranial features within the Carnivora.

All carnivorans possess a well-developed osseus tentorium, with a strong medial projection, in the cranial cavity (FER 2). Although an ossified tentorium is present in other eutherians, it is generally small and has a relatively restricted distribution among groups. Clark and Guensburg (1982) note a massive tentorium in at least two species of *Hyaenodon*, so that creodonts and carnivorans appear to share this derived feature.

Early Tertiary carnivorans and some creodonts (e.g. hyaenodontids) have a strong processus hyoideus ('mastoid tubercle', see Novacek 1986) of the petrosal that arches towards, and closely approaches, the promontorium (FER 3). Few eutherians exhibit this morphology, and it may be a derived feature shared by carnivorans and creodonts.

Van der Klaauw (1931) and Mellett (1977) identified an ossified entotympanic bulla in several specimens of *Hyaenodon*. Although Van der Klaauw (1931) thought the bulla was completely entotympanic, it is apparent that the bulla was composed of both a caudal entotympanic and a petrosal component (Mellett 1977; and *pers. obs.*). These specimens also preserve, lateral to the bulla, a free ectotympanic (a dorsally incomplete, narrow ring), whose plane is relatively steeply inclined (dorsoventrally) and oriented anteromedially to posterolaterally. An entotympanic contribution to the bulla may be a derived feature uniting carnivorans and creodonts (FER 4), although it would not be uniquely derived in that many other eutherians also have an entotympanic contribution to the bulla.

Many specimens of creodonts show grooves on the promontorium of the petrosal whose pattern is similar to that found in early carnivorans; a groove present across the lateral surface of the promontorium may indicate a transpromontorial path of an internal carotid artery. Creodonts also preserve a groove running dorsolaterally across the promontorium (towards the fenestra ovalis) that may have carried a stapedial artery. All living carnivorans (and most, or all, fossil carnivorans) lack a stapedial artery. If the primitive eutherian course of the internal carotid artery was transpromontorial (as suggested by Wible 1986), the carnivoran/creodont pattern may not indicate a close phylogenetic relationship. However, if a more medial position is primitive for eutherians (as would be suggested by outgroup comparison with monotremes, marsupials, and edentates), the transpromontorial position of this artery in early carnivorans (and some living feloids; see Wible 1986, and below) and in creodonts may be derived for both groups (FER 5).

In the tarsus, carnivorans and creodonts appear to share a uniquely derived condition in which the medial border of the astragalar tibial trochlea is much longer than the lateral border (FER 6). In the calcanea of both taxa, the anterior plantar tubercle is relatively far posterior and the peroneal process is reduced relative to the primitive eutherian condition (Szalay 1977) (FER 7).

Phylogeny of the living Carnivora

Principal subdivisions

The modern concept of phylogeny of the living Carnivora is based on the work of Turner (1848) and Flower (1869), who emphasized structures of the basicranium, especially the auditory bulla and surrounding area, to define morphological groups. More comprehensive and detailed reviews of the osteological structure of the basicranium were produced by Van Kampen (1905) and Van der Klaaw (1931) in descriptive anatomies virtually devoid of systematic conclusions. Hunt's (1974*a*) recent reappraisal of the structure and ontogeny of the bulla in carnivorans from a comparative viewpoint yields data amenable to the methods of phylogenetic systematics. Additional important descriptive work on the auditory ossicles (Doran 1877) and a comparative study of the arterial circulation in the skull by Tandler (1899) and Bugge (1978), add further data useful in phylogeny reconstruction.

The division of the Carnivora into terrestrial (Fissipeda) and pelagic (Pinnipedia) types was retained by Turner (1848) and Flower (1869), although at different ranks. They saw the close similarity of the pinnipeds to the bear-like division of the fissipeds, a point further elaborated by Mivart (1885; our Fig. 4.1E), but they did not group the pinnipeds with the ursids in the classification as their comparisons and statements clearly allowed (Fig. 4.1A and B). Assumption of pinniped monophyly and separation from terrestrial carnivores continues to be the most widely held view, as expressed in such influential classifications as Simpson (1945; our Fig. 4.1C), although diphyly has also been accepted by many workers.

The structure of the auditory bulla provided the clearest evidence for subdivision of the fissipeds. Both Turner (1848) and Flower (1869) discerned three types of bullae:

(1) those lacking a longitudinal septum within the bulla: Arctoidea (Flower 1869), including the families Ursidae, Ailuridae, Procyonidae, and Mustelidae;

(2) those possessing a complete septal separation of the tympanic space into two chambers: Aeluroidea (Flower 1869), including the families Felidae, Viverridae, and Hyaenidae;

(3) those having a low septum in the anterior part of the tympanic cavity: Cynoidea (Flower 1869), including a single family, Canidae.

Ontogenetic studies (Hunt 1974*a*) show that the auditory bulla is composed of ectotympanic (tympanic ring) and entotympanic (ossifying from connective tissue) components and that the entotympanic is

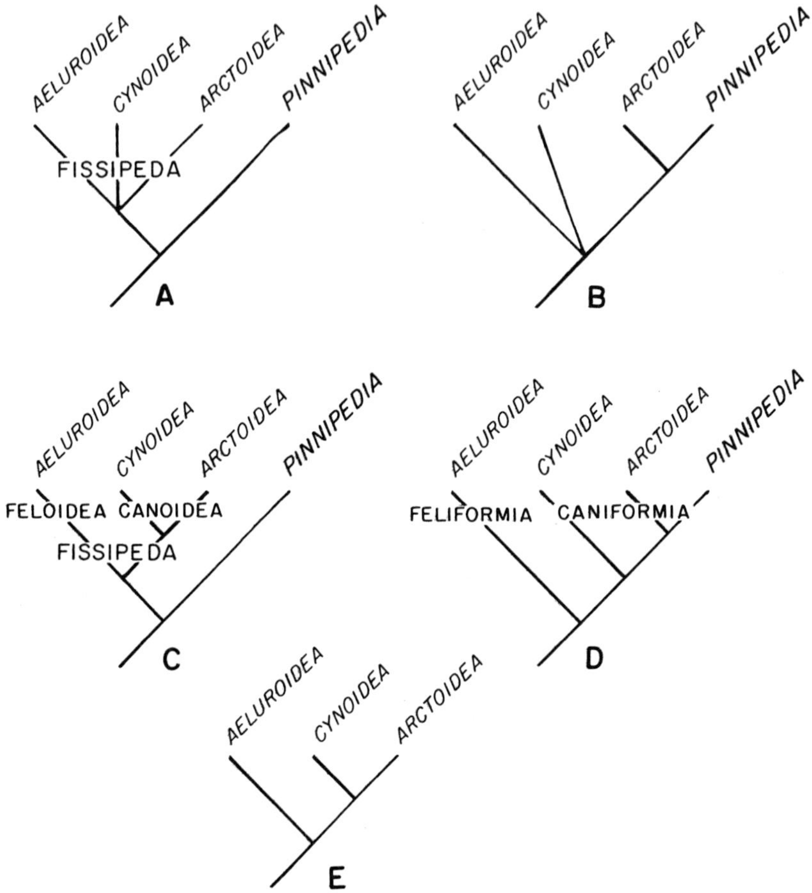

Fig. 4.1. Historical propositions for the relationships of the major carnivoran clades. A, cladogram from the classification of Flower (1869); B, phyletic relationships of pinnipeds and arctoids discussed in the text of Flower (1869); C, phyletic consequences of Simpson's (1945) classification; D, pinnipeds (families Otariidae, Odobenidae, and Phocidae) as a monophyletic group related to arctoids (Wyss 1987); E, pinnipeds diphyletic, and subsumed within Arctoidea (Mivart 1885; Tedford 1976).

composed of at least two centres of ossification: a rostral entotympanic surrounding the internal carotid artery within the anteromedial wall of the bulla, and a caudal entotympanic forming the medial and posterior part of the tympanic cavity.

The septum bulla in the Aeluroidea forms by an inflection dorsally of the edges of the ectotympanic and caudal entotympanic where they abut across the floor of the developing bulla. The inturned edges grow dorsally to form a bilaminar septum that nearly touches the promontorium, thus nearly completely dividing the hypotympanic sinus into two chambers. The bilaminar nature of the septum bullae has been observed in felids, herpestids, and viverrids, and inferred for hyaenids (Hunt 1974*a*, 1987). The ectotympanic contribution to the hyaenid septum bullae has been confirmed (Hunt 1974*a*); therefore if the septum is unilaminar, it is likely to be a specialized condition within the Aeluroidea by having reduced or lost the entotympanic contribution.

In the Canidae, in contrast, the partial septum is formed entirely by the entotympanic curving in at its anterior edge. Evidence from ontogeny suggests that the primitive condition of the bulla in the Carnivora is that of a little expanded tympanic ring nearly covering the petrosal promontorium. The entotympanic is narrow medially, and little expanded posteriorly. This condition is found little modified in adults in certain living Ursidae.

Bullar morphology clearly should be analysed as a series of discrete characters, rather than a single unit character. Although the carnivoran bullar types of Turner (1848), Flower (1869), and Hunt (1974*a*) are valid descriptive morphologic groupings, their use in phylogenetic analysis is limited because they represent variable combinations of primitive and derived character states. Components of bullar morphology on which these morphotypes are based include development of septae and details of their construction, geometry and degree of inflation of entotympanic and ectotympanic elements, variation in development of hypotympanic space and excavation of surrounding bones, bullar topology relative to the internal carotid artery, position of the tympanic cavity relative to other basicranial landmarks, and details of the lateral expansion of the ectotympanic. For example, Hunt's (1974*a*) morphotypes D (felids, viverrids) and E (hyaenids) share such derived characters as an ectotympanic (and probably bilaminar) septum, and ventral and posterior inflation of the caudal entotympanic. These characters of morphotype D have been used to unite the Aeluroidea; therefore morphotype D is primitive within the Aeluroidea and does not unite only felids and viverrids in a smaller monophyletic group. Morphotype E is in several features autapomorphous for hyaenids. Bullar morphotype B of Hunt (1974*a*) and the arctoid bulla of Turner (1848) and Flower (1869), however, are a combination of primitive (e.g. single chamber) and derived (e.g. ventral and posterior expansion of the caudal entotympanic) characters. Ventral and posterior expansion of the caudal entotympanic is found in felids, viverrids, hyaenids, canids, phocids, procyonids (except *Ailurus*), mustelids (not ventrally expanded in lutrines, mephitines, or *Potamotherium*), and possibly otariids. The caudal entotympanic is not expanded in the feliform *Nandinia*, the arctoid Ursidae, the pinniped Odobenidae (and possibly Otariidae), and the procyonid *Ailurus*. The wide distribution of the derived component of the 'type B' bulla within the Carnivora, and its absence primitively in many carnivoran clades, suggests that this condition has independently evolved as many as five times within the Carnivora. We conclude that useful phylogenetic inferences can only be drawn from detailed study of the distribution of derived components of bullar morphology, not from the general categorization of the entire bulla.

Historically, these features of the auditory bullae were combined with other characters of the basicranium (Flower 1869) and soft anatomy (Turner 1848), to distinguish the major subdivisions of the Carnivora. Outgroup comparison establishes the following derived character states as a basis for recognition of the following monophyletic groups: six synapomorphies support the Arctoidea and Cynoidea as sister-taxa—lack of Cowper's glands [CAN 1]; maxilloturbinals large and branching, excluding nasoturbinals from narial opening [CAN 2]; upper molars with complete lingual cingulae [CAN 3]; and lack of parastyle on the P^4 [CAN 4]; Flynn and Galiano (1982) added loss of fibular facet on the calcaneum [CAN 5] and presence of a large teres process of the scapula [CAN 6], as additional support for the monophyly of Caniformia. These and other phenetic similarities have been used as bases for grouping the cynoids and arctoids as Hypomycteri Cope 1882, Arctoidei Winge 1895, Ursoidae Hay 1930, Canoidea Simpson 1931, or Caniformia Kretzoi 1943.

Autapomorphies defining the bear-like and dog-like clades are: (1) Arctoidea: external auditory meatus tubular, prolonged laterally [CAN 7]; mastoid process of petrosal prominent, salient [CAN 8]; lack of a caecum [CAN 9]; prostate glands contained within the wall of the urethra [CAN 10]; and (2) Cynoidea: bulla with partial septum formed by entotympanic [CAN 11]; internal carotid medial in position in shallow groove in caudal entotympanic, carotid canal very short and confined to rostral entotympanic [CAN 12]; base of paroccipital process buttressing the caudal entotympanic (only in Neogene and living canoids) [CAN 13].

The Aeluroidea are united by their unique double-chambered bulla with its bilaminar septum (FEL 1] (perhaps modified to retain only the ectotympanic portion in hyaenids); paroccipital process short, broadened, cupping the caudal entotympanic [FEL 2, distinct from CAN 13]; posterior opening of the carotid canal farther anterior to the posterior lacerate foramen than in primitive caniforms [FEL 3]; condyloid foramen usually in common fossa with posterior lacerate foramen, except in *Nandinia* and some felids [FEL 4]; postglenoid foramen absent [FEL 5]; alisphenoid canal absent [FEL 6, as in CAN 15]; baculum small or absent [FEL 7]; and reduction of molars to M^{1-2}, M_1 [FEL 8, paralleled in several caniform taxa]. This group has been termed the Epimycteri Cope 1882; Herpestoidei Winge

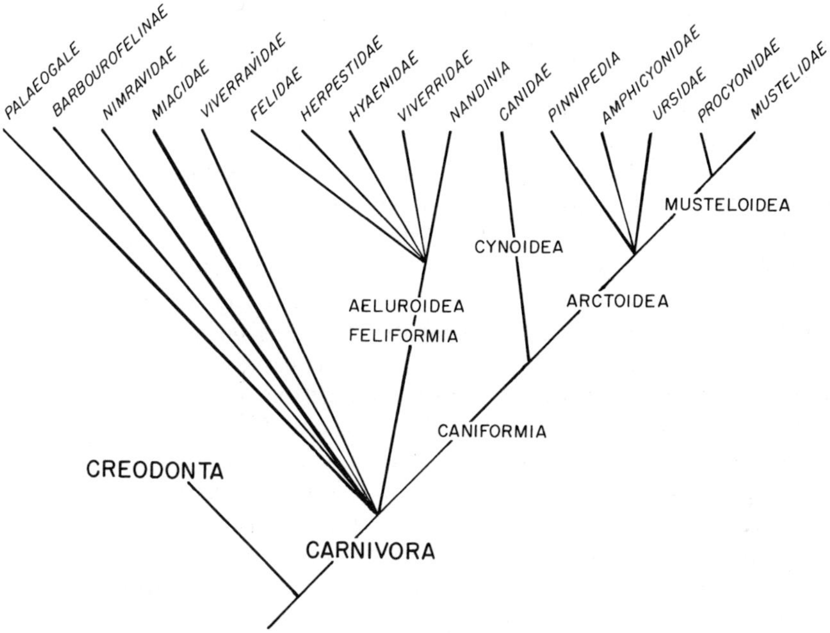

Fig. 4.2. Consensus phylogeny of the Carnivora. A conservative approach is adopted in this phylogeny, and the phylogenetic position of controversial taxa is left unresolved at the highest possible level of resolution. Major clades include: (1) Carnivora plus Creodonta (Ferae, Characters FER 1–7); (2) Carnivora (CAR 1–10); (3) Feliformia (FEL 1–14); (4) Caniformia (CAN 1–6); (5) Aeluroidea, Cynoidea, and Arctoidea (including Musteloidea). Taxa of controversial relationships include the 'Miacoidea' ('Miacidae', 'Viverravidae'), Nimravidae (see Fig. 4.6), Barbourofelinae, *Palaeogale*, *Nandinia*, Pinnipedia (see Fig. 4.5), and Amphicyonidae (see Fig. 4.7).

1895; Feloidea Hay 1930 (followed by Simpson 1930); and Feliformia Kretzoi 1945.

The data presented in this section and further information below yields the consensus phylogeny depicted in Fig. 4.2.

Phylogeny of the Feliformia

In addition to the cranial characters just cited, the Feliformia share a number of derived dental features, including:

(1) large P^4 parastyle [FEL 9];

(2) deep, narrow carnassial notch in P^4 [FEL 10];

(3) M^3_3 lost [FEL 11];

(4) M^2 reduced, simple [FEL 12];

(5) M_2 talonid extremely elongate, oval [FEL 13];

(6) M_2 hypoconulid larger than, or equal to, hypoconid [FEL 14].

The four extant aeluroid families—the Viverridae, Herpestidae, Felidae, and Hyaenidae—form a very well-defined group, as described above. However, no clear consensus exists about relationships within the Aeluroidea. Because of the degree of controversy, the contents of the families must be established before reviewing the confusion about family relationships.

The boundaries of the extant Felidae are unquestioned, although considerable variation exists in the generic-level assignments, perhaps in part because of the morphological uniformity that makes the family so easy to delimit. Derived features characterizing the family include: cheek teeth reduced to P_2^{3-2} and M_1^1; small chisel-like incisors arranged in a straight transverse line; the crushing function of molars entirely lost, and shearing function of the carnassials emphasized, with reduction of M^1 to a small transverse bar and the loss of the talonid and metaconid on the M_1; skull with short rostrum and facial area and short, inflated braincase, implying powerful bite (Radinsky 1981); large orbits and well-developed visual abilities (Radinsky 1981), with stereoscopic vision; and tongue covered with numerous, posteriorly directed, horny papillae. The postcranial skeleton is generalized, retaining retractile claws and sharing digitigrade adaptations with hyaenids and herpestids.

Within the family, the pantherines share the ligament in the hyoid associated with the ability to roar. The precise relationships of the cheetah, clouded leopard, and puma to the rest of the cats are not established, and it is far from clear whether the remaining species form a natural group of 'small cats', or what the relationships are within that remainder.

The extant hyaenids (including *Proteles*) all share a characteristic disproportion wherein the hindlimbs are shorter than the forelimbs, reduction of the molars to M_1^1, with a loss of postcarnassial crushing function, great reduction and loss of the M_1 talonid and reduction of the metaconid, all of which we interpret as paralleling felids; enlargement of the premolars for crushing (except in the highly modified, anteating aardwolf, *Proteles*); auditory bulla with greatly enlarged ectotympanic chamber and reduced entotympanic chamber; septum bullae either fusing the two layers very early or losing entotympanic contribution entirely (character state uncertain for all hyaenids); digestion modified to process all organic material from bone, producing faeces of almost pure calcium, phosphate, and carbonate. The most specialized genus, *Proteles*, is clearly a hyaenid with extensive adaptations for a diet of ants and termites, and thus presents no real phylogenetic problem.

The viverrids and herpestids exhibit a much greater diversity than do the living felids or hyaenids. The viverrids retain a probably primitive arboreal habitus (and retractile claws covered by fleshy sheaths), whereas the mongooses display more terrestriality (and the concomitant digitigrady and protracted claws). Wozencraft (1984) has presented an extensive survey of the morphology of these two families and, on that basis, strongly supports previous workers' recognition of these two groups as quite separate families (e.g. Pocock 1916*b* [Herpestidae called Mungotidae]; Gregory and Hellman 1939). We know of no clearly derived features defining even a restricted family Viverridae, nor did Wozencraft's (1984) extensive analysis produce any. Certainly the family is diverse morphologically, even when the Herpestidae are recognized as a distinct group, suggesting that the Viverridae *s.s.* may not be monophyletic, but may instead contain taxa from a variety of lineages that branched off independently from the early aeluroid stock. For example, Petter (1974) concludes that the civets on Madagascar, *Cryptoprocta*, *Fossa*, and *Eupleres*, are an early branch, together with the herpestids, from the ancestral lineage leading to the viverrids. However, the Galidiinae share with the Herpestidae a specialized bullar morphology with a tubular auditory meatus, and only share primitive features with the viverrids *s.s.*, and so are best assigned to the Herpestidae. *Cryptoprocta*, on the other hand, has a typical viverrid bulla, elongate but without the specializations seen in Herpestidae, and no derived characters such as the expansion of the middle-ear cavity into the mastoid or the extensive perineal glands of the viverrines, which would firmly unite them with that family.

The familial assignment of several genera has been the centre of some confusion: first, both *Eupleres* and *Cryptoprocta* are extremely specialized dentally (and even cranially in *Eupleres*), and their ear regions closely resemble the Viverridae *s.s.* only in primitive features. As a result of their morphological distance and lack of shared derived features, their inclusion within the Viverridae should be considered provisional.

The African palm civet, *Nandinia*, shows the opposite situation: dentally, cranially, and postcranially, it is quite an unexceptional viverrid. However, in the basicranium it exhibits a cartilaginous entotympanic that several authors have argued was primitive relative to the fully ossified entotympanic portion of the bulla in all other living Carnivora. However, cartilage is not necessarily the evolutionary precursor of bone: as Novacek (1977) has convincingly argued, the middle ear chamber was more probably primitively floored by a fibrous membrane. Such a fibrous membrane is present in many modern eutherians without a fully ossified bulla, and in the early developmental stages of those forms with an ossified entotympanic bulla. There are statements in the literature that the entotympanic ossifies within cartilage in various taxa, but the precursor has not been shown to be cartilage rather than membrane in these cases. Thus, *Nandinia*, like the two megachiropterans with a cartilaginous bulla, is uniquely specialized within the Carnivora in its development of cartilage instead of bone, but in other respects its anatomy is unexceptionally viverrid.

However, recently, Hunt (1987) has come to the opposite conclusion. He argues that *Nandinia*'s bulla is primitive relative to other aeluroids in seven other features that he judges are not associated with the cartilaginous composition of the caudal entotympanic. For example, the backwardly-directed paroccipital process not cupping the bulla, and the separation of the condyloid foramen from the posterior lacerate (or jugular) foramen are primitive features compared to the condition in other aeluroids. The absence of a true bilaminar septum bullae is less clearly unrelated to the cartilaginous nature of the entotympanic: perhaps normal ossification is necessary for the interaction between ecto- and entotympanic to produce the bilaminar septum. However, because of the significant question raised by Hunt about *Nandinia*'s phylogenetic position, we place *Nandinia incertae sedis* within the Feliformia, reflecting its status as a separate taxon within the feliform multifurcation (Fig. 4.2). This suggests that, if the ear region is truly a primitive retention, then *Nandinia* may be a sister-group to the rest of the Feliformia, rather than a derived viverrid.

The relationships among these four families are disputed. Most possible arrangements have been proposed (Fig. 4.3). Although all of the features in Fig. 4.3 have been argued to be derived, some are clearly more complex than others, and most occur elsewhere within the Carnivora. For example, the condition of the perineal glands and anal sacs is more diverse than can be fully described here, and deserves additional analysis. Ewer (1973) notes that the perineal glands are absent in the Felidae, Hyaenidae, Herpestinae, Cryptoproctinae, and the Prionodontinae within the Viverrini. They are present, but of relatively simple construction, in the rest of the viverrids and herpestids. The condition in *Nandinia* is unusual in that the gland is not divided bilaterally as in other taxa. Finally, the perineal glands are large and complexly developed in the Viverrini, and absent elsewhere in the Carnivora.

Anal sacs are widespread within the Carnivora, reduced or lost only within the bears and in the aquatic taxa—otters and pinnipeds. They are paired, relatively small, and open next to the anus in almost all other carnivorans. In *Cryptoprocta* (as the name implies), the herpestines, and hyaenids, the skin around the anus forms a pouch into which the anal sacs empty.

Another character complex that has provided several features used in phylogenetic analysis is the cephalic arterial system, studied exten-

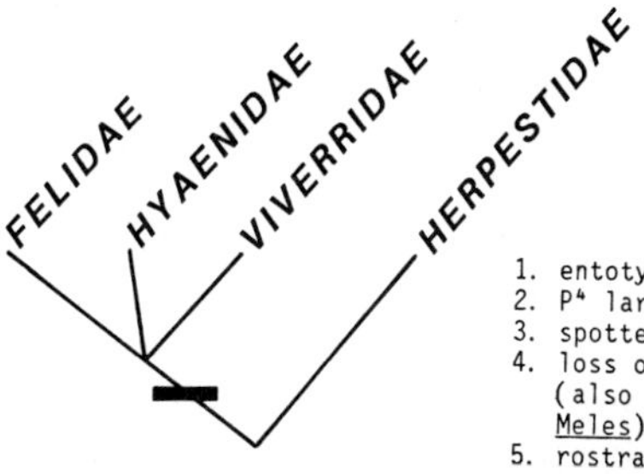

1. entotympanic elongated
2. P^4 larger than M^1
3. spotted coat pattern
4. loss of supraorbital ramus of medial meningeal artery (also lost in Ursus, Ailurus, Procyon, Mustela, and Meles)
5. rostral entotympanic tube short; groove or trough in caudal entotympanic for internal carotid artery

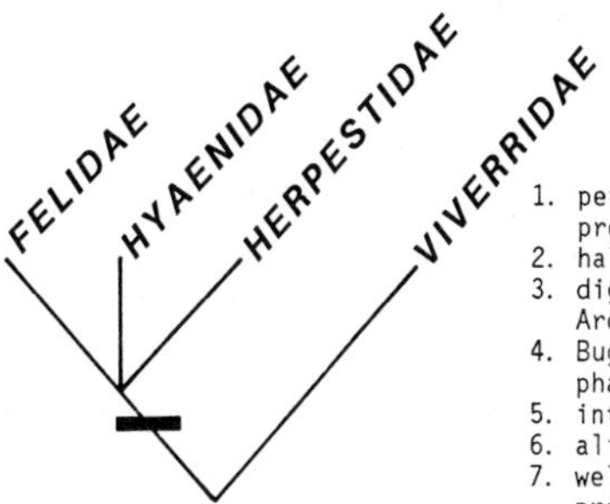

1. perineal scent glands reduced or lost (reversal; absence probably primitive for carnivorans)
2. hallux and pollex reduced or lost (also in Canidae)
3. digitigrade, with compressed metapodials (also in Arctoidea)
4. Bugge's 'anastomosis Y' present, between ascending pharyngeal artery and internal carotid artery
5. internal opthalmic artery reduced
6. alisphenoid canal lost or absent
7. well-developed cruciate sulcus present in brain (also present in extant arctoids)

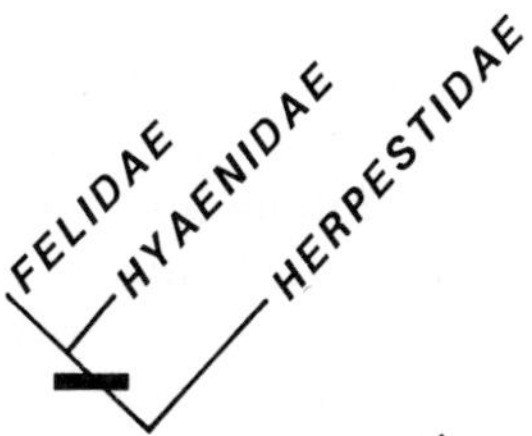

1. M^1 and M_2 very reduced or absent
2. internal carotid artery reduced or lost
3. elaborate intra- and extra-cranial rete present in anastomic artery and external opthalmic artery, connecting the infraorbital ramus and the circle of Willis (intracranial portion only present also in Canidae)
4. postglenoid foramen lost (also in otariids and phocids)

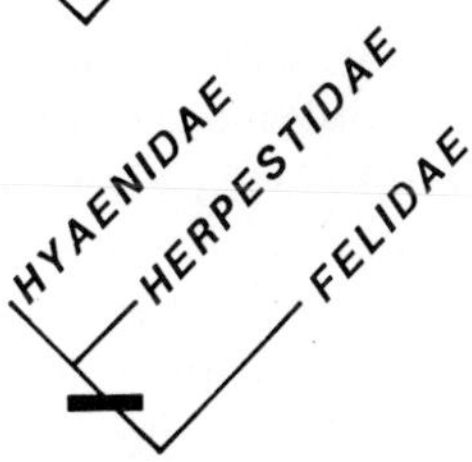

1. ectotympanic enlarged (much more so in hyaenids)
2. anal glands enclosed within pouch
3. claws non-retractile (permanently protracted; also found in several arctoid families)
4. bursa absent (also in ursids, most mustelids)

Fig. 4.3. Hypotheses of relationships within the Aeluroidea. The conflicting distribution of derived characters led to the presentation of the families of Aeluroidea as an unresolved multichotomy in Fig. 4.2.

sively by Bugge (1978). He found the hyaenid and felid patterns to be markedly different from those of other families, in that the external carotid arterial system has annexed the stapedial area, taken over supply of the bulbar part of the orbit, and feeds into the intracranial circulation via large extra- and intracranial arterial rete. The internal carotid artery is lost, and the vertebral-basilar artery is reduced in both families.

Phylogeny of the Caniformia

We have already listed some synapomorphies that unite the living members of the fissiped families Canidae, Ursidae, Procyonidae, and Mustelidae with the pinnipeds, into a monophyletic group for which we take the term Caniformia Kretzoi 1943, at subordinal rank. We also detailed features distinguishing the monophyletic clades Cynoidea and Arctoidea. The monophyly of the living families, and their phyletic relationships with one another have been supported from investigations of several features.

Auditory bulla Hunt (1974*a*) distinguishes two caniform morphotypes. In Type A, the caudal entotympanic is very small relative to the rostral entotympanic and to the ectotympanic, the internal carotid artery lies completely within a tube formed mostly by the rostral entotympanic, and the posterior carotid foramen lies at the posteromedial corner of the bulla. The Type A bulla is present at an early ontogenetic stage of all carnivorans, and in adult Ursidae (including *Ailuropoda*), the lesser panda (*Ailurus*), lutrines, mephitines, and the pinnipeds Otariidae and Odobeniidae, although details of bulla morphology differ among these groups. A ventrally and posteriorly expanded caudal entotympanic is present in Canidae, Procyonidae (except *Ailurus*), the remaining Mustelidae, and Phocidae (Hunt's 'Type B'). The Canidae also have an entotympanic septum. Expansion of the caudal entotympanic is usually accompanied by relative movement of the posterior carotid foramen which results in shortening of the carotid canal [CAN 14], separating the posterior opening of the canal from the posterior lacerate foramen. In the canids only, the carotid canal is confined to the very small rostral entotympanic; consequently the canal is very short, and restricted far anteriorly. The internal carotid runs in a shallow groove on the medial surface of the caudal entotympanic, ventral to the inferior petrosal sinus, which is more similar to the situation in juvenile *Felis* than in other Caniformia.

Basicranium As noted by Turner (1846) and Flower (1869), the canids, ursids, lesser panda, otariids, and odobenids (primitively) retain the alisphenoid canal, whereas the procyonids, mustelids, and phocids have lost this structure [CAN 15].

Pocock (1920, 1921, 1929) sought to extend Flower's observations, and Segall (1943) and Hough (1948) investigated cynoid and arctoid auditory regions in some detail. Their work suffered from limited sampling, but it again confirmed the unity of structures cited above for the Canidae. Segall (1943) grouped the ursids, procyonids, and 'Ailuridae' (including both lesser and giant pandas) as a sister-taxon to the Mustelidae. Analysis of his characters shows that few are synapomorphies supporting a special relationship of bears and racoons.

Segall (1943, p. 39) was the first to describe the pit in the mastoid process of the squamosal that forms the posterodorsal roof of the external auditory meatus in procyonids. Davis (1964, p. 320) described a shallower, but homologous, structure in the pandas. Segall named this the suprameatal fossa, and both he and Davis (1964) discovered that it receives the pars flaccida of the tympanic membrane and so communicates with the tympanic cavity through a small fleshy opening near the neck of the malleus. Recently, Schmidt-Kittler (1981) has shown that the suprameatal fossa in living and fossil arctoids is also present in mustelids, but is usually floored by the developing external auditory tube in adult mustelids and is sometimes eliminated entirely since this tube is fused to the mastoid process. In the pandas and bears the fossa is shallow. This structure thus seems to be a synapomorphy for the arctoids as a whole [CAN 16], but its character states seem to typify specific families:

(1) shallow, open ventrally in ursids, pandas, and primitive Musteloidea (primitive arctoid condition);

(2) deep, open ventrally in procyonids [CAN 17];

(3) shallow, floored by the meatus, and thus opening medially, or closed in mustelids [CAN 18].

Pocock (1920, 1921, 1929) and Segall (1943) described the distribution of the expanded tympanic cavity or hypotympanic sinus of arctoids. In arctoids, even in the ursids and the pandas, which have primitive, little expanded bullae, the hypotympanic space is extended into the base of the auditory tube [CAN 19] so that the crista tympani appears to be produced into the tympanic cavity. The procyonids have extended the cavity further into the roof of the meatus through expansion of the suprameatal fossa. Details of patterns of enlargement of the typanic cavity may provide synapomorphies useful in recognizing lower-level clades within the Arctoidea.

Auditory ossicles Doran's (1877) classic study of the auditory ossicles of the Mammalia described a representative suite of carnivoran ossicles, but drew few systematic conclusions. He did contrast the carnivoran malleus with that of other mammals and showed that the development of the thin lamina between the head and neck supported by the processus gracilus in Carnivora was approximated only by the condition in certain ruminants and murid rodents. Within the caniforms, Doran (1877) noted that the canid malleus was distinctive in the size and anterior curvature of its manubrium. In this, and in the general massiveness and form of the bone [CAN 20], the overall similarity is with the felids and hyaenids, rather than with the arctoids.

Arctoid mallei are diverse in form, as also reported by Segall (1943), making generalizations difficult. Perhaps the most distinctive mallei are those of the bears and giant panda, which have a very reduced or absent muscular process [CAN 21] for insertion of the tensor tympani at the base of the neck. Procyonids and *Ailurus* have a small, but consistently present, processus muscularis [CAN 22], in contrast with the canids and mustelids in which the process is primitively large. Segall (1943) grouped the ursids and procyonids together, on the basis of the reduced muscular process, the shape of the head of the malleus, the presence in some taxa of an anterior prominence on the head just above the lamina, and the large size of the lamina itself. However, interspecific variation within this group and within the mustelids blurs some of these generalizations and decreases their usefulness in phylogenetic analyses.

Dentition The dentition is reduced in the caniforms from a primitive M_3^3 (although M^3 is unknown in living caniforms) to M_3^2 [CAN 23] in the ursids, to M_2^2 [CAN 24] in the procyonids (and *Ailurus*), and to M_2^1 [CAN 25] in the living mustelids. Highly derived canids (*Speothos, Cuon*, and *Lycaon*) also reduce the posterior molars, producing an ultimate dental formula of $M_2^{1,2}$, but this is easily seen to be a parallelism given the primitive dental formula for the monophyletic Canidae and the morphological differences of the remaining molars.

The morphology of the molar dentition offers further clues to relationship. The canids uniquely enlarge the posterior limb of the buccal cingulum of the upper molars [CAN 26] to form a hypocone, which occludes with the prominent entoconid on the lower molars. Only the hypercarnivorous canids (*Speothos*, *Cuon*, and *Lycaon*), with their reduced dental formulae, suppress the internal cingulum on the upper molars and achieve, a monocuspidate talonid on the lower molars. The ursids and procyonids enlarge the upper molars by expansion of the metaconule [CAN 27] as well as the cingulum. In the bears, the protocone and metaconule of the upper molars form a longitudinal crest [CAN 28] to occlude with the enlarged talonid cusps of the lower molars, the M^2 develops a posteriorly projecting platform [CAN 29] to occlude with the large M_3, and the M_2 trigonid loses its paraconid [CAN 30]. The procyonids have lost M_3, but have elongated the talonid of M_2 [CAN 31] to occlude with the talons of the upper molars with their enlarged metaconules. In addition, the enlarged hypoconulid [CAN 32] becomes a more important cusp on the M_2 in procyonids, even in *Bassariscus* in which the upper molars otherwise retain a primitive form and the M^2 is reduced in size. The dental morphology and formula of *Ailurus* seems most comparable with that of the procyonids, despite its complexity of form. The morphology of the procyonid M_2 seems an important dental synapomorphy for this group. Living mustelids are all characterized by loss of the carnassial notch on P^4 [CAN 33], as discussed by Schmidt-Kittler (1981); this is a unique feature in the Carnivora, and surely a synapomorphy.

Cephalic arteries Bugge (1978) affirmed Tandler's (1899) finding that the arctoids have the most primitive arrangement of cephalic arteries, especially ursids and pandas in which the internal carotid and vertebral arteries exclusively supply the brain, and there are no anastomoses between the extracranial and intracranial circulations. In the procyonids and the mustelids there are limited anastomoses with the external carotid, but none of these is as well developed as in the feliforms, and no retia are formed. Ursids (including the giant panda) fold the internal carotid artery intracranially [CAN 34] within the inferior petrosal sinus (Hunt 1974*a*). The bony groove in the basioccipital containing the sinus is enlarged and extended behind the posterior lacerate foramen to accommodate this loop of the carotid.

Bugge (1978) agreed that the arctoids were characterized by their, albeit primitive, cephalic arterial pattern, but he believed that the canids showed special resemblance to the felids in the extent to which the external carotid system had taken over supply to the brain and also because of the development of a rudimentary intracranial arterial rete. Hough (1953) also proposed uniting the Canidae and Felidae (including the Nimravidae) into a superfamily 'Cynofeloidea' on superficial and largely primitive features in which the nimravids (thought to be felids) were considered closely related to canids. Ginsburg (1982) also proposed such a union, based on the bulla septum, which has been shown not to be homologous in the Canidae and Feliformia by Hunt (1974*a*).

Figure 4.4 presents the phylogenetic relationships of the families as developed from the data discussed above.

Problem Caniformia A number of living caniforms have posed difficulties to systematists, particularly those having characters considered diagnostic of different families. Among those cited in the literature, Flower (1869) refuted the supposed viverrid affinities of the cacomistles (*Bassariscus*), Segall (1943) questioned the status of the kinkajou (*Potos*) as a procyonid, and Radinsky (1973) explored the relationship of the stink badgers (*Mydaus*, *Suillotaxus*) and the skunks. None of these problems has attracted so much attention as those posed by the pandas and pinnipeds, partly because these two latter groups represent extreme adaptations for terrestrial and aquatic carnivores.

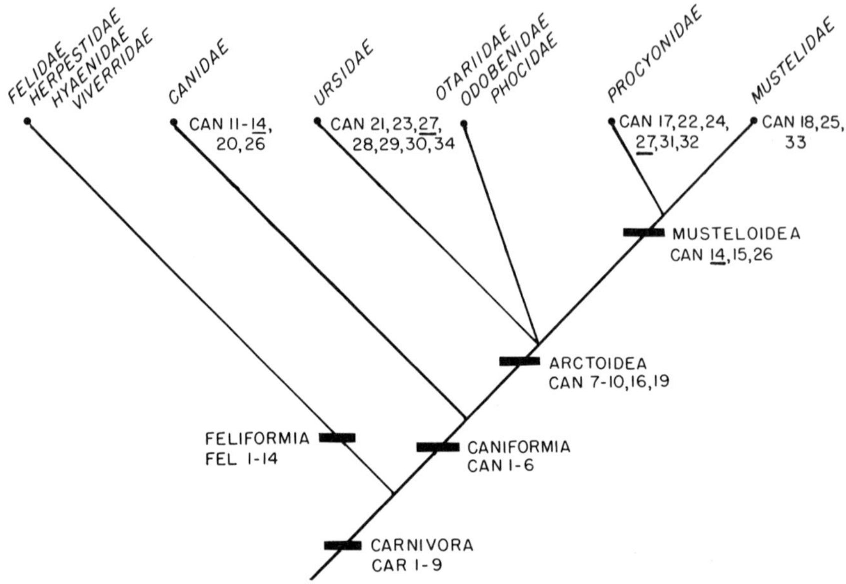

Fig. 4.4. Cladogram of living Caniformia, best supported by the evidence discussed in this paper. Numbered characters and character states are identified and discussed in the text. Parallel acquisition of some characters in this hypothesis is indicated by underlining. Monophyly of the pinniped families follows the evidence of Wyss (1987; see text).

Pandas In the 19th century the lesser or red panda (*Ailurus*) was recognized as an arctoid, but was usually relegated to a monotypic family standing between the bears (based on possession of an alisphenoid canal) and the procyonids (based on reduction of the molars to M^2_2). After discovery of the giant panda (*Ailuropoda*), it was common to group the pandas as a Family Ailuridae (Segall 1943), to relegate them both to the Procyonidae (Gregory 1936), or to separate them, assigning the giant panda to the Ursidae (Davis 1964) and leaving the lesser panda in the Procyonidae. Davis's (1964) monograph on the comparative anatomy of the giant panda firmly allied that form with the Ursidae, a position unchallenged by biochemical systematists (O'Brien *et al.* 1985). The question of the affinities of the red panda remain. Ginsburg (1982) most recently placed *Ailurus* as a sister-taxon to *Ailuropoda*, and proposed the so-constituted Ailuridae as a sister-family of the Ursidae. His evidence for this relationship was dental, and included no synapomorphies. Thenius (1979) and Schmidt-Kittler (1981) both removed the lesser panda from its special relationship to the giant panda (an ursid in their view), regarding it as a procyonid (Thenius), or as a living representative of primitive Musteloidea and a sister-taxon to the Procyonidae plus Mustelidae (Schmidt-Kittler). Certainly, the structure of the basicranium indicates the primitive arctoid nature of *Ailurus*, but the dental evidence more closely allies *Ailurus* with the procyonids, especially the shared-derived enlargement of upper molar conules (significantly the metaconule) and the elongation of the M_2 talonid and enlargement of its hypoconulid.

Pinnipeds The pinnipeds have been a persistent systematic problem, not because their essentially arctoid relationships have been in doubt, but because their monophyly has been questioned (Fig. 4.5). Flower (1869) was convinced that the pinnipeds were most closely related to the arctoids, specifically the ursids, even though he did not express this in his classification. Mivart (1885) extended Flower's observations and concluded that 'the Otaries might have been derived from Bear-like animals, while the Phocidae had another, possibly a Lutrine, origin' (p. 497), listing nine osteological, principally cranial features supporting this view. Where mentioned, the walrus was more closely allied with the 'otaries'. Twentieth century systematists have most consistently

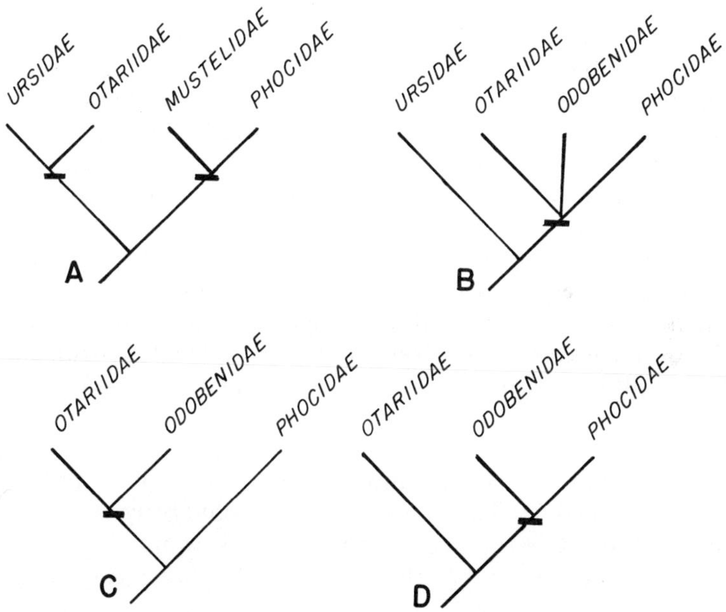

Fig. 4.5. Phylogeny of the Pinnipedia. The pinnipeds clearly are arctoids, although previously proposed relationships within the Arctoidea include conflicting hypotheses of diphyly (upper left) or monophyly (upper right). The hypothesis of a monophyletic Pinnipedia seems to be the most strongly supported (see text). Two alternative three-taxon hypotheses have been suggested for the living pinniped families; (C) walruses as sister-group to the sea-lions (traditional); and (D) walruses as sister-group to the seals (Wyss 1987). These hypotheses can be considered independently of evaluations of pinniped monophyly. We support Wyss' (1987) hypotheses of a sister-group relationship between the Odobenidae and Phocidae (see text).

considered the pinnipeds as monophyletic, although diphyly has recently become more widely accepted. Diphyly was revived by McLaren (1960) and given a positive statement by Tedford (1976), who concluded that the Otariidae and Odobenidae share the 'Type A' bulla and other primitive arctoid features with ursids, while the Phocidae, with their 'Type B' bulla, resemble the mustelids in some aspects of bullar morphology. Tedford (1976, p. 370) admitted that these differences, and others related to the adaptation of the ear-region for underwater hearing, did not preclude monophyly of the pinniped families, but he did not find synapomorphies unrelated to aquatic adaptation that could unite them.

Wyss (1987) concluded that the walrus (*Odobenus*) is most closely related to the phocids, and amassed convincing morphological evidence for pinniped monophyly. Walruses have been largely relegated to an autapomorphous sideline in phylogenetic thinking because of their aberrant feeding mechanism and modifications of structure thought to be related to it. Doran (1877) called attention to the hypertrophy of the head of the malleus and the body of the incus in the walrus, as a peculiar similarity with the condition in the phocids. Otariids, on the other hand, retain small, but modified, ossicles in which the malleus lacks the muscular process, has a small lamina, and has a pit in the anterior face of the head, as in the other pinnipeds. Wyss (1987) used these derived ossicle characters, and other aspects of cranial and postcranial osteology and soft anatomy, to build a case for a phocid–odobenid sister-relationship. He used 43 derived characters to support a cladogram of pinniped (monophyletic) relationships involving both living and fossil forms, a clade that has long been favoured by biochemical systematists and karyologists. Wyss (1987) summarized available biochemical data (serum albumins, DNA hybridization, α-lens crystallin A sequences, a unique bile acid), karyology, and the ectoparasite distributions, all of which support the monophyly of pinnipeds. Although there is still some controversy, we feel that pinniped monophyly is the most strongly supported phylogenetic hypothesis (Fig. 4.5). This pinniped clade would be rooted at the base of the Arctoidea, either as an ursid sister-group (Fig. 4.5) or in unresolved multichotomy with other arctoid families (Fig. 4.2).

Biochemical phylogeny

Phenetic analyses (protein electrophoresis, serum immunology, karyotype) have been applied widely to higher level carnivoran systematics (see the following papers, and references cited therein: serum immunology: Leone and Wiens 1956; Pauly and Wolfe 1957; Sarich

1969*a,b*; Farris 1972; protein electrophoresis: Seal 1969; Feng *et al.* 1985; karyotype: Fay *et al.* 1967; Wurster and Benirschke 1968; Arnason 1974; Couturier and Dutrillaux 1986). We reject the use of these phenetic techniques in phylogeny reconstruction, in favour of cladistic analysis of character distributions. Data amenable to cladistic analysis have been generated by a number of biochemical techniques, including protein sequencing (e.g. Romero-Herrera *et al.* 1978; de Jong 1986; de Jong and Goodman 1982; Goodman *et al.* 1982; Miyamoto and Goodman 1986) and some karyotype studies (e.g. Arnason 1974; Wurster-Hill and Gray 1975; Wurster-Hill and Bush 1980; Wurster-Hill and Centerwall 1982; O'Brien *et al.* 1985). Although few in number, such studies will provide important new information for understanding carnivoran phylogeny. The phylogenetic significance of DNA–DNA hybridization studies is currently being debated; such studies of Carnivora have been presented by O'Brien *et al.* (1985) and Arnason and Widegren (1986).

Cladistic analysis of non-phenetic biochemical information has yielded the following conclusions about the higher-level phylogeny of the Carnivora:

(1) The Pinnipedia are monophyletic (Arnason 1974; Romero-Herrera *et al.* 1978; Arnason and Widegren 1986; de Jong and Goodman 1982; Goodman *et al.* 1982; Miyamoto and Goodman 1986).

(2) The giant panda (*Ailuropoda*) is most closely related to the Ursidae (O'Brien *et al.* 1985).

(3) The lesser panda (*Ailurus*) is most closely related to the Procyonidae, although the evidence is weak (O'Brien *et al.* 1985).

(4) The monophyletic Pinnipedia are most closely related to the Mustelidae (Arnason and Widegren 1986), or to the Musteloidea (Mustelidae plus Procyonidae; Miyamoto and Goodman 1986).

(5) The Procyonidae and Mustelidae are sister-groups (de Jong 1986; Miyamoto and Goodman 1986).

(6) The Caniformia and Feliformia are monophyletic (Miyamoto and Goodman 1986), although sampling is sparse within the Feliformia.

(7) The Canidae and Ursidae are sister taxa (Miyamoto and Goodman 1986), and therefore the Arctoidea are not a monophyletic group.

Most of these conclusions (except number 7, and possibly number 4) are consistent with the phylogenies, based on morphology, proposed within our paper and in many previous studies.

Phylogenetic relationships of fossil Carnivora

Early Tertiary carnivorans ('Miacoidea'): 'Miacidae', Viverravidae, Didymictidae

Since Cope (1880*a*) first erected the 'Miacidae', many authors have included the early Tertiary carnivorans within this family. Although most workers have considered the 'Miacidae' to be 'ancestral' to the later Tertiary and living groups of terrestrial carnivorans (Fissipeda), the 'Miacidae' variably have been classified as Creodonta or basal Fissipeda (Miacoidea, Simpson 1945). Flynn and Galiano (1982) suggested that the 'Miacidae' (or 'Miacoidea') were not a monophyletic group, and that many of the dental and skeletal features supporting the dichotomy recognized in the living Carnivora (Feliformia/Caniformia) were also present in the early Tertiary carnivorans ('Miacoidea'). As such, Flynn and Galiano (1982) assigned two monophyletic groups of early Tertiary 'Miacoidea', the Didymictidae and Viverravidae, to the Feliformia (FEL 9–14; as sister-taxa to the living families of Feloidea), and the miacid 'miacoids' to the Caniformia (CAN 2–6).

Although these Early Tertiary carnivorans share many derived features with either the Feliformia or Caniformia, there are some features that do not support inclusion of the 'Miacoidea' within these clades. All members of the clades of living carnivoran families have ossified bullae (CAR 7) (except *Nandinia*), lose the stapedial artery (CAR 8), and fuse the scaphoid and lunar bones (CAR 10).

Although 'miacoids' are not known to have possessed an ossified bulla, this is negative evidence: 'miacoids' could have had a fully ossified but weakly attached bulla. At least one group of arctoid caniforms (the daphoenine Amphicyonidae) also included members that do not preserve an ossified entotympanic, even in specimens with a well preserved ectotympanic, thus indicating a perhaps secondarily cartilaginous bulla. Another, although slightly less parsimonious, hypothesis is that there was independent acquisition of a completely ossified bulla in the Feliformia, cynoid Caniformia, and once or twice within the arctoid Caniformia.

Although all living carnivorans lose the stapedial artery, the loss or extent of reduction of this artery in individual 'miacoids' is uncertain (see discussion above, in the section on the Definition of the order).

Some 'miacoid' skeletons possess an independent scaphoid and lunar. However, many 'miacoids' (*Vassacyon*, *Procynodictis*, and some *Vulpavus*) do fuse the scaphoid and lunar into a single scapholunar bone, as in all other Caniformia and Feliformia, and some other eutherians. The wide and variable distribution of this character sug-

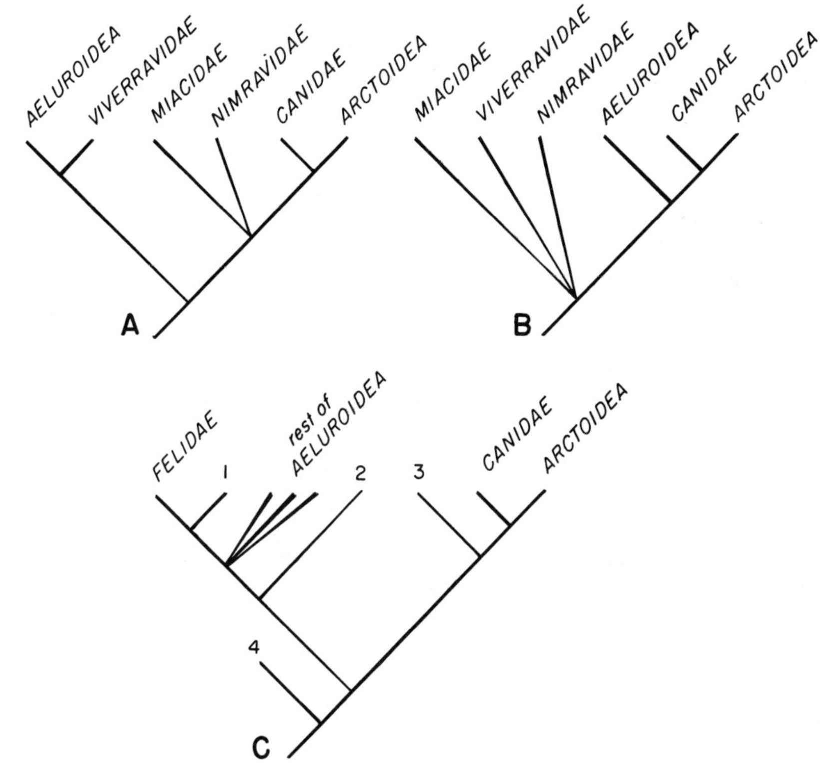

Fig. 4.6. Phylogenetic hypotheses for the 'Miacoidea' and Nimravidae. (A) Flynn and Galiano (1982) proposed that the bipartite split recognized in living Carnivora was also present in the Early Tertiary 'Miacoidea' and in the Nimravidae; they assigned the Viverravidae to the Feliformia and the Miacidae and Nimravidae to the Caniformia. (B) An alternative hypothesis recognizes several characters that support monophyly of the living families of Carnivora, to the exclusion of the 'Miacoidea' and Nimravidae. (C) The phylogenetic relationships of the Nimravidae are particularly perplexing (lower diagram); placement of this family at positions 1 and 2 does not seem to be well supported, while placement at positions 3 or 4 seems better supported. Given these conflicting arrangements, the Nimravidae are placed *incertae sedis* within the Carnivora (Fig. 4.2).

gests multiple independent acquisition of a fused scapholunar, and does not strongly support the exclusion of the Early Tertiary 'miacoids' from the Caniformia and Feliformia clades.

Alternative hypotheses of relationships of the 'miacoids' to other carnivorans are shown in Fig. 4.6.

Palaeogale

The phylogenetic relationships of *Palaeogale*, an Oligocene taxon from Eurasia and North America, are uncertain. Although many previous workers had considered *Palaeogale* to be a mustelid, Hunt (1974*b*) and Flynn and Galiano (1982) presented evidence that indicated it might be a feliform. It lacks many caniform-derived features (e.g. lacks a teres major process on the scapula, and lacks complete lingual cingulum on the upper molars) and mustelid-derived features (e.g. retains a carnassial notch on P^4). *Palaeogale* also possesses many derived features of the Feliformia or Aeluroidea (of Flynn and Galiano 1982), including a relatively large P^4 parastyle, slit-like carnassial notch, open trigonid on M_1, absence of molar conules, reduction of molars to M^2_2 (also in some arctoids), elongate M_2, and loss of M_1 metaconid (as in some Feloidea). Further, within the Aeluroidea, *Palaeogale* appears to be most closely related to *Viverravus*; both taxa share reduction of the paroccipital and supraoccipital processes, and location of the large postglenoid foramen far posterolaterally on the postglenoid process of the squamosal. *Palaeogale* is excluded from a position within the Feloidea because of the absence of any internal septae (particularly ectotympanic) in its inflated bulla, and retention of a large postglenoid foramen. It also lacks an alisphenoid canal, and has the posterior carotid foramen far posterior, the condyloid foramen separate from the posterior lacerate foramen, and the median lacerate foramen clearly visible anteriorly.

Nimravidae and fossil Felidae

The Oligocene sabre-toothed carnivorans pose another currently controversial question. Cope (1880*b*) named a Family Nimravidae for the Oligocene cat-like carnivorans, defining it and the Felidae with basicranial characters, without postulating how close the relationship was between the two families. Matthew (1910) proposed a phylogeny in which the modern cats evolved from the non-sabre-toothed Oligocene nimravids, and the sabre-toothed felids arose from the Oligocene sabre-toothed forms, based primarily on the overall dental similarities. This hypothesis was accepted in North America with very little question, except for Hough's (1953) analysis of the auditory region. She proposed a classification in which all sabre-toothed forms were collected in the superfamily Machairodontinae, and the Aeluroidea were restricted to Viverridae and questionably Hyaenidae, with the addition of Daphoenidae. A third superfamily, the Cynofeloidea, contained the Canidae and Felidae (the latter including the Felinae and Nimravinae). The basicranial morphology of the Canidae was recognized as being primitive in many respects: the nimravine ear was interpreted to be very similar, and the canid–nimravid condition was argued to be ancestral to the derived Felidae. The problem here was partly a willingness to unite taxa on the basis of primitive characters, but even more so a failure to analyse correctly the character conditions in the nimravids.

Meanwhile, Teilhard de Chardin (1945) proposed two successive pairs of radiations within the cat-like carnivorans—a lineage of sabre-toothed and a lineage of non-sabre-toothed forms in the Oligocene, and a second pair in the Miocene to Pliocene. He did not specify the degree of relationship between the two pairs of radiations. Although Teilhard's hypothesis was largely adopted by the European palaeontological community, when statements of familial-level relationship were made, the Nimravidae were considered to be most closely related to the Felidae (e.g. Beaumont 1964), an hypothesis espoused in North America as recently as 1980 by Martin (position '1' in Fig. 4.6C).

When derived characters were sought for various groupings within the Carnivora, however, it was quickly recognized that few, if any, characters directly supported a sister-group relationship between the Felidae and the Nimravidae, and placement of the Nimravidae in that position contradicted the strong suite of characters otherwise defining the Aeluroidea. This led to several informal suggestions that the nimravids were the sister-group to the aeluroids (position '2' in Fig. 4.6C): this left the synapomorphies which united the Aeluroidea, and it was suggested by the presence of retractile claws in the Nimravidae when that feature was thought to be more narrowly distributed. Flynn and Galiano's (1982) analysis of the 'miacoids' produced the first cladogram in Fig. 4.6A, in which the Nimravidae were placed within the Caniformia (position '3') on the basis of their lack of a calcaneal fibular facet in the pes; reduced parastyle on P^4; reduced protocone on P^4; and large teres major process on the scapula. The maxilloturbinals of *Dinictis* are aeluroid in form, simple, unexpanded, and unbranched (Hunt, pers. comm.), a condition which is probably primitive for the Carnivora.

A detailed analysis of the basicranial anatomy of the Nimravidae was undertaken to choose between the hypotheses '1', '2', and '3' (Fig. 4.6), with the disconcerting result that none was supported by the auditory region, and a fourth hypothesis (position '4') suggested instead (Neff 1983). The nimravids were found to have a very specialized basicranial structure (largely misinterpreted by Hough) that excluded them from placement within the Aeluroidea—they lacked a completely ossified bulla, lacked a bilaminar septum bullae, or indeed any extensive septum at all, and were autapomorphous in their peculiar thick-walled entotympanic, a promontorial position of the internal carotid artery (condition 5 of the internal carotid artery, see above), a vertical flange on the sides of the basioccipital (as in ursids), and a bony separation between the exits of the sigmoid sinus and the inferior petrosal sinus. The lateral position of the grooves across the promontory, from which a 'promontory artery' was inferred, was accompanied by the presence of a small posterior carotid foramen through the posterior wall of the bulla in a position farther lateral than seen in any other carnivoran.

If the 'miacids' and viverravids were excluded from consideration, the rest of the Carnivora shared the synapomorphies of ossification of the bulla (with daphoenines and *Nandinia* being exceptions), and capture of a medial internal carotid artery in a tube within the rostral entotympanic for some portion of its basicranial traverse (condition 2 or 3). No basicranial features were found to support Flynn and Galiano's (1982) placement of the 'miacoids', although they were not the primary subject of the analysis: Neff's (1983) hypothesis suggests that the 'miacoids' do not primarily fall within the Aeluroidea/Arctoidea split. Because of the high taxonomic level of this controversy, the multichotomy shown in Figs. 4.2 and 4.6 encompasses the variety of hypotheses.

The barbourofelines turned out to be a larger puzzle than first realized. *Barbourofelis* and *Sansanosmilus* in Europe, unlike the Oligocene nimravids, have a completely ossified, thin-walled entotympanic, lacking a septum bullae, and with a very reduced canal presumably for an internal carotid artery in a promontorial position. The barbourofelines also presented some contradictions to Flynn and Galiano's analysis of the nimravid condition: a large parastyle is present in *Barbourofelis*, for example. Again, because of the amount of apparently contradictory information, the barbourfelines are placed with the 'miacids', viverravids, and nimravids, *incertae sedis* within the order (Fig. 4.2).

That the nimravids have nothing to do with the ancestry of the felids is now clear, however. Instead, the relatives of the modern felids may be sought among the miscellaneous forms lumped together in the genus *Pseudaelurus* and related taxa. In both Europe and North America, *Proailurus* may offer even more generalized morphology within the lineage leading to the felids—the European form apparently lacks any ossification of the entotympanic portion of the bulla, for example. A transitional form leading to nimravid morphology is lacking so far, unless Adams' (1896) *Aelurotherium* and Chow's (1975) *Chailicyon* turn out to be primitive relatives of the Nimravidae.

Canidae

Hough (1948) showed that the canid bulla was present in North American Early Oligocene *Hesperocyon* and *Mesocyon*, differing only in that there is no tubular meatus and the paroccipital process is salient rather than cupping the inflated caudal ectotympanic. They possessed the derived upper molars of living canids, and were digitigrade, although the first metacarpal and metatarsal was not so reduced as in modern canids. These criteria allow recognition of a diverse assemblage of fossil canids in North America; the lack of fossil canids outside North America suggests that the family did not occur outside that continent until the Late Tertiary. The fossil record does not alter the phylogenetic affinities of the Family Canidae derived from the living forms, although it demonstrates the antiquity of this major caniform clade and slightly modifies our concept of the morphotypical canid.

Amphicyonidae

The Family Amphicyonidae has most consistently been regarded as a canid, or a canid sister-taxon, but in fact their osteology is a close approximation to the primitive arctoid morphotype (Fig. 4.7A). They are relatively large animals with elongated skulls, massive skeletons, long tails, and plantigrade pentadactyl feet. Dentally, they are primi-

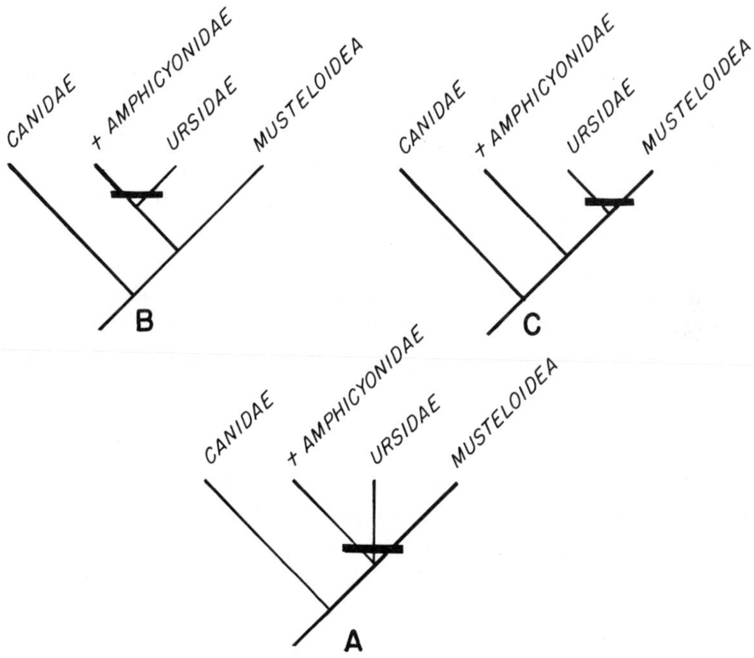

Fig. 4.7. Alternative hypotheses for the relationships of the Amphicyonidae. A, amphicyonids as arctoids, united by possession of carotid loop and postscapular fossa (present in some primitive musteloids); B, amphicyonids as ursid sister-taxon supported by few synapomorphies beyond those probably primitive for arctoids (e.g. ventral bending of the lateral margins of the basioccipital; Ginsburg 1982); C, amphicyonids as primitive arctoids: ursids retain carotid loop and postscapular fossa, but add suprameatal fossa as synapomorphy with musteloids; these features are modified or lost in musteloids.

tive caniforms in possessing M^3 (lost in some later genera); the upper molars have a strong lingual cingulum nearly symmetrically disposed about the protocone; the upper carnassial has an anteriorly placed protocone; the M_2 has a paraconid; the lower molar talonids have a low entoconid, and the hypoconid is a longitudinal crest more centrally located on the talonid; and the premolars are simple in form, low-crowned, elongated, and often separated by diastemata in the long jaws. The bulla is remarkable in the small size of the entotympanic, which was apparently unossified in the most primitive Oligocene forms (Daphoeninae), but when ossified it mostly consists of the carotid canal. The bulla thus is dominated by the ectotympanic, which just covers the promontorium and does not extend posterior to the mastoid process ('Type A' of Hunt 1974*a*). The external auditory meatus is inflated by extension of the hypotympanic sinus into its walls to a larger extent than in other arctoids. Hunt (1977) showed that the basioccipital of the primitive amphicyonid *Cynelos* is deeply grooved medially, as in the living bears. In bears, the groove accommodates the inferior petrosal venous sinus and a unique posterior loop of the internal carotid, as mentioned above. The morphology of this groove, and the pit in the basisphenoid anterior to the carotid canal for the first loop of the carotid, convinced Hunt that the amphicyonids (including the Daphoeninae) had an internested carotid–venous system, as in the bears. This feature, and the remarkable similarity of the scapula, complete with postscapular fossa (Davis 1949), constitute synapomorphies with the Ursidae, despite the overall primitiveness of the amphicyonid skeleton (Fig. 4.7B). An alternative interpretation (Fig. 4.7C) is that the amphicyonids occupy a more primitive position within the Arctoidea and that the folded internal carotid and scapular features are primitive for arctoids, retained in bears, and reduced in procyonids and mustelids.

Ursidae

Skulls available for the European Oligocene genera *Cephalogale, Amphicynodon,* and *Pachycynodon* show all the basicranial and dental features of later ursids, including the primitive 'Type A' bulla, excavated basioccipital, and a shallow suprameatal fossa. *Cephalogale* is digitigrade, yet retains the ursid postscapular fossa. The M^3 is absent, and the upper carnassial has a shelf-like protocone that is more medial in position. The upper molars lack parastyles, their protocones are elongated, the lower molar talonids are correspondingly enlarged transversely and have peripheral cuspids, the M_2 lacks a paraconid, and the M_3 is a large tooth. This type of dentition characterizes the primitive (and paraphyletic) Amphicynodontinae and the digitigrade Hemicyoninae, while the plantigrade Tertiary Ursinae further modify the dentition as described above for the living bears.

Musteloidea

Fossil representatives of the Procyonidae and Mustelidae having osteological features characteristic of their living counterparts are recognizable in the Early Miocene of the New and Old World. Oligocene musteloids give evidence of lineages other than those that may constitute stem groups for the living taxa. Primitive Musteloidea, like their sister-taxon the Ursoidea, have shallow suprameatal fossae open ventrally, an alisphenoid canal, and a posterior carotid foramen posteriorly placed, near the posterior lacerate foramen. Such taxa (e.g. *Amphictis* and *Mustelictis*) have primitive dentitions that retain the primitive conditions of variable development of a parastyle on M^1; a small metaconule on M^1; and a paraconid on M_2. Besides the loss of M_3, *Amphictis* has a long M_2, with a talonid bearing a large hypoconulid, whereas *Mustelictis* has a small, but still double-rooted, M_2, less elongate than in *Amphictis*. The structure of the M_2 in *Amphictis* is a procyonid synapomorphy. *Mustelictis* has small M^2_2, which by virtue of their small size relative to M^1_1, could represent a trend in the mustelid direction. Thus, early in the Oligocene, primitive musteloids show initial stages of dental divergence that later characterized the procyonids and mustelids.

Additional diversity is encountered in the Late Oligocene European *Plesictis* whose deeply excavated, open suprameatal fossa, and inflated 'Type B' bulla with its shortened carotid tube all closely resemble the condition in living procyonids. Dentally, *Plesictis* resembles a mustelid with greatly reduced M^2_2 with one or two roots, loss of metaconule on M^2, and entoconid on M^1.

By the Early Miocene, forms with procyonid auditory regions, large $M^{1,2}$ with metaconules, and M_2 with elongated talonids occur in Europe (*Broiliana* and *Stromeriella*,) and North America ('*Bassariscus*'). In the Late Cenozoic, procyonids whose auditory regions preserve a primitive arctoid condition are represented by two divergent types: the hypocarnivorous lesser panda (*Ailurus*) and the hypercarnivorous *Simocyon* whose superficially canid-like upper dentition and short-faced, massive skull made this taxon the name-bearer of a polyphyletic array of taxa in conventional canid classifications (e.g. Simpson 1945). The diagnostic elongated M_2 talonid and arctoid bulla demonstrates the true procyonid affinities of *Simocyon*.

Schmidt-Kittler's (1981) work has shown that a deep suprameatal fossa occurs in a variety of mustelids, as can be seen in early ontogenetic stages before further development of the posterior limb of the tubular external auditory meatus floors the fossa or fills it in. *Plesictis*, which lacks the tubular meatus, has an auditory region corresponding to an early ontogenetic stage in living forms. That genus, and other Late Oligocene and Early Miocene forms in Europe (*Paragale*, *Plesiogale*) and North America (*Promartes*, *Aelurocyon*, *Zodiolestes*), retain a tiny M^2 and the carnassial notch on P^4, but have lost the alisphenoid canal and show variable covering of the suprameatal fossa depending on the relative size of the tubular meatus. This paraphyletic assemblage must include the progenitor of the living mustelids that appear in the Early Miocene.

Pinnipeds

During the last 15 years there has been an enormous increase in the number of fossil pinniped specimens (summarized by Barnes *et al.* 1985). Although much remains undescribed, significant new information is available on the oldest-known pinnipeds (Mitchell and Tedford 1973; Barnes 1979) and the diversity of Late Tertiary odobenid-like forms (Repenning and Tedford 1977). Early Miocene pinnipeds from California (*Enaliarctos* and *Pinnarctidion*) were assigned to a paraphyletic Enaliarctinae, but it seems clear from comparative osteology that these represent separate clades (Barnes 1979; Wyss 1987). Derived features in the heterodont dentitions of these earliest-known forms support the relationship of pinnipeds to terrestrial arctoids. The shelf-like and anteromedially placed P^4 protocone, and the transversely narrow form of M^1, with its longitudinally elongated protocone, are ursid synapomorphies. The presence of a wide, deeply impressed basioccipital sulcus for the inferior petrosal sinus in *Enaliarctos* and *Pinnarctidion* strongly suggests a carotid loop within the venous plexus as in ursids or, alternatively, as in primitive arctoids as suggested above. Most of the other structures cited by Mitchell and Tedford (1973) in support of an ursid ('Hemicyoninae') relationship are primitive arctoid features, as Wyss (1987) contends. The skull and dentition of *Enaliarctos* are those of a primitive arctoid (most probably an ursid) adapted to pelagic life (Mitchell and Tedford 1973, pp. 261–3). This view is reinforced by the undescribed postcranial skeleton of *Enaliarctos* figured in Barnes *et al.* (1985, fig. 5) in which the characteristic otariid scapula with its large supraspinatus fossa can be seen. Other otariid synapomorphies of *Enaliarctos* are the small-diameter crista tympanica and the periotic paroccipital process. *Pinnarctidion*, on the other hand, shows a number of odobenid and phocid synapomorphies, as pointed out by Wyss (1987).

By Middle and Late Miocene, a number of odobenid-like taxa occur, some known from most of the skeleton (e.g. *Allodesmus, Imagotaria*).

These animals are nearly homodont, but lack the dental specialization of living walruses. They possess a number of odobenid and phocid cranial synapomorphies, as discussed by Wyss (1987). Middle Miocene deposits on the eastern seaboard of North America provide the oldest cranial material (attributed to '*Monotherium*') showing the inflated bulla, large auditory ossicles and petrosal pachyostosis typical of phocids (Ray 1976). Wyss (1987) cites the posterior widening of the palate and a mortised squamosal–jugal union in *Pinnarctidion* and *Allodesmus* as phocid synapomorphies not shared with odobenids. Fossil odobenids confirm the sister-group relationship to phocids indicated by the comparative anatomy of the living forms.

Knowledge of the fossil record of pinnipeds thus serves to expand the diversity of the group and indicate the minimum ages of the principal clades. It does not modify our conception of the phylogenetic relationships of the living families that was deduced from their present-day representatives, but it does provide both additional evidence that pinnipeds are rooted in primitive arctoids (with ursids as their most likely sister-taxon) and a demonstration of the extent of parallel development of some features relative to other clades (e.g. the lack of homology between the 'Type B' bulla of phocids and some mustelids).

Potamotherium/Semantor

Despite our knowledge of the complete osteology of *Potamotherium*, this early arctoid remains phylogenetically elusive. Morphologically it is a mosaic of both primitive and derived musteloid and pinniped features. Any specific phylogenetic conclusion presents character conflicts that are difficult to resolve.

Tedford (1976), Schmidt-Kittler (1981), and Muizon (1982) contended that the late Oligocene *Potamotherium* was a musteloid and a sister taxon to the Phocidae, a relationship alluded to by Savage (1957) in his analysis of that genus. Tedford (1976, p. 370) explicitly set aside the ossicle and petrosal similarities among pinnipeds as parallel adaptations to common pelagic life, although Wyss (1987) showed that more detailed analysis of these features can be useful in phylogeny reconstruction.

Potamotherium possesses features similar to those found in some phocids, some of which are also found in musteloids and/or other caniforms. These features include an inflated bulla, large-diameter crista tympanica, lack of an alisphenoid canal, well-separated paroccipital and mastoid processes, a scapula with enlarged teres process, and evidence for phocid-like swimming detailed by Savage (1957). Enlargement of the auditory ossicles may support a relationship to odobenids and phocids (but see Wyss 1987), and variable development of a round-window fossula resembles the pinniped condition.

Except for the presumed aquatic adaptations, the ear region of *Potamotherium* is primitive for an arctoid. The bulla, although having an important caudal entotympanic component, is a relatively uninflated 'Type A' with a well-developed meatal tube. The malleus retains a small muscular process. The anterior half of the sulcus for the inferior petrosal sinus is markedly widened, possibly for a short carotid loop reflected from a pit in the basisphenoid at the anterior opening of the long carotid canal. Primitively, the crista tympanica is not produced into the tympanic cavity and the paroccipital process is formed by the exoccipital alone.

The dentition includes a P^4 with carnassial notch and M^1 with a parastyle as in primitive arctoids. The M_3 is lacking, and the M^2_2 are tiny and single rooted. The M^1 is musteloid-like in being transversely elongate without a metaconule and having a paracone larger than the metacone. The expanded, shelf-like protocone of P^4 is similar to that in otters, but also to ursids and primitive pinnipeds. The premolars are large and have strong anterior and posterior cusps as in primitive pinnipeds, and some other arctoids.

Three phylogenetic alternatives seem supported by the evidence cited above. The first alternative, that *Potamotherium* is another independently aquatically adapted taxon of musteloid affinity, implies parallel evolution of a number of features similar to those in phocids, or odobenids plus phocids, reflecting a specific mode of aquatic adaptation. The other two alternatives regard these similarities as synapomorphies with pinnipeds, with *Potamotherium* being either a sister taxon of the odobenids plus phocids, or the phocids alone. The musteloid features would then be parallel developments in *Potamotherium*. If *Potamotherium* is a sister taxon to odobenids plus phocids, the pinniped synapomorphies not present in *Potamotherium* would be independently acquired in otariids, and phocids plus odobenids, and the phocid-like similarities would be independently derived homoplasies. If *Potamotherium* is a sister taxon to phocids, the pinniped synapomorphies not present in *Potamotherium* would be independently acquired in three taxa (otariids, odobenids, and phocids).

Potamotherium is geologically older than known pinnipeds. It survived, apparently little changed morphologically, until medial Miocene time; its sister-group *Semantor* appeared in the Late Miocene. The latter genus is even more phocid-like. This rather sterotyped lineage is an isolated clade no matter what its ultimate phylogenetic relationship is determined to be. Both *Potamotherium* and *Semantor* presently are best placed *incertae sedis* within the Arctoidea.

References

Adams, G. I. (1896). The extinct Felidae of North America. *Am. J. Sci.* **1**, 419–44.
Archibald, J. D. (1977). Ectotympanic bone and internal carotid circulation of Eutherians in reference to anthropoid origins. *J. Hum. Evol.* **6**, 609–22.
Arnason, U. (1974). Comparative chromosome studies in Pinnipedia. *Hereditas* **76**, 179–226.
Arnason, U. and Widegren, B. (1986). Pinniped phylogeny enlightened by molecular hybridizations using highly repetitive DNA. *Mol. Biol. Evol.* **3**, 356–65.
Barnes, L. G. (1979). Fossil enaliarctine pinnipeds (Mammalia: Otariidae) from Pyramid Hill, Kern County, California. *Contrib. Sci., Nat. Hist. Mus. Los Angeles Co.* **318**, 1–41.
Barnes, L. G., Domning, D. P., and Ray, C. E. (1985). Status of studies on fossil marine animals. *Mar. Mamm. Sci.* **1**, 15–53.
Beaumont, G. de (1964). Remarques sur la classification des Felidae. *Eclogae Geol. Helv.* **57**, 837–45.
Bugge, J. (1978). The cephalic arterial system in carnivores, with special reference to the systematic classification. *Acta Anat.* **101**, 45–61.
Chow, M. C. (1975). Some carnivores from the Eocene of China. *Vertebr. Palasiat.* **13**, 165–9.
Clark, J. and Guensburg, T. E. (1972). Arctoid genetic characters as related to the genus *Parictis*. *Fieldiana, Geol.* **26**, 1–76.
Cope, E. D. (1875). On the supposed Carnivora of the Eocene of the Rocky Mountains. *Proc. Acad. Nat. Sci. Philadelphia* **27**, 444–9.
Cope, E. D. (1880*a*). On the Genera of the Creodonta. *Proc. Am. Philos. Soc.* **1882**, 76–82.
Cope, E. D. (1880*b*). On the extinct cats of America. *Am. Nat.* **14**, 833–58.
Couturier, J. and Dutrillaux, B. (1986). Evolution chromosomique chez les Carnivores. *Mammalia* **50**, 128–62.
Davis, D. D. (1949). The shoulder architecture of bears and other carnivores. *Fieldiana, Zool.* **31**, 285–305.
Davis, D. D. (1964). The giant panda: a morphological study of evolutionary mechanisms. *Fieldiana, Zool. Mem.* **3**, 1–339.
de Jong, W. W. (1986). Protein sequence evidence for monophyly of the carnivore families Procyonidae and Mustelidae. *Mol. Biol. Evol.* **3**, 276–81.
de Jong, W. W. and Goodman, M. (1982). Mammalian phylogeny studied by sequence analysis of the eye lens protein α-crystallin. *Zeit. Saug.* **47**, 257–76.
Doran, A. H. G. (1877). Morphology of the mammalian Ossicula Auditûs. *Trans. Linn. Soc. London* **1**, 371–497.
Eldredge, N. and Cracraft, J. (1980). *Phylogenetic Patterns and the Evolutionary Process: Method and Theory in Comparative Biology*, viii+349 pp. Columbia University Press, New York.
Ewer, R. F. (1973). *The Carnivores*, xv+494 pp. Cornell University Press, Ithaca.

Farris, J. S. (1972). Estimating phylogenetic trees from distance matrices. *Am. Nat.* **106**, 645–68.

Fay, F. H., Rausch, V. R., and Feltz, E. T. (1967). Cytogenetic comparisons of some pinnipeds (Mammalia: Eutheria). *Can. J. Zool.* **45**, 773–8.

Feng, W., Luo, C., Ye, Z., Zhang, A., and He, G. (1985). The electrophoresis comparison of serum protein and LDH isoenzyme in 5 Carnivora animals—giant panda, red panda, asiatic black bear, cat and dog. *Acta Ther. Sinica* **5**, 151–5.

Flower, W. H. (1869). On the value of the characters of the base of the cranium in the classification of the Order Carnivora and the systematic position of *Bassaris* and other disputed forms. *Proc. Zool. Soc. London* **1869**, 4–37.

Flynn, J. J. and Galiano, H. (1982). Phylogeny of Early Tertiary Carnivora, with a description of a new species of *Protictis* from the Middle Eocene of northwestern Wyoming. *Am. Mus. Novit.* **2725**, 1–64.

Ginsburg, L. (1982). Sur la position systématique du petit panda, *Ailurus fulgens* (Carnivora, Mammalia). *Géobios, Mém. Spéc.* **6**, 247–58.

Gonyea, W. and Ashworth, R. (1975). The form and function of retractile claws in the Felidae and other representative carnivorans. *J. Morphol.* **145**, 229–38.

Goodman, M., Romero-Herrera, A. E., Dene, H., Czelusniak, J., and Tashian, R. E. (1982). Amino acid sequence evidence on the phylogeny of the primates and other eutherians. In *Macromolecular Sequences in Systematic and Evolutionary Biology* (ed. M. Goodman) pp. 3–51. Plenum Press, New York.

Goodman, M., Czelusniak, J., and Beebers, J. E. (1985). Phylogeny of primates and other Eutherian orders: a cladistic analysis using amino acid and nucleotide sequence data. *Cladistics* **1**, 171–85.

Gregory, W. K. and Hellman, M. (1939). On the evolution and major classification of the civets (Viverridae) and allied fossil and recent Carnivora: a phylogenetic study of the skull and dentition. *Proc. Am. Philos. Soc.* **81**, 309–92.

Gregory, W. R. (1936). On the phylogenetic relationships of the giant Panda (*Ailuropoda*) to other arctoid Carnivora. *Am. Mus. Novit.* **878**, 1–29.

Hough, J. R. (1948). The auditory region in some members of the Procyonidae, Canidae and Ursidae: its significance in the phylogeny of the Carnivora. *Bull. Am. Mus. Nat. Hist.* **92**, 67–118.

Hough, J. R. (1953). Auditory region in North American fossil Felidae: its significance in phylogeny. *U.S. Geol. Surv., Prof. Pap.* **243**, 95–115.

Hunt, R. M., Jr. (1974*a*). The auditory bulla in Carnivora: an anatomical basis for reappraisal of carnivore evolution. *J. Morphol.* **143**, 21–76.

Hunt, R. M., Jr. (1974*b*). *Daphoenictis*, a cat-like carnivore (Mammalia, Amphicyonidae) from the Oligocene of North America. *J. Paleontol.* **48**, 1030–47.

Hunt, R. M., Jr. (1977). Basicranial anatomy of *Cynelos* Jourdan (Mammalia: Carnivora), an Aquitanian amphicyonid from the Allier Basin, France. *J. Paleontol.* **51**, 826–43.

Hunt, R. M., Jr. (1987). Evolution of the aeluroid Carnivora: significance of auditory structure in the nimravid cat *Dinictis*. *Am. Mus. Novit.* **2886**, 1–74.

Leone, C. A. and Wiens, A. L. (1956). Comparative serology of carnivores. *J. Mammalogy* **37**, 11–23.

Linneaus, C. (1758). *Systema Naturae per regna tri naturae, secundum classis, ordines, genera, species cum characteribus, differentiis, synonymis locis*, 10th edn., Vol. 1, 824 pp. Laurentii Salvii Stockholm.

McKenna, M. C. (1975). Toward a phylogenetic classification of the Mammalia. In *Phylogeny of the Primates* (eds. W. P. Luckett and F. S. Szalay), pp. 21–46. Plenum, New York.

McLaren, I. A. (1960). Are the Pinnipedia biphyletic? *Syst. Zool.* **9**, 18–28.

MacPhee, R. D. E. (1981). Auditory regions of primates and eutherian insectivores: morphology, ontogeny, and character analysis. In *Contributions to Primatology*, Vol. 18, 282 pp. Karger, Basel.

Martin, L. D. (1980). Functional morphology and the evolution of cats. *Trans. Nebraska Acad. Sci.* **8**, 141–54.

Matthew, W. D. (1909). The Carnivora and Insectivora of the Bridger basin, Middle Eocene. *Mem. Am. Mus. Nat. Hist.* **9**, 289–567.

Matthew, W. D. (1910). The phylogeny of the Felidae. *Bull. Am. Mus. Nat. Hist.* **28**, 289–316.

Mellett, J. S. (1977). Paleobiology of North American *Hyaenodon* (Mammalia, Creodonta). In *Contributions to Vertebrate Evolution*, Vol. 1, 134 pp. Karger, Basel.

Mitchell, E. D. and Tedford, R. H. (1973). The Enaliarctinae, a new group of extinct aquatic Carnivora and a consideration of the origin of the Otariidae. *Bull. Am. Mus. Nat. Hist.* **151**, 201–84.

Mivart, St. G. (1885). Notes on the Pinnipedia. *Proc. Zool. Soc. London* **1885**, 484–500.

Miyamoto, M. M. and Goodman, M. (1986). Biomolecular systematics of eutherian mammals: phylogenetic patterns and classification. *Syst. Zool.* **35**, 230–40.

Muizon, C. de (1982). Les relations phylogénétiques des Lutrinae (Mustelidae, Mammalia). *Géobios, Mém. Spéc.* **6**, 259–77.

Neff, N. A. (1982). *The Big Cats: The Paintings of Guy Coheleach*, 244 pp. Abrams, New York.

Neff, N. A. (1983). The Basicranial Anatomy of the Nimravidae (Mammalia: Carnivora): Character Analyses and Phylogenetic Inferences, Ph.D. Thesis, 643 pp. City University, New York.

Neff, N. A. (1986). A rational basis for *a priori* character weighting. *Syst. Zool.* **35**, 110–23.

Novacek, M. J. (1977). Aspects of the problem of variation, origin, and evolution of the eutherian auditory bulla. *Mammal Rev.* **7**, 131–49.

Novacek, M. J. (1986). The skull of leptictid insectivorans and the higher-level classification of eutherian mammals. *Bull. Am. Mus. Nat. Hist.* **183**, 1–112.

Novacek, M. J. and Wyss, A. R. (1986). Higher-level relationships of the recent eutherian orders: morphological evidence. *Cladistics* **2**, 257–87.

O'Brien, S. J., Nash, W. G., Wildt, D. E., Bush, M. E., and Benveniste, R. E.

(1985). A molecular solution to the riddle of the giant panda's phylogeny. *Nature (London)* **317**, 140–4.

Pauly, L. K. and Wolfe, H. R. (1957). Serological relationships among members of the Order Carnivora. *Zoologica* **42**, 159–66.

Petter, G. (1974). Rapports phyletiques des viverridés (Carnivores Fissipèdes). Les formes de Madagascar. *Mammalia* **38**, 605–36.

Pocock, R. I. (1916*a*). On some of the cranial and external characters of the hunting leopard or cheetah (*Acinonyx jubatus*). *Ann. Mag. Nat. Hist.* **18**, 419–29.

Pocock, R. I. (1916*b*). On some of the external characters of *Cryptoprocta*. *Ann. Mag. Nat. Hist.* **17**, 413–25.

Pocock, R. I. (1920). On the external and cranial characters of the European badger (*Meles*) and American badger (*Taxidea*). *Proc. Zool. Soc. London* **1920**, 423–36.

Pocock, R. I. (1921). The auditory bulla and other cranial characters in the Mustelidae. *Proc. Zool. Soc. London* **1921**, 473–86.

Pocock, R. I. (1929). The structure of the auditory bulla in the Procyonidae and Ursidae, with a note on the bulla of *Hyaena*. *Proc. Zool. Soc. London* **1928**, 963–74.

Powell, R. A. (1982). *The Fisher*. University of Minnesota Press, Minneapolis.

Presley, R. (1979). The primitive course of the internal carotid artery in mammals. *Acta Anat.* **103**, 238–44.

Radinsky, L. B. (1973). Are stink badgers skunks? Implications of neuroanatomy for mustelid phylogeny. *J. Mammal.* **54**, 585–93.

Radinsky, L. B. (1981). Evolution of skull shape in carnivores. 1. Representative modern carnivores. *Biol. J. Linn. Soc. London* **15**, 369–88.

Ray, C. E. (1976). *Phoca cuymani* and other Tertiary seals (Mammalia: Phocidae) described, from the eastern seaboard of North America. *Smithson. Contrib. Paleobiol.* **28**, 1–36.

Repenning, C. A. and Tedford, R. H. (1977). Otarioid seals of the Neogene. *U. S. Geol. Surv., Prof. Pap.* **992**, (i–vi) 1–93.

Romero-Herrera, A. E., Lehmann, H., Joysey, K. A., and Friday, A. E. (1978). On the evolution of myoglobin. *Phil. Trans. R. Soc. Lond. B* **283**, 61–163.

Sarich, V. M. (1969*a*). Pinniped origins and the rate of evolution of carnivore albumins. *Syst. Zool.* **18**, 286–95.

Sarich, V. M. (1969*b*) Pinniped phylogeny. *Syst. Zool.* **18**, 416–22.

Savage, R. J. G. (1957). The anatomy of *Potamotherium*, an Oligocene lutrine. *Proc. Zool. Soc. London* **129**, 151–244.

Schmidt-Kittler, N. (1981). Zur stammegeschichte der marderverwandten Raubtiergruppen (Musteloidea, Carnivora). *Eclog. Geol. Helv.* **74**, 753–801.

Seal, U. S. (1969). Carnivora systematics: a study of hemoglobins. *Comp. Biochem. Physiol.* **31**, 799–811.

Segall, W. (1943). The auditory region of the arctoid carnivores. *Zool. Ser. Field Mus. Nat. Hist.* **29**, 33–59.

Shoshani, J. (1986). Mammalian phylogeny: comparison of morphological and molecular results. *Mol. Biol. Evol.* **3**, 222–42.

Simpson, G. G. (1945). The principles of classification and a classification of mammals. *Bull. Am. Mus. Nat. Hist.* **85**, (i–xvi) 1–350.

Szalay, F. S. (1977). Phylogenetic relationships and a classification of the eutherian Mammalia. In *Major Patterns in Vertebrate Evolution* (eds. M. K. Hecht, P. C. Goody, and B. M. Hecht), pp. 315–74. Plenum, New York.

Tandler, J. (1899). Zur vergleichenden Anatomie der Kopfarterein bei den Mammalia. *Denkschr. Akad. Wissens. Wien, Math-Naturwiss. Kl.* **67,** 677–784.

Tedford, R. H. (1976). Relationship of pinnipeds to other carnivores (Mammalia). *Syst. Zool.* **25**, 363–74.

Teilhard de Chardin, P. (1945). Les Félides de Chine: Première Partie. Les formes fossiles. *Publ. Inst. Geobiol.* **11**, 1–35.

Thenius, E. (1979). Zur systematischen und phylogenetischen Stellung des Bambusbären: *Ailuropoda melanoleuca* David (Carnivora, Mammalia). *Z. Saugetierk.* **44**, 286–305.

Turner H. N., Jr. (1848). Observations relating to some of the foramina at the base of the skull in Mammalia, and on the classification of the Order Carnivora. *Proc. Zool. Soc. London* **1848**, 63–88.

Van der Klaaw, C. J. (1931). The auditory bulla in some fossil mammals, with a general introduction to this region of the skull. *Bull. Am. Mus. Nat. Hist.* **52**, 1–352.

Van Kampen, P. N. (1905) Die Tympanalgegend des Säugetierschädels. *Morphol. Jb.* **34**, 321–722.

Wible, J. R. (1983). The internal carotid artery in early eutherians. *Acta Palaeontol. Polon.* **28**, 281–93.

Wible, J. R. (1984). The ontogeny and phylogeny of the mammalian cranial arterial pattern, Unpublished Ph.D. Thesis, 705 pp. Duke University, North Carolina.

Wible, J. R. (1986). Transformations in the extracranial course of the internal carotid artery in mammalian phylogeny. *J. Vertebr. Paleontol.* **6**, 313–25.

Wozencraft, W. C. (1984). A phylogenetic reappraisal of the Viverridae and its relationship to other Carnivora, Unpublished Ph.D. Thesis, University of Kansas.

Wurster, D. H. and Benirschke, K. (1968). Comparative cytogenetic studies in the Order Carnivora. *Chromosoma (Berl.)* **24,** 336–82.

Wurster-Hill, D. H. and Bush, M. (1980). The interrelationships of chromosome banding patterns in the giant panda (*Ailuropoda melanoluca*), hybrid bear (*Ursus middendorffi* × *Thalarctos maritimus*), and other carnivores. *Cytogenet. Cell Genet.* **27**, 147–54.

Wurster-Hill, D. H. and Centerwall, W. R. (1982). The interrelationships of chromosome banding patterns in canids, mustelids, hyaena, and felids. *Cytogenet. Cell Genet.* **34**, 178–92.

Wurster-Hill, D. H., and Gray, C. W. (1975). The interrelationships of chromosome banding patterns in procyonids, viverrids and felids. *Cytogenet. Cell. Genet.* **15**, 306–31.

Wyss, A. R. (1987). The walrus auditory region and the monophyly of pinnipeds. *Am. Mus. Novit.* **2871**, 1–13.

5. Phylogeny of the insectivores

P. M. BUTLER
Department of Zoology, Royal Holloway and Bedford New College, Englefield Green, Surrey, UK

Abstract

The Lipotyphla (or Insectivora *sensu stricto*) are one of the main products of the Late Cretaceous eutherian radiation. By the Middle Palaeocene they had divided into the suborders Erinaceomorpha and Soricomorpha, of which the most primitive genera are *Adunator* and *Leptacodon*, respectively.

The classification of Erinaceomorpha is revised. Dormaaliinae (including *Macrocranion*), Litocherinae (including *Litolestes*), and Amphilemuridae are removed from the Lipotyphla. The sister-group of Erinaceidae is the North American family Scenopagidae. The centre of erinaceid evolution was in Asia. A new subfamily Tupaiodontinae is proposed for *Tupaiodon*, *Ictopidium*, and the North American *Entomolestes*. *Adapisorex* and *Leipsanolestes* are not erinaceids.

The most primitive soricomorph family is Nyctitheriidae. An early African branch gave rise to the Tenrecidae and Chrysochloridae. *Solenodon* is not closely related to Soricidae. It may be the only survivor of a North American branch that includes *Centetodon* and *Nesophontes*, and possibly *Apternodus*. Talpidae and Soricidae are sister-groups, probably of Asiatic origin.

Palaeoryctes is not a soricomorph. The Cretaceous ancestry of the Lipotyphla is unknown: evidence supporting a leptictid or a palaeoryctid relationship is unconvincing. The Order may have existed as an independent group since the Campanian, but whether *Paranyctoides* is an early member has yet to be established.

The Phylogeny and Classification of the Tetrapods, Volume 2: Mammals (ed. M. J. Benton), Systematics Association Special Volume No. 35B, pp. 117–41. Clarendon Press, Oxford, 1988.

Introduction

The earlier conception of an Order Insectivora, ancestral to the other eutherian orders and defined by primitive characters (e.g. Simpson 1945) has now been abandoned. Although there is still no consensus on the relationships of many of the fossil families, the living insectivores are now generally divided into the three orders Scandentia, Macroscelidea, and Lipotyphla (Butler 1972). This paper deals only with the Lipotyphla, which contains six out of the eight living families. It has been proposed to use Insectivora as the name of this group (McKenna 1975; Novacek 1980), but the term is still much employed with its older, wider meaning, and to avoid confusion I will continue to use Lipotyphla.

Far from being a relict group, the Lipotyphla is the fifth largest living order of Eutheria, exceeding the Primates in genera and species; yet comparatively little attention has been paid to its phylogeny. Although the fossil record goes back to the Palaeocene, few genera prior to the Oligocene are known by more than teeth and jaws.

The living lipotyphlan families are: Erinaceidae (hedgehogs and moon-rats), Talpidae (moles and desmans), Soricidae (shrews), Solenodontidae (*Solenodon*), Tenrecidae (tenrecs and otter-shrews), and Chrysochloridae (golden moles). The Nesophontidae (*Nesophontes*) are subfossils in the West Indies. On the basis of skull characters, the Erinaceidae are the sister-group of the remaining families, and the suborders Erinaceomorpha and Soricomorpha were distinguished by Butler (1956); the Soricomorpha possess derived features of the skull, lacking in Erinaceidae (Fig. 5.1).

Lipotyphlan characters

Monophyly of the Lipotyphla rests upon derived characters shared by Erinaceidae and Soricomorpha.

A characteristic lipotyphlan apomorph is the mobile snout or proboscis, used in probing for insect prey among ground litter, in soil, etc. It is associated with an enhanced dependence on olfaction and the tactile sense of the vibrissae; the eye is reduced, the olfactory organs are enlarged, and the infraorbital canal is of large diameter for the passage of sensory nerves. The muscles that move the snout are also enlarged. These characters are more advanced in soricomorphs than in erinaceids, which are at an intermediate level of specialization. The erinaceid eye and optic foramen, though proportionately smaller than in most mammals, are larger than in soricomorphs. In most sori-

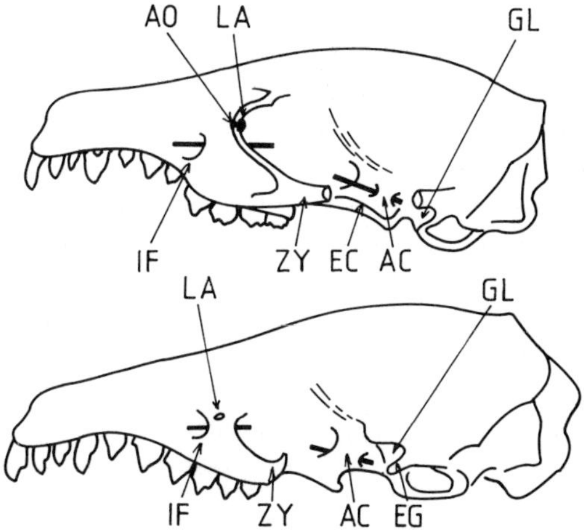

Fig. 5.1. Skulls of an erinaceomorph (*Galerix*, above) and a soricomorph (*Microgale*, below). AC, alisphenoid canal; AO, antorbital crest; EC, ectopterygoid process; GL, glenoid fossa; IF, infraorbital foramen; LA, lacrimal foramen; ZY, zygomatic process (arch of *Galerix* removed).

comorphs, the infraorbital foramen has retreated to below the eye, reducing the infraorbital canal to a bridge. In erinaceids, and in primitive soricomorphs such as *Solenodon*, the nasolabial muscles originate from a depressed area of the maxilla, below, anterior to, and above the orbit, and limited posteriorly by an antorbital crest. In most living soricomorphs the muscles are longer, passing over and under the eye to originate from the frontal and squamosal, and the antorbital crest has disappeared (Fig. 5.1).

In the medial wall of the orbit of all lipotyphlans, a large wing of the maxilla has displaced the palatine, which is far removed from the lacrimal and confined to the floor of the orbit. Novacek (1980) suggested that the composition of the orbital wall was related to the relative development of the olfactory capsule, eye, and brain.

In most mammals, the jugal forms the lower edge of the orbit and much of the zygomatic arch, overlying the maxilla, but in lipotyphlans it is reduced or absent, and the lower border of the orbit is formed entirely from the maxilla. Possibly, reduction of the jugal is related to the weakness of the zygomatic arch, which has been lost in most soricomorphs.

McDowell (1958) gave as a character of Lipotyphla the absence of a true postglenoid process, but there is much diversity of the jaw joint within the group. Primitively (e.g. didelphids, leptictids, tupaiids), the postglenoid process articulates with the medial part of the posterior surface of the condyle, the lateral part being occupied by a ligament.

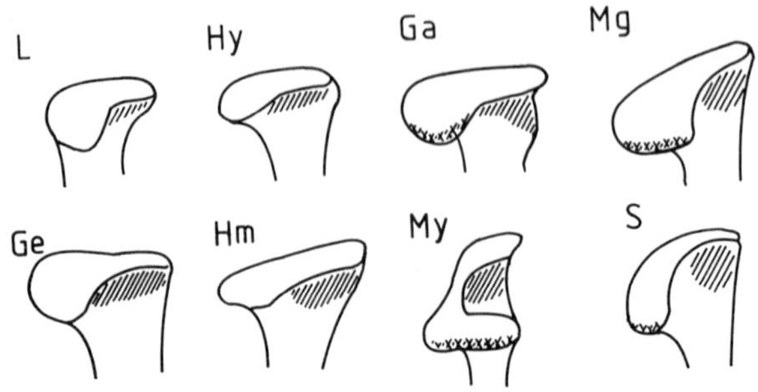

Fig. 5.2. Right mandibular condyles, posterior view. Primitive: *Leptictis* (L), *Gesneropithex* (Amphilemuridae) (Ge). Erinaceomorphs: *Hylomys* (Hy), *Hemiechinus* (Hm). Soricomorphs: *Galemys* (Talpidae) (Ga), *Myosorex* (Soricidae) (My), *Microgale* (Tenrecidae) (Mg), *Setifer* (Tenrecidae) (S).

The articular facet of the condyle extends on to the posterior surface towards the medial side. In erinaceids, with reduction of the postglenoid process, the facet is mainly dorsal, but early lipotyphlans such as *Leptacodon* and *Adapisorex* retain the primitive condition. In most soricomorphs, the condyle is extended medially, together with the postglenoid process (then called entoglenoid process), and the well-developed posteriorly-facing facet can spread even to the ventral surface, as in shrews (McDowell 1958) (Fig. 5.2).

The tympanic membrane is more horizontal in lipotyphlans than in other eutherians (e.g. inclined at about 20° in *Erinaceus*). This has generally been considered a primitive character, but Kielan-Jaworowska (1981) found an elevation of about 45° in the Cretaceous eutherian *Asioryctes*, where the tympanic lies near the plane of the dentary, and she considered the more horizontal orientation to be derived (see also MacPhee 1981).

The floor of the tympanic cavity, medial to the tympanic membrane, is ossified in erinaceids and in some soricomorphs (Talpidae, Tenrecidae, Chrysochloridae) from a tympanic wing of the basisphenoid, supplemented posteriorly by a wing of the petrosal. The tympanic may extend laterally in the wall of the meatus, but it never expands medially to the membrane, and there is no entotympanic. The basisphenoid bulla is peculiar to lipotyphlans. However, in other soricomorphs (Soricidae, *Solenodon*, *Centetodon*, *Apternodus*) it is absent, as well as in the Eocene erinaceomorph *Diacodon* (Novacek, McKenna *et al.* 1983). These have a large pyriform fenestra (an unossified area in the roof of the tympanic chamber between the petrosal and the alisphenoid). The pyriform fenestra is an embryonic character of all insectivores (MacPhee 1981) and is probably primitive. It persists in most tenrecids and is largest in *Geogale* where the basisphenoid wing is small, and absent or nearly so in Potamogalinae and Tenrecinae where the wing is large. In the Talpidae, the fenestra is present in *Uropsilus* but not in the more fossorial genera in which the ear is highly ossified. Closure of the fenestra and development of the basisphenoid wing can be interpreted as two aspects of a greater ossification of the middle-ear region. It seems probable that the bulla was primitively absent and that it developed more than once within the Lipotyphla.

The internal carotid, after giving off the stapedial artery, crosses the tympanic chamber in a groove on the promontorium (secondarily closed over in some erinaceids and most talpids). The transpromontorial course is more primitive than the medial course followed in some mammals, including marsupials and monotremes (Wible 1986). The earlier view that promontorial and medial carotid arteries were both primitively present, is rejected following the work of Presley (1979), who showed that both types of carotid are derived from the embryonic dorsal aorta. The absence of a promontorial groove in *Kennalestes* and *Asioryctes* (Kielan-Jaworowska 1981), the erinaceomorph *Diacodon* (Novacek, McKenna *et al.* 1983), and palaeoryctoids (McKenna *et al.* 1984) is probably to be explained by a failure of the artery to make close contact with the promontorium; this might be a primitive character.

McDowell (1958) mentions a process of the periotic immediately behind the fenestra rotunda, that defines a pit into which the fenestra opens. This is the caudal tympanic process of MacPhee (1981). It is absent in leptictids (Novacek 1986) and in *Palaeoryctes* (McDowell 1958), but present in *Asioryctes* and in *Kennalestes* (Kielan-Jaworoska 1981), in *Diacodon*, and in the palaeoryctoid *Prosarcodon* (McKenna *et al.* 1984). It is regarded as primitive.

Lipotyphlan dentitions are very varied, especially the anterior teeth. The primitive eutherian condition, with small incisors and larger canines, occurs in some Palaeocene lipotyphlans, and it survived in some galericine erinaceids, perhaps also in *Nesophontes* and some tenrecids, if their large canines are not secondary. The enlarged upper canine of *Talpa* is surely derived, as it functions with a caniniform P_1. There are primitively four premolars, here numbered 1–4, though 1,2,4,5 would probably be strictly correct (Novacek 1985*a*). *Leptacodon tener* might be exceptional in having five upper premolars (McKenna 1968), but there are only four in the lower jaw; possibly tooth B is an incisor, pushed back *post mortem* beyond the canine tooth (A). P^4_4 are semimolariform in Palaeocene lipotyphlans such as *Adapisorex*, *Leptacodon*, and *Adunator*, and in Eocene Nyctitheriidae; P^4 has a metacone, and P_4 has a basined talonid with two or three cusps. All later lipotyphlans have lost the P^4 metacone and have simplified the P_4 talonid; molariform premolars survive only in zalambdodonts, in which these changes have taken place on the molars as well. In

Talpidae and *Nesophontes* the metaconid of P_4 is also lost, and it is reduced in Soricidae.

The molars of all modern lipotyphlans have departed more or less widely from the primitive tribosphenic pattern, but this is approached in Palaeocene forms such as *Leptacodon*. McDowell (1958) gives as a lipotyphlan character the interradicular crests on the basal surface of the molars, which fit into corresponding grooves in the interalveolar bone (Butler 1948). On upper molars three crests meet in the centre of the tooth, where in Soricidae, Talpidae, and *Nesophontes* there is a short supplementary root. On lower molars the crest is median in Erinaceidae and in Plesiosoricidae, and lateral in Talpidae; in Soricidae a boss of alveolar bone fits into the concave base of the tooth and is crossed at the top by a groove for the interradicular crest. Weaker crests, which do not clearly mark the bone, are present in many mammals, and they are probably primitive; they have been seen in *Adapisorex* and *Macrocranion*. Their further development might have taken place independently in different lipotyphlan families. They are absent in tenrecids, chrysochlorids, and *Centetodon*, perhaps secondarily.

Lipotyphlans are basically small ambulatory mammals and, apart from special adaptations to fossorial (Talpidae, Chrysochloridae) or aquatic life (Potamogalinae, Desmaninae), their postcranial skeletons are mostly generalized (Novacek 1980). A clearly derived character is the greatly reduced pubic symphysis, situated far posteriorly. It is lost in Talpidae (except *Desmana* and *Condylura*) and in Soricidae, although it is present in the embryo (Leche 1883). The significance of reduction of the symphysis is unknown; it is otherwise found in bats, dermopterans, and some rodents and edentates.

The absence of the caecum in lipotyphlans is a derived character, possibly associated with their animal diet. The caecum is reduced or lost within several eutherian orders (Stevens 1980).

Some of the resemblances between extant lipotyphlans may be the result of parallel evolution, as palaeontological evidence seems to suggest. During their long independent history, lipotyphlans have no doubt undergone much change. Some have remained remarkably primitive, especially the Tenrecidae, isolated in Madagascar, which are at a low stage in the evolution of homoeothermy (Crompton *et al.* 1978); the tenrecid *Echinops* has the smallest neocortex among living mammals (Stephan 1967). Lipotyphla are more diverse than other orders in early embryonic development (Strauss 1942). Although most have an advanced haemochorial placenta, that of the mole *Scalopus* is epitheliochorial like that of lemurs (Mossman 1939).

Erinaceomorpha

Erinaceomorpha and Erinaceoidea have been used in the past in a wide, poorly defined sense. Thus, Gregory (1910) included the Lepticidae and Dimylidae, and McDowell (1958) included the Talpidae. Van Valen (1967) proposed Adapisoricidae for the basal family of Erinaceota (i.e. Lipotyphla minus the zalambdodonts), and included in it many Early Tertiary genera of obscure relationships. In a recent revision, Novacek *et al.* (1983) divided the Palaeocene and Eocene erinaceomorphs into three families, the Erinaceidae, Dormaaliidae, and Amphilemuridae. Following Krishtalka (1976*a*), they transferred *Adapisorex* and some other genera to the Erinaceidae.

The two existing subfamilies of Erinaceidae, the Erinaceinae and Galericinae, were already distinct in the Oligocene, when *Amphechinus* and *Tetracus* entered Europe, presumably from Asia. Both subfamilies invaded North America and Africa in the Miocene. The Galericinae are the more primitive in the skull, dentition and postcranial skeleton (Butler 1948, 1956, 1984). Some fossil forms have been put into other subfamilies: Neurogymnurinae (Oligocene, Europe and Asia); Brachyericinae (*Exallerix*, Oligocene, Asia, and *Brachyerix*, Miocene, North America; Rich and Rich 1971; Rich 1981); Protericinae (Oligocene, North America; Gawne 1968; Bjork 1975; there is a probable Late Eocene North American ancestor, as yet undescribed). Except the Protericinae, all the erinaceid subfamilies seem to have originated in Asia, probably early in the Eocene.

The erinaceid dentition is characterized by an emphasis on the P^4–M_1 shear, which bears some resemblance to the carnassial function of carnivores; P^4 has a high paracone and a long metacrista; M_1 is enlarged, with an extended trigonid due to the forwardly placed paraconid, which takes the form of a high paracristid. Shear between P^4 and M_1 is primitive, occurring in soricomorphs and Cretaceous eutherians, and it has been enlarged in soricids in the same way as in erinaceids. In both families, M^1 also has an extended metacrista, which shears against the trigonid of M_2, and in talpids this shear is more important than that between P^4 and M_1. Erinaceids differ, however, in that on $M_{2,3}$ the paraconid does not take part in the shear; it is reduced to a transverse crest, at the level of the talonid of the preceding tooth, and it wears on the top against the enlarged hypocone of the upper molar. With emphasis on P^4 and M^1, the posterior molars are reduced, the M^2 metacrista is very short, and the zygomatic process of the maxilla is shifted forwards to bring the adductor jaw muscles nearer to the shearing teeth. Similar changes have taken place in the Carnivora.

Tupaiodon and *Ictopidium*, from the Oligocene of Mongolia and China

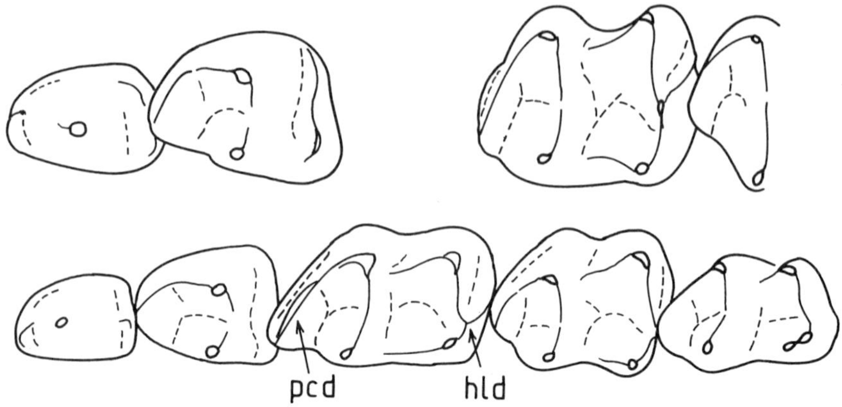

Fig. 5.3. Right lower cheek teeth of *Ictopidium lechei* (above) and *I. tatalgolensis* (below). hld, hypoconulid; pcd, paracristid.

(Matthew and Granger 1924; Zdansky 1930; Sulimski 1969), are more primitive than Galericinae and I propose for them a new subfamily **Tupaiodontinae**. M_1 is unknown in *I. lechei*, but from the space available in the type it was subequal in length to M_2, whereas in *I. tatalgolensis* and *T. minutus* M_2 is smaller than M_1 as in other erinaceids (Fig. 5.3). However, P_4 and M_2 of *I. lechei* closely resemble the other species, and the relationship is undeniable (*contra* Novacek *et al.* 1983). The upper dentition, known only in *T. morrisi*, is distinguished from that in Galericinae by the more transverse molars, with small hypocones (Fig. 5.4,T). The lower molars are also proportionately wider, and the M_1 trigonid is shorter than it is wide, though the paracristid forms a high shearing crest as in other erinaceids. The hypoconulid, absent in other erinaceids, is present as a rudiment on most specimens of M_1 and M_2. On M_3 it projects posteriorly and is connate with the entoconid (the two cusps are fused in *I. tatalgolensis*). There is a wide posterior cingulum, ending at the hypoconulid. The anterior dentition is specialized in that the canines are small and the premolar series anterior to P_3 is shortened: in *I. tatalgolensis* P_1 is absent and P_2 has one root: the other species have two alveoli between P_3 and the canine, either for a two-rooted P_2 or two one-rooted premolars. The maxilla of *Tupaiodon* has a strong antorbital crest as in *Galerix*.

Some isolated teeth, very similar to *Tupaiodon* and *Ictopidium*, have been reported from the Early Oligocene of Yunnan (Rich *et al.* 1983). *Entomolestes grangeri*, from the Middle Eocene of North America, seems to be tupaiodontine. In the lower dentition it closely resembles the Asiatic species, except that the buccal and posterior cingula of the molars are narrower, and the anterior premolars are less reduced: P_1 is present, and P_2, though small, is two-rooted (Krishtalka and West 1977).

The mainly Eocene family Dormaaliidae was divided by Novacek (1985*b*) into the subfamilies Scenopaginae, Dormaaliinae, and Sespedectinae. It is distinguished from Erinaceidae in lacking the specialization of the M_1 trigonid: the paraconid forms a low transverse paracristid, as on the other molars. M_2^2 are not reduced, and the zygomatic arch arises farther back (over M^2). The hypoconulids are better developed, and the talonid of P_4, simple in erinaceids, shows signs of a former semimolariform condition. There are typically two mental foramina, whereas one is normally absent in erinaceids. A derived character, paralleled in tupaiodontines, is the undifferentiated lower anterior dentition, with small canine and short, one-rooted P_2.

McKenna and Simpson (1959) put *Scenopagus* in the Erinaceidae on account of the resemblance of its upper cheek teeth to *Tupaiodon*. *Ankylodon*, its Oligocene descendant, shows exaggeration of the shearing crests, reduced hypoconulids, and an enlarged hypocone, joined to the trigon by a crest as in most erinaceids (Lillegraven *et al.* 1981). *Scenopagus* and *Ankylodon* agree with erinaceids in that the protocone is in line with the reduced paraconule and the paracone, and the metaconid is lingual or slightly forward from the protoconid (Fig. 5.4,Sc). The skull of *Ankylodon* has a strong antorbital crest as in *Galerix* and *Tupaiodon* (Fox 1983), and the mandibular condyle has a dorsal articulating surface, implying a weak postglenoid process (Lillegraven *et al.* 1981).

On the other hand, the dormaaliine *Macrocranion* has less acute molar cusps, shorter shearing crests, paraconule subequal to the metaconule, more posteriorly placed protocone and metaconid (Fig. 5.4,M), and reduced metaconid on P_4. The skull (Maier 1979) is markedly different from that of the *Ankylodon*, with weak antorbital crest and strong postglenoid process. The upper canine is enlarged. *Macrocranion* departs from erinaceids and other lipotyphlans in the jugal, which extends over the maxilla at the edge of the orbit, as in leptictids; also the braincase is larger, the hindlimbs are longer (Koenigswald and Storch 1983), and the distal phalanges are very short, stout and triangular. The basis for including *Macrocranion* in the Lipotyphla is weak, and until more information is available it is best left as Insectivora *incertae sedis* (i.e. Proteutheria *sensu* Butler 1972). *Scenopagus* and *Ankylodon*, which share derived characters with Erinaceidae, are therefore removed from the Dormaaliidae into a new family Scenopagidae (Novacek 1985*b*, new rank). The Sespedectinae share some characters with scenopagids rather than *Macrocranion* (small upper canine, large P^3 as in *Ankylodon*, reduced paraconules, lingual P_4 metaconid), and they may tentatively be classified as a subfamily of Scenopagidae.

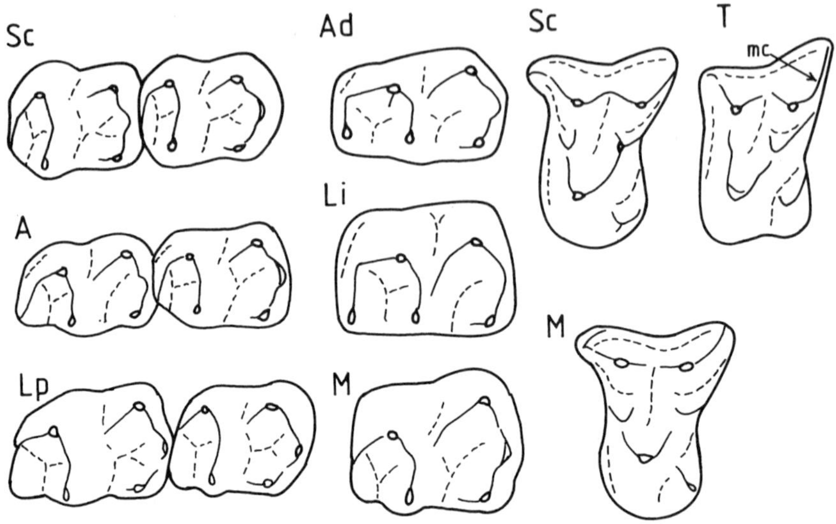

Fig. 5.4. Right $M_{1,2}$ of *Scenopagus edenensis* (Sc), *Adunator minutus* (A), and *Leptacodon tener* (Lp); right M_1 of *Adapisorex abundans* (Ad), *Litolestes ignotus* (Li), and *Macrocranion nitens* (M); left M^1 of *Scenopagus edensis* (Sc), *Tupaiodon morrisi* (T), and *Macrocranion nitens* (M). mc, metacrista.

The common ancestor of Scenopagidae and Erinaceidae may be conceived as having an M_1 as in *Leptacodon*, with a moderately high paracristid that sheared against P^4, and was inclined forward at an angle of about 20° from the transverse direction (as seen in crown view). In Erinaceidae it became higher and more oblique (primitively about 45°): in Scenopagidae it became lower and more transverse, and shearing was confined to the preprotocristid. Some poorly-known genera may represent stages in the evolution of the scenopagid paracristid. *Talpavus* has a low paracristid that is more divergent than in *Scenopagus*. *Dartonius* has a higher paracristid, much like *Leptacodon*. It differs from the Erinaceidae, where it was assigned by Novacek *et al.* (1983) in having higher molar trigonids, a partly molariform P_4 talonid, and a one-rooted P_2, agreeing in these characters with *Scenopagus*.

Adunator (including *Diacocherus* and *Mckennatherium*; see Bown and Schankler 1982), from the Palaeocene of Europe and North America, is probably the most primitive erinaceomorph genus. It is very similar to *Leptacodon*, from which it differs in the lower, transverse, non-sectorial paraconids of $M_{2,3}$. In *A. ladae*, M_1 also has this type of paraconid, and Krishtalka (1976*a*) thought that this species was near the ancestry of *Scenopagus*. In *A. lehmani* (Storch and Lister 1985, fig. 35) and *A. minutus*, the paracristid of M_1 is higher and projects forward at an angle of about

30°, as would be expected in a potential erinaceid (Figs. 5.4,A). *Adunator* is more primitive than *Scenopagus* but resembles *Galerix* in the enlarged lower canine and unreduced, two-rooted P_2. P_4 is submolariform, and the upper dentition of *A. minutus* (Gingerich 1983) shows a metacone on P^4 and P^3. The leptictid-like upper teeth assigned to *A. lehmani* by Russell (1964) and Storch and Lister (1985) are probably misidentified.

Diacodon, from the Early Eocene of North America (Novacek 1982), resembles *Adunator* in most dental characters, and is probably a direct descendant. The paracristid of M_1 is transverse, as in *A. ladae*. Some characters of the skull are known (Novacek, McKenna *et al.* 1983): there is a strong postglenoid process, articulating with the back of the condyle; the bulla is absent; a pyriform fenestra is present; and there is no groove for the promontory artery. *Adunator* presumably also possessed these primitive characters.

The Palaeocene genera *Litolestes, Adapisorex* and *Leipsanolestes*, included in the Erinaceidae by Novacek *et al.* (1983), do not resemble tupaiodontines or galericines in the M_1 trigonid. *Leipsanolestes* has a short paracristid that does not reach the lingual side of the tooth. In *Adapisorex* the trigonid is extended, but not due to the paracristid, which is transverse, with a lingual swelling representing the paraconid (Fig. 5.4,Ad). In *Litolestes* the paracristid has a similar structure, and the metaconid is backwardly displaced (Fig. 5.4,Li). In *Adapisorex* and *Leipsanolestes* the posterior premolars are submolariform as in *Adunator*, *Leipsanolestes* has forwardly placed metaconids, low trigonids, and reduced hypoconulids as in Erinaceidae, and it could be an offshoot from the scenopagid–erinaceid stem. *Adapisorex* has blunt-cusped, quadrate upper molars with well-developed hypocones. It has been considered as a hyopsodontid, but it may provisionally be left in the Erinaceomorpha, *incertae sedis*.

Litolestes may be a dormaaliid; Bown and Schankler (1982) noted resemblances to *Macrocranion*. It has low molar cusps, backwardly placed metaconids on P_4 and M_1, reduced P_4 talonid, one-rooted P_2 and reduced lower canine (regarded as a first premolar by Schwarz and Krishtalka 1976). Two characters—enlargement of P_4^4 and M_1^1, and diversion of the oblique crest on the lower molars towards the metaconid—are shared with *Litocherus*, *Cedrocherus*, and the Early Eocene *Eolestes*; the four genera can be grouped as a subfamily Litocherinae (Gingerich 1983), provisionally included in the Dormaaliidae. *Litocherus* is primitive in having a larger lower canine, a less reduced P_4 talonid and a two-rooted P_2.

The Amphilemuridae (Eocene, Europe) are probably not lipotyphlan. *Pholidocercus* has a large jugal (Koenigswald and Storch 1983, fig. 2). A relationship to Dormaaliidae is possible, but there are many differences. Meaningful discussion must await Storch's report on the skull of *Pholidocercus*.

Soricomorpha

Gregory (1910) included in the Soricomorpha only the Soricidae and Talpidae. Butler (1956) and McDowell (1958) added *Nesophontes* and the zalambdodonts (*Solenodon*, Tenrecidae, Chrysochloridae), but McDowell removed the Talpidae, which he regarded as erinaceomorph. Butler (1972) added a number of fossil families (Nyctitheriidae, Geolabididae, Plesiosoricidae, Micropternodontidae, and (?) Apternodontidae). He removed the Tenrecidae and Chrysochloridae into separate suborders. Removal of the Tenrecidae was based mainly on the resemblance to erinaceids of a cranium referred to the Miocene tenrecid *Erythrozootes*, but new material shows that this specimen was misidentified and belongs to the erinaceid *Amphechinus* (Butler 1984); the Tenrecidae are here regarded as soricomorph. Schmidt-Kittler (1973) showed that the Dimylidae, formerly assigned to the Erinaceomorpha, are related to the Talpidae. *Palaeoryctes* and the Micropternodontidae were classified as Soricomorpha by McKenna *et al.* (1984), but are excluded here for reasons given in the following section.

Soricomorph skulls are, in general, more derived than those of erinaceids, in such characters as smaller eyes, shorter infraorbital canal, elimination of the midcranial constriction, and loss of ectopterygoid processes. However, erinaceid characters survive as plesiomorphs in several families, indicating parallel evolution. Thus, the Miocene tenrecid *Protenrec* has a long infraorbital canal opening anteriorly above P^3, an antorbital crest, and a midcranial constriction (Butler 1984); ectopterygoid processes are large in the talpid Desmaninae, and present as rudiments in the talpid *Uropsilus* and some soricids, e.g. *Myosorex*. The zygomatic arch has been lost within the Soricidae, as it was retained in the oldest subfamily Heterosoricinae; it is present in Talpidae, Chrysochloridae, and Geolabididae.

An apparently diagnostic soricomorph character is the long alisphenoid canal whose medial wall is unossified so that the internal maxillary artery passes medially to the alisphenoid, leaving through the sphenorbital fissure. In the erinaceids, it is external to the alisphenoid, in a groove that is bridged over for a short distance (Fig. 5.1).

The transversely extended mandibular condyle of *Solenodon*, *Nesophontes*, and most tenrecids is probably an autapomorph of the suborder (McDowell 1958). It is present in the nyctitheriids *Saturninia* and *Amphidozotherium* (Sigé 1976), but not in *Leptacodon tener* (McKenna

1968), which is probably primitive. This type of condyle occurs in the Desmaninae (Fig. 5.2,Ga), and in a more reduced form in *Urotrichus*, but in other Talpidae the condyle is not extended, the glenoid cavity being restricted medially by the bulla: the latter condition is regarded as secondary. The same probably applies to the simple condyle of the Chrysochloridae. The knob-like condyle of the Tenrecinae (Fig. 5.2,S) is adapted for rotation of the jaw around its long axis in chewing (Oron and Crompton 1985).

If the absence of the basisphenoid bulla is primitive (witness *Diacodon*), its presence in the Talpidae and also in the Tenrecidae and Chrysochloridae must be a parallelism with the Erinaceidae. Other characters by which McDowell (1958) supported a talpid–erinaceid relationship are primitive (ectopterygoid processes) or they are also found in Soricidae (crossed mm. rectus abdominis). Likewise, the lack of a bulla does not support a special relationship between the Soricidae and *Solenodon*.

It has long been held that shrews and moles are closely related. This is partly because of the shrew-like appearance of the most primitive moles, which resemble shrews in the limb skeleton (Campbell 1939; Reed 1951; Hutchison 1968). This might be convergence or plesiomorphy, but the two families share some unusual characters. Thus, in the chondrocranium the auditory capsule has an opercular process unknown in other insectivores (de Beer 1937); the cartilaginous pubic symphysis is resorbed, except in Desmaninae (Leche 1883); m. quadratus femoris is fused with m. obturator externus, uniquely among mammals (Leche 1883). Both families differ from *Solenodon* in loss of the antorbital crest, dilambdodont molars, ventral margin of the foramen magnum reaching beyond the level of the jugular foramina, two papillae circumvallatae on the tongue (three in *Solenodon* and Tenrecidae), crossed mm. rectus abdominis, and more extensive fusion of the tibia and fibula; the last two characters are shared with the Erinaceidae.

McDowell (1958) listed 17 characters in which *Solenodon* resembled the Soricidae and differed from the Tenrecidae, but several of these are shared also with the Talpidae: postpalatine torus, absence of frenum, reduction of milk dentition, inguinal testes, and conus inguinalis; to these might be added interradicular crests and upright coronoid process. *Solenodon* resembles the Soricidae in embryological characters (Wislocki 1940): large yolk-sac, persistent yolk-sac placenta, small allantoic vesicle, antimesometrial implantation. The Talpidae are more advanced in having a small yolk-sac, but otherwise they share these characters (Strauss 1942). There remain as shared–derived characters only the lateral glands, and the behavioural character of 'chattering'. McDowell compared the latter with the 'twittering' of shrews. It might be associated with echolocation, like the tongue-clicking of tenrecids (Gould 1965). The view taken here is that the Soricidae and Talpidae are sister-groups, *Solenodon* is related to their common ancestor, and the Tenrecidae branched off earlier.

The molar teeth of the soricomorphs have specialized in two divergent directions, towards zalambdodonty and towards dilambdodonty (Fig. 5.5) In zalambdodonts, the stylar shelf of the upper molars is hypertrophied, with large styles B and D (nomenclature of Simpson 1929), the metacone is lost, the protocone is reduced, and the paracone retreats lingually. On the lower molars, the trigonid is more elevated and the talonid is reduced and simplified. In dilambdodonts, the stylar shelf develops a mesostyle (style C), which is joined by crests to the metacone and paracone, and is frequently doubled. The metacone tends to be enlarged, and it is separated from the paracone by a deep groove through which the hypoconid passes during chewing. On the lower molars, the trigonid and talonid are more equal in area and

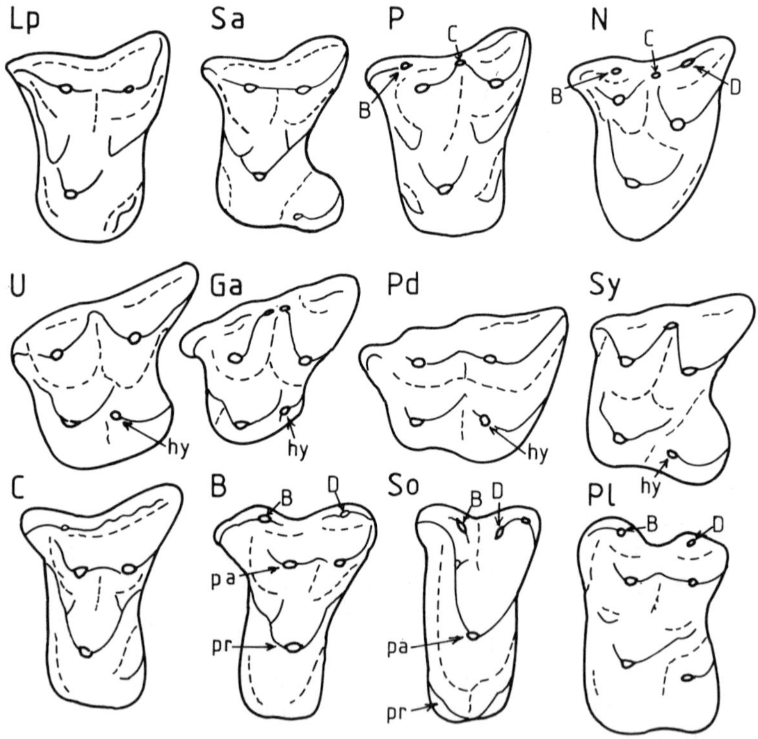

Fig. 5.5. Right M^1 of *Leptacodon* (Lp), *Saturninia* (Sa), *Pontifactor* (P), *Nesophontes* (N), *Uropsilus* (U), *Galemys* (Ga), *Plesiodimylus* (Pd), *Sylvisorex* (Sv), *Centetodon* (C), *Batodonoides* (B), *Solenodon* (So), and *Plesiosorex* (Pl). B, C, D, stylar cusps; hy, hypocone; pa, paracone; pr, protocone.

height. Both these types of molar were presumably derived from the erinaceoid type found in *Leptacodon*, in which the stylar shelf is merely a cingulum. An early stage in the trend towards dilambdodonty is seen in *Nyctitherium velox*, in which the groove between paracone and metacone is deep and the cingulum is elevated to form a mesostyle at the end of the groove. In *Pontifactor*, the stylar shelf is better developed, and it bears a complete series of styles, of which C is joined to the paracone and metacone. A still stronger stylar shelf, with an enlarged style C, is present in *Nesophontes*.

The trend towards zalambdodonty may perhaps be seen in the nyctitheriid *Saturninia*, in which the metacone is lower than the paracone, and the intervening groove is at a more ventral level than the buccal cingulum; at the same time, the trigonid and talonid are more disparate in height. This trend is more advanced in the Geolabididae (Lillegraven *et al.* 1981). The styles are rudimentary in *Centetodon*, but in *Batodonoides* (Novacek 1976) they are larger, especially B (Fig. 5.5,B). Styles B and D are well developed in the Plesiosoricidae, and in *Butselia* the metacone is reduced and the trigonid is elevated (Quinet and Misonne 1965; Butler 1972). In the Tenrecidae, *Potamogale* has a small metacone, and *Protenrec* a rudimentary one (Butler 1984). Zalambdodonty and dilambdodonty have evolved in the Soricomorpha more than once.

McDowell (1958) homologized the high lingual cusp of *Solenodon* with the high metacone of *Nesophontes* and maintained that the paracone, rather than the metacone, had been lost. This does not accord with the occlusal relations: in the bat *Harpiocephalus*, where the paracone is reduced, the lower molars are modified quite differently from *Solenodon*.

It is commonly stated that the Talpidae differ from the Soricidae in the absence of the hypocone, which is replaced by an enlarged metaconule. Comparison of *Uropsilus* and dimylids with soricids (Fig. 5.5) indicates, however, that it is the metaconule that has disappeared: *Uropsilus* resembles shrews in having a wide posterior cingulum extending beyond the hypocone; the cingulum is reduced in other talpids. Expansion of this cingulum also occurs in Nyctitheriidae (but not in *Leptacodon* or *Pontifactor*) and (especially on P^4) in *Plesiosorex*. In *Nesophontes* the hypocone has been lost, a parallelism to some tupaiids (Butler 1980).

Speculation on the relationships of the fossil soricomorph families can be made only in a general way. Current studies of the skulls of *Centetodon* and the Apternodontidae by McKenna and co-workers will no doubt clarify the picture. A tentative scheme is presented in Fig. 5.6.

The oldest and most primitive soricomorphs are the Palaeocene—Early Oligocene Nyctitheriidae (reviewed by Sigé (1976) and Krishtalka (1976*b*)). They are known only by teeth and jaws. The family is

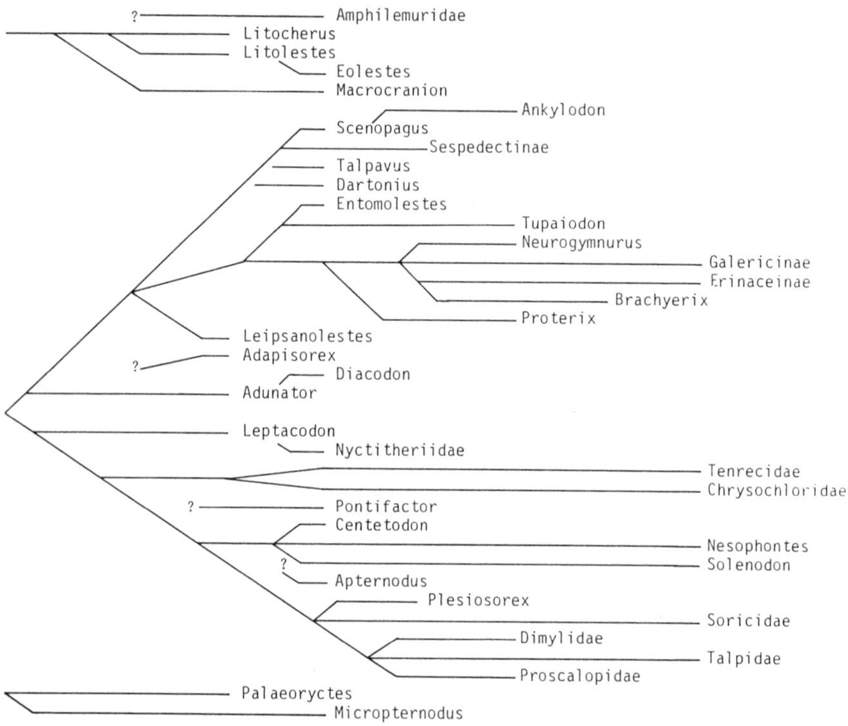

Fig. 5.6. Postulated phylogenetic relations of insectivores.

probably paraphyletic, and it has been made rather artificial by a tendency to use it as a 'waste-basket'. I would, for instance, exclude *Pontifactor* and the Eocene genera from Mongolia described by Russell and Dashzeveg (1986), and I would include only those genera that have submolariform last premolars, i.e. *Leptacodon, Plagioctenodon* (Bown and Schankler 1982), *Nyctitherium, Saturninia, Scraeva,* and *Amphidozotherium.* Advances that can be detected within the family are shortening of the infraorbital canal (the foramen is above P^3 in *Leptacodon catulus,* above P^4 in later genera), de-differentiation of the canines (largest in *Leptacodon* and *Plagioctenodon*), evolution of multicuspid incisors, widening of the hypocone cingulum (narrow in *Leptacodon*), widening of the mandibular condyle (rounded in *Leptacodon tener*), loss of a mental foramen (in the European genera), and increased verticality of the coronoid process (*Amphidozotherium*).

The geolabidid *Centetodon* (McKenna 1960; Lillegraven *et al.* 1981) is more advanced in the loss of the P^4 metacone and reduction of the P_4 talonid, and probably also in reduction of the milk dentition. The skull agrees with that in *Solenodon* and *Nesophontes* in many respects, but apparently retains the zygomatic arch. There is a rather close resemblance to *Amphidozotherium* in the maxilla and the lower jaw. It is not impossible that *Centetodon, Solenodon*, and *Nesophontes* had a common North American ancestor close to the Nyctitheriidae. If so, a considerable diversification of the dentitions would be implied.

McDowell (1958) removed *Apternodus* from the Lipotyphla, but his account of the skull contained some misinterpretations (McKenna 1975). The skull, though specialized, has soricomorph characters. Despite its zalambdodont molars, *Apternodus* is probably not especially related to *Solenodon*: the cusp patterns differ in certain details, as McDowell noted.

Plesiosorex resembles moles in the extremely short, bridge-like infraorbital canal (Wilson 1960). It resembles shrews in the large incisors (I_2^1; the homology of the lower incisor of shrews is uncertain) and reduction of canines and anterior premolars, in the expanded talon of P^4, the emphasis on the P^4 and M_1 shear, and loss of the zygomatic arch. Unlike shrews and moles, the hypoconulid is median, and the metaconid of P_4 is well developed and lingual. The Plesiosoricidae are probably of Asiatic origin; the oldest known genus is *Pakilestes*, from the Eocene of Pakistan, which resembles *Butselia* (Russell and Gingerich 1981). They may be related to Talpidae and Soricidae, which also probably originated in Asia.

Of the three living subfamilies of Talpidae, the Uropsilinae contains only *Uropsilus* from China, and it has a poor palaeontological record. Primitive members of the Desmaninae and Talpinae, however, approach *Uropsilus* in structure (Hutchison 1968, 1974). The oldest talpid, *Eotalpa* (Late Eocene, UK; Sigé, *et al.* 1977), appears to be a talpine. The Talpinae did not reach North America till the Miocene. They were preceded there by the Proscalopidae (Oligocene—Early Miocene), fossorial animals in which the hands moved backwards beneath the body instead of laterally as in the Talpinae, and which resemble chrysochlorids in the form of the skull (Barnosky 1981). Their teeth share derived characters with the Talpidae and reveal their relationship.

The Dimylidae, from the Oligocene and Miocene of Europe, are also related to the Talpidae. Their cheek teeth, though low-crowned and blunt-cusped, are derivable from the primitive talpid pattern preserved in *Uropsilus*. The root pattern is also similar. Schmidt-Kittler (1973) inferred from the blindly ending zygomatic process that dimylids had no zygomatic arch; however, the process might be the equivalent of the subzygomatic process of *Uropsilus*, from which the arch arises dorsally.

The record of the African families, Tenrecidae and Chrysochloridae, goes back only to the Miocene (Butler 1984). They share a number of primitive characters, such as the cloaca, undescended testes, and large yolk-sac, but also some derived characters, including basisphenoid bulla, zalambdodont molars, and orthomesometrial implantation of the blastocyst (Strauss (1942) regards this as primitive, Luckett (1975) considers it derived). They diverge widely, however, in the male reproductive system: thus, in the Chrysochloridae the penis is small, backwardly pointing and without the accessory erectile tissue present in other soricomorphs (Kaudern 1910). The chrysochlorid skull is highly modified in adaptation to burrowing: there is a marked cranial flexure, and the chondrocranium is very distinctive (Roux 1947). Jacobson's organ resembles that of macroscelidids (Broom 1915), but some of its characters are present in the Soricidae (du Toit 1942). A common African origin of chrysochlorids and tenrecids is tentatively accepted, but the differences indicate an early separation.

The origin of the Lipotyphla

It is beyond the scope of this article to consider the relationships of the Lipotyphla to other orders (see Chapter 3), but it is desirable to comment briefly on their possible relation to some other insectivoran groups. The leptictids have long been held to be related to the erinaceoids (Simpson 1945; Clemens 1973), and Novacek (1986) has recently found support for a lipotyphlan–leptictid relationship from characters of the skull. On the other hand, a soricomorph–palaeoryctid relationship is favoured by McKenna, who regards *Batodon* as a probable geolabid (in Novacek 1976) and classifies *Palaeoryctes* in the Soricomorpha (McKenna *et al.* 1984). A third hypothesis, proposed by Fox (1984), is that lipotyphlans are derived from neither leptictoids nor palaeoryctoids but from a third group of Cretaceous eutherians, represented by *Paranyctoides*.

Palaeocene lipotyphlans such as *Leptacodon* and *Adunator* have some resemblance to leptictids in the semimolariform premolars, but in leptictids the molarization is more advanced, and P^3_3 are enlarged in comparison to P^4_4; these characters can be seen also in *Gypsonictops* as far back as the Judithian (near the Campanian–Maastrictian boundary; Lillegraven and McKenna 1986). Resemblances in the bones of the face, found by Novacek (1986), could be parallelisms: reduced posterior width of the nasals occurs in many eutherians; reduction of the facial wing of the lacrimal, and shortening of the intraorbital canal (which is longer in *Gypsonictops*) could be the result of a parallel enlargement of the nasolabial musculature; the orbital wing of the maxilla is much smaller in leptictids than in lipotyphlans. The common recess for the sphenopalatine and dorsal palatine foramina occurs in

Asioryctes (Kielan-Jaworowska 1981). Most shared characters of the ear region are probably primitive, as some are found in *Asioryctes*, but there are significant differences in the presence in leptictids of an entotympanic bulla and a high clinoid process, and the absence of a caudal tympanic process, while in lipotyphlans the postglenoid part of the skull is lengthened so that the mastoid exposure is lateral rather than posterior. Some resemblances, such as the ectopterygoid processes and the groove for the promontorial artery, are shared with tupaiids and macroscelideans.

The skull of *Palaeoryctes* (McDowell 1958; Van Valen 1966) agrees with soricomorphs in possessing a pyriform fenestra. The absence of ectopterygoid processes is probably a primitive character shared with *Asioryctes*, but in the Soricomorpha it is secondary. McDowell excluded *Palaeoryctes* from the Lipotyphla because it has no groove for a promontorial artery, and no caudal tympanic process. The bulla is formed from the ectotympanic, as in *Asioryctes*. Other characters of the skull are more primitive than in soricomorphs, but shared with erinaceids: midcranial constriction; a groove for the maxillary artery on the external surface of the alisphenoid; mandibular condyle not extended transversely; infraorbital foramen above P^3 and well separated from the lacrimal foramen, which is above M^1. The dentition, with well-developed canines, non-molariform last premolars, molar metacones appressed to the paracones, no hypocone or posterior cingulum, and elevated trigonids, can easily be derived from *Cimolestes propalaeoryctes* (Lillegraven 1969), but it is very distinct from primitive Lipotyphla. The presence of a pyriform fenestra, unsupported by other evidence, seems an insufficient reason to include *Palaeoryctes* in the Soricomorpha.

A stronger case can be made for inclusion of the Micropternodontidae. *Micropternodus* appeared in North America in the Late Eocene, but three genera are known from the Palaeocene of Asia (McKenna *et al.* 1984). *Micropternodus* differs from the Palaeoryctidae in having enlarged hypocones, a semimolariform P_4 with a metaconid, a short infraorbital canal, and no zygomatic arch (as in *Solenodon*). In the Palaeocene *Prosarcodon*, the hypocone is represented by a cingulum and the infraorbital canal is longer. The basicranium of *Prosarcodon* shows a caudal tympanic process, but otherwise it is very like *Palaeoryctes*, with a pyriform fenestra and no groove for a promontorial artery. The canines are enlarged, the trigonids are high, and the metacone of M^1 is lower than and close to the paracone (M^3 has been lost). Thus, *Micropternodus* appears to have been derived from a form near to *Palaeoryctes*. However, it is much too specialized to be considered as a primitive soricomorph; more probably it is a palaeoryctoid that has paralleled the Soricomorpha in some respects. The characters listed by McKenna *et al.* (1984) in support of lipotyphlan affinities, such as the resemblance of the tympanic region to *Diacodon*, and the weak sella turcica, are plesiomorphous and not diagnostic.

Evidence that *Batodon* is a geolabidid rather than a palaeoryctid is weak. *Batodon* differs from *Centetodon* in having an incipient metacone on P^4, like *Cimolestes simpsoni*, and a large lower canine and a one-rooted P_1, as in all species of *Cimolestes*. It differs from most species of *Cimolestes* in having cingula near the protocone of the upper molar, and a metaconid on P_4; in these respects, it resembles micropternodontids. The high trigonids of *Centetodon*, which give the lower cheek teeth a palaeoryctoid appearance, are probably derived in comparison with the less-elevated trigonids of the Palaeocene lipotyphlans.

The discovery of *Paranyctoides* (Fox 1970, 1979, 1984) in the Aquilan and Judithian faunas of North America showed the existence, at an early date, of eutherians that did not have the high trigonids of leptictids and palaeoryctoids, but resembled *Leptacodon* in the lower molars. Though Fox classified *Paranyctoides* as a lipotyphlan, he recognized a similarity also to *Protungulatum*; the teeth in fact have a structure that could be ancestral to many types of molar. The upper molars have low cusps, with subequal and well-separated paracone and metacone, and well-developed conules. As Fox stated, there are a number of differences from *Leptacodon* and other early lipotyphlans: the metaconid of P_4 is absent in *P. sternbergi* and very rudimentary in *P. maleficus*; the P_4 talonid is not semimolariform; the upper molars are not transverse, and they have a rather wide stylar shelf with cuspules. These could be primitive features. Fox (1984) derived the Condylarthra and Primates from erinaceoids, but apart from a resemblance in the molars which could reasonably be attributed to parallel evolution, the evidence that these orders are of lipotyphlan origin is slight. *Paranyctoides* might belong to any or none of these groups. What it shows is how little we know about eutherian evolution in the Cretaceous.

Acknowledgements

In preparing this paper I have drawn upon observations of fossil insectivores made over many years in numerous museums, and I take this opportunity to thank the many curators who have provided access to specimens and facilities to study them. Special thanks should go to Drs M. C. McKenna and M. J. Novacek at the American Museum of Natural History, to Dr J. J. Hooker at the British Museum (Natural History), and to Dr D. R. Russell at the Museum d'Histoire Naturelle, Paris, for stimulating discussions and much help and encouragement.

References

Barnosky, A. D. (1981). A skeleton of *Mesoscalops* (Mammalia, Insectivora) from the Miocene Deep River Formation, Montana, and a review of the proscalopid moles: evolutionary, functional, and stratigraphic relationships. *J. Vertebr. Paleontol.* **1**, 285–239.

Bjork, P. R. (1975). Observations on the morphology of the hedgehog genus *Proterix* (Insectivora: Erinaceidae). *Univ. Mich. Pap. Paleontol.* **12**, 81–8.

Bown, T. M. and Schankler, D. (1982). A review of the Proteutheria and Insectivora of the Willwood Formation (lower Eocene), Bighorn Basin, Wyoming. *U.S. Geol. Surv. Bull.* **1523**, 1–79.

Broom, R. (1915). On the organ of Jacobson and its relations in the Insectivora. *Proc. Zool. Soc. London* **1915**, 157–62, 347–54.

Butler, P. M. (1948). On the evolution of the skull and teeth in the Erinaceidae, with special reference to fossil material in the British Museum. *Proc. Zool. Soc. London* **118**, 446–500.

Butler, P. M. (1956). The skull of *Ictops* and the classification of the Insectivora. *Proc. Zool. Soc. London* **126**, 453–81.

Butler, P. M. (1972). The problem of insectivore classification. In *Studies in Vertebrate Evolution* (eds. K. A. Joysey and T. S. Kemp), pp. 253–65. Oliver and Boyd, Edinburgh.

Butler, P. M. (1980). The tupaiid dentition. In *Comparative Biology and Evolutionary Relationships of Tree Shrews* (ed. W. P. Luckett), pp. 171–204. Plenum, New York.

Butler, P. M. (1984). Macroscelidea, Insectivora and Chiroptera from the Miocene of East Africa. *Palaeovertebrata* **14**, 117–200.

Campbell, B. (1939). The shoulder anatomy of the moles. A study in phylogeny and adaptation. *Am. J. Anat.* **64**, 1–39.

Clemens, W. A. (1973). Fossil mammals of the type Lance Formation, Wyoming. Part III. Eutheria and summary. *Univ. Calif. Publ. Geol. Sci.* **94**, 1–102.

Crompton, A. W., Taylor, C. R. and Jagger, J. A. (1978). Evolution of homeothermy in mammals. *Nature (London)* **272**, 333–6.

de Beer, G. R. (1937). *The Development of the Vertebrate Skull.* Clarendon Press, Oxford.

du Toit, A. E. (1942). The nasal region of the chondrocranium of *Elephantulus. S. Afr. J. Med. Sci.* **7**, 33–49.

Fox, R. C. (1970). Eutherian mammal from the Early Campanian (Late Cretaceous) of Alberta, Canada. *Nature (London)* **227**, 630–1.

Fox, R. C. (1979). Mammals from the Upper Cretaceous Oldman Formation, Alberta. III. Eutheria. *Can. J. Earth Sci.* **16**, 114–25.

Fox, R. C. (1983). New evidence on the relationships of the Tertiary insectivoran *Ankylodon* (Mammalia). *Can. J. Earth Sci.* **20**, 968–77.

Fox, R. C. (1984). *Paranyctoides maleficus* (new species), an early eutherian mammal from the Cretaceous of Alberta. *Carnegie Mus. Spec. Publ.* **9**, 9–20.

Gawne, C. E. (1968). The genus *Proterix* (Insectivora, Erinaceidae) of the Upper Oligocene of North America. *Am. Mus. Novit.* **2315**, 1–26.

Gingerich, P. D. (1983). New Adapisoricidae, Pentacodontidae and Hyposodontidae (Mammalia, Insectivora and Condylarthra) from the Late Paleocene of Wyoming and Colorado. *Contrib. Mus. Paleontol. Univ. Mich.* **26**, 227–55.

Gould, E. (1965). Evidence for echolocation in the Tenrecidae of Madagascar. *Proc. Am. Philos. Soc.* **109**, 352–60.

Gregory, W. K. (1910). The orders of mammals. *Bull. Am. Mus. Nat. Hist.* **27**, 3–524.

Hutchison, J. H. (1968). Fossil Talpidae (Insectivora, Mammalia) from the later Tertiary of Oregon. *Bull. Mus. Nat. Hist. Univ. Oregon* **11**, 1–117.

Hutchison, J. H. (1974). Notes on type specimens of European Miocene Talpidae and a tentative classification of the Old World Tertiary Talpidae (Insectivora, Mammalia). *Géobios* **7**, 211–56.

Kaudern, W. (1910). Studien über die männlichen Geschlechtsorgane von Insectivoren und Lemuriden. *Zool. Jahrb. Abt. Anat.* **31**, 1–106.

Kielan-Jaworowska, Z. (1981). Evolution of the therian mammals of the late Cretaceous of Asia. Part IV. Skull structure in *Kennalestes* and *Asioryctes*. *Palaeontol. Polon.* **42**, 25–78.

Koenigswald, W. von and Storch, G. (1983). *Pholidocercus hassiacus*, ein Amphilemuride aus dem Eozän der 'Grube Messel' bei Darmstadt (Mammalia, Lipotyphla). *Senckenbergiana Lethaea* **54**, 447–95.

Krishtalka, L. (1976*a*). Early Tertiary Adapisoricidae and Erinaceidae (Mammalia, Insectivora) of North America. *Bull. Carnegie Mus. Nat. Hist.* **1**, 1–40.

Krishtalka, L. (1976*b*). North American Nyctitheriidae (Mammalia, Insectivora). *Ann. Carnegie Mus.* **46**, 7–28.

Krishtalka, L. and West, R. M. (1977). The Bridgerian insectivore, *Entomolestes grangeri*. *Contrib. Biol. Geol. Milwaukee Publ. Mus.* **14**, 1–11.

Leche, W. (1883). Zur Beckenregion bei Insectivora, mit besonderer Berücksichtigung ihrer morphologische Beziehungen zu derjenigen anderer Säugethiere. *K. Sven. Vet. Akad. Handb.* **20**, 1–113.

Lillegraven, J. A. (1969). Latest Cretaceous mammals of upper part of Edmonton Formation of Alberta, Canada, and review of marsupial–placental dichotomy in mammalian evolution. *Univ. Kans. Paleontol. Contrib.* **50**, 1–122.

Lillegraven, J. A. and McKenna, M. C. (1986). Fossil mammals from the 'Mesaverde' Formation (Late Cretaceous, Judithian) of the Bighorn and Wind River Basins, Wyoming, with definitions of Late Cretaceous North American land-mammal ages. *Am. Mus. Novit.* **2840**, 1–68.

Lillegraven, J. A., McKenna, M. C., and Krishtalka, L. (1981). Evolutionary relationships of Middle Eocene and younger species of *Centetodon* (Mammalia, Insectivora, Geolabididae), with a description of the dentition of *Ankylodon*. *Univ. Wyo. Publ.* **45**, 1–115.

Luckett, W. P. (1975). Ontogeny of the fetal membranes and placenta. Their bearing on primate phylogeny. In *Phylogeny of the Primates* (eds. W. P. Luckett and F. S. Szalay), pp. 157–82. Plenum, New York.

MacPhee, R. D. E. (1981). Auditory regions of primates and eutherian insectivores. Morphology, ontogeny and character analysis. *Contrib. Primatol.* **18**, 1–284.

Maier, W. (1979). *Macrocranion tupaiodon*, an adapisoricid(?) insectivore from the Eocene of 'Grube Messel' (Western Germany). *Paläontol. Z.* **53**, 38–62.

Matthew, W. D. and Granger, W. (1924). New insectivores and ruminants from the Tertiary of Mongolia, with remarks on the correlation. *Am. Mus. Novit.* **105**, 1–7.

McDowell, S. B. (1958). The Greater Antillean insectivores. *Bull. Am. Mus. Nat. Hist.* **115**, 113–214.

McKenna, M. C. (1960). The Geolabidinae, a new subfamily of early Cenozoic erinaceoid insectivores. *Univ. Calif. Publ. Geol. Sci.* **37**, 131–64.

McKenna, M. C. (1968). *Leptacodon*, an American Paleocene nyctithere (Mammalia, Insectivora). *Am. Mus. Novit.* **2317**, 1–12.

McKenna, M. C. (1975). Toward a phylogenetic classification of the Mammalia. In *Phylogeny of the Primates* (eds. W. P. Luckett and F. S. Szalay), pp. 21–46. Plenum, New York.

McKenna, M. C. and Simpson, G. G. (1959). A new insectivore from the Middle Eocene of Tabernacle Butte, Wyoming. *Am. Mus. Novit.* **1952**, 1–12.

McKenna, M. C., Xue, X. and Zhow, M. (1984). *Prosarcodon lonanensis*, a new Paleocene micropternodontid palaeoryctoid insectivore from Asia. *Am. Mus. Novit.* **2780**, 1–17.

Mossman, H. W. (1939). Comparative morphogenesis of the fetal membranes and accessory uterine structures. *Contrib. Embryol. Carnegie Inst. Wash.* **26**, 129–246.

Novacek, M. J. (1976). Insectivora and Proteutheria of the later Eocene (Uintan) of San Diego County, California. *Spec. Publ. Nat. Hist. Mus. Los Angeles County.* **283**, 1–52.

Novacek, M. J. (1980). Cranioskeletal features in tupaiids and selected Eutheria as phylogenetic evidence. In *Comparative Biology and Evolutionary Relationships of Tree Shrews* (ed. W. P. Luckett), pp. 35–93. Plenum, New York.

Novacek, M. J. (1982). *Diacodon alticuspis*, an erinaceomorph insectivore from the early Eocene of northern New Mexico. *Contrib. Geol. Univ. Wyo.* **20**, 135–49.

Novacek, M. J. (1985*a*). The primitive eutherian dental formula. *J. Vertebr. Paleontol.* **6**, 191–6.

Novacek, M. J. (1985*b*). The Sespedectinae, a new subfamily of hedgehog-like insectivores. *Am. Mus. Novit.* **2822**, 1–24.

Novacek, M. J. (1986). The skull of leptictid insectivorans and the higher-level classification of eutherian mammals. *Bull. Am. Mus. Nat. Hist.* **183**, 1–112.

Novacek, M. J., Bown, T. M. and Schankler, D. (1983). On the classification of the early Tertiary Erinaceomorpha (Insectivora, Mammalia). *Am. Mus. Novit.* **2813**, 1–22.

Novacek, M. J., McKenna, M. C., Neff, N. A. and Cifelli, R. L. (1983). Evidence from the earliest known erinaceomorph basicranium that

insectivorans and primates are not closely related. *Nature (London)* **306**, 683–4.
Oron, U. and Crompton, A. W. (1985). A cineradiographic and electromyographic study of mastication in *Tenrec ecaudatus*. *J. Morphol.* **185**, 155–82.
Presley, R. (1979). The primitive course of the internal carotid artery in mammals. *Acta Anat.* **103**, 238–44.
Quinet, G. E. and Misonne, X. (1965). Les insectivores zalambdodontes de l'Oligocène inférieur belge. *Bull. Inst. R. Sci. Nat. Belg.* **41**, (19), 1–15.
Reed, C. A. (1951). Locomotion and appendicular anatomy in three soricoid insectivores. *Am. Midl. Nat.* **45**, 513–671.
Rich, T. H. V. (1981). Origin and history of the Erinaceinae and Brachyericinae (Mammalia, Insectivora) in North America. *Bull. Am. Mus. Nat. Hist.* **171**, 1–116.
Rich, T. H. V. and Rich, P. V. (1971). *Brachyerix*, a Miocene hedgehog from western North America, with a description of the tympanic regions of *Paraechinus* and *Podogymnura*. *Am. Mus. Novit.* **2477**, 1–58.
Rich, T. H. V., Zhang, Y.-P. and Hand, S. J. (1983). Insectivores and a bat from the early Oligocene Caijiachong Formation of Yunnan, China. *Aust. Mammal.* **6**, 61–75.
Roux, G. H. (1947). The cranial development of certain Ethiopian 'insectivores' and its bearing on the mutual affinities of the group. *Acta Zool. (Stokholm)* **28**, 165–397.
Russell, D. E. (1964). Les mammifères paléocènes d'Europe. *Mem. Mus. Natl. Hist. Nat. Paris. Ser. C* **13**, 1–324.
Russell, D. E. and Dashzeveg, D. (1986). Early Eocene insectivores (Mammalia) from the People's Republic of Mongolia. *Palaeontology* **29**, 269–91.
Russell, D. E. and Gingerich, P. D. (1981). Lipotyphla, Proteutheria(?) and Chiroptera (Mammalia) from the Early-Middle Eocene Kuldana Formation of Kohat (Pakistan). *Contrib. Mus. Paleontol. Univ. Mich.* **25**, 277–87.
Schmidt-Kittler, N. (1973). *Dimyloides*-Neufunde aus der oberoligocänen Spaltenfüllung 'Ehrenstein 4' (Süddeutschland) und die systematische Stellung der Dimyliden (Insectivora, Mammalia). *Mitt. Bayer. Staatssamml. Paläontol. Hist. Geol.* **13**, 115–39.
Schwarz, J. H. and Krishtalka, L. (1976). The lower antemolar teeth of *Litolestes ignotus*, a late Paleocene erinaceid (Mammalia, Insectivora). *Ann. Carnegie Mus.* **46**, 1–6.
Sigé, B. (1976). Insectivores primitifs de l'Eocène supérieur et Oligocène inférieur d'Europe occidentale. Nyctitheriidés. *Mem. Mus. Natl. Hist. Nat., Paris, Ser. C* **34**, 1–140.
Sigé, B., Crosset, J. Y. and Insole, A. (1977). Les plus vieilles taupes. *Géobios, Mém. Spéc.* **1**, 141–57.
Simpson, G. G. (1929). American Mesozoic mammals. *Mem. Peabody Mus.* **3**, 1–235.
Simpson, G. G. (1945). The principles of classification and a classification of mammals. *Bull. Am. Mus. Nat. Hist.* **85**, 1–350.
Stephan, H. (1967). Zur Entwicklungshöhe der Insektivoren nach Merkmalen des Gehirns und die Definition der 'Basalen Insektivoren'. *Zool. Anz.* **179**, 177–99.

Stevens, C. E. (1980). The gastrointestinal tract of mammals: major variations. In *Comparative Physiology: Primitive Mammals* (eds. K. Schmidt-Nielsen, L. Bolis and C. R. Taylor), pp. 52–62. Cambridge University Press.

Storch, G. and Lister, A. M. (1985). *Leptictidium nasutum*, ein Pseudorhynchocyonide aus dem Eozän der 'Grube Messel' bei Darmstadt (Mammalia, Proteutheria). *Senckenbergiana Lethaea* **86**, 1–37.

Strauss, F. (1942). Vergleichende Beurteilung der Placentation bei den Insektivoren. *Rev. Suisse Zool.* **49**, 269–82.

Sulimski, A. (1969). On some Oligocene insectivore remains from Mongolia. *Palaeontol. Polon.* **21**, 53–70.

Van Valen, L. (1966). Deltatheridia, a new order of mammals. *Bull. Am. Mus. Nat. Hist.* **132**, 1–126.

Van Valen, L. (1967). New Paleocene insectivores and insectivore classification. *Bull. Am. Mus. Nat. Hist.* **135**, 217–84.

Wible, J. R. (1986). Transformations in the extracranial course of the internal carotid artery in mammalian phylogeny. *J. Vertebr. Paleontol.* **6**, 313–25.

Wilson, J. R. (1960). Early Miocene rodents and insectivores from north eastern Colorado. *Univ. Kans. Paleontol. Contrib. Vertebr.* **7**, 1–92.

Wislocki, G. B. (1940). The placentation of *Solenodon paradoxus*. *Am. J. Anat.* **66**, 497–531.

Zdansky, O. (1930). Die Alttertiären Säugetiere Chinas nebst stratigraphischen Bemerkungen. *Paleontol. Sinica* **6**(2), 1–87.

6. A phylogenetic analysis of the Primates

PETER ANDREWS
Department of Palaeontology, British Museum (Natural History), London, UK

Abstract

A definition of the Order Primates is attempted, based on 54 characters of the living species of lemurs, lorises, tarsiers, platyrrhine monkeys, catarrhine monkeys, and apes. With reference to this definition, evidence is presented for referring omomyoids and adapiforms to the order, but there is insufficient evidence for the inclusion of plesiadapiforms and tupaioids. Adapiforms share mainly primitive characters with lemuriforms; phylogenetic analysis using the parsimony (PAUP) program separates them in all most-parsimonious analyses (Consistency Index, CI=0.636), but omomyoids are consistently grouped with tarsiers in the infraorder Haplorhini. Omomyoidea is paraphyletic in all most-parsimonious trees, with microchoerids more closely related to tarsiers than are omomyids, and the evidence supporting a tarsiiform clade is slightly stronger than that linking tarsiers more closely with anthropoids. The most-parsimonious branching patterns (branch length 99) differ from the generally accepted phylogeny of Szalay and Delson (branch length 111, CI=0.606 in the user-tree option of PAUP) in the consistent grouping of plesiadapiforms as outgroup to tupaioids plus primates and in the separation of adapiforms from lemuriforms. Alternative phylogenies are less parsimonious, i.e. grouping tarsiers with plesiadapiforms and lemuriforms with anthropoids (branch length 125, CI=0.504), or grouping tarsiers with lorisiforms (branch length 115, CI=0.548).

The Phylogeny and Classification of the Tetrapods, Volume 2: Mammals (ed. M. J. Benton), Systematics Association Special Volume No. 35B, pp. 143–75. Clarendon Press, Oxford, 1988.

Introduction

Classifications of the Order Primates attract attention because of its special relevance to human evolution. Although the basic groups of living primates have long been recognized, there are still uncertainties as to what constitutes the order and how the groups are related to each other. There has been much controversy, for instance, over the inclusion of tree shrews in the order, and over the recognition as primate of such early Paleogene groups as the Plesiadapiformes. In this paper, I first provide a definition of the Order Primates based on the living species only, and the 28 defining characters will be combined with a further 26 in a phylogenetic analysis to determine the relatedness of key fossil groups to the order and to find the most-parsimonious sets of relationships of the main groups of primates. This will focus on the relationships of the strepsirhines and haplorhines, and the position of tarsiers and other problem groups with respect to this basic dichotomy within the Primates.

Definition of Primates

Primates as a group have often been considered difficult to define: for instance, Linneaus included bats and colugos in Primates. Mivart (1873) first defined the Primates, and his definition was still standard until a few years ago. He included two suborders, Prosimii and Anthropoidea, with the following definition:

> Unguiculate, claviculate placental mammals, with orbits encircled by bone; three kinds of teeth, at least at one time of life; brain always with a posterior lobe and calcarine fissure; the innermost digit of at least one pair of extremities opposable; hallux with a flat nail or none; a well-developed caecum; penis pendulous; testes scrotal; always two pectoral mammae.

In 1929, Wood Jones reviewed these characters, and concluded that not one of them is peculiar to the Order Primates exclusively. He had earlier (1916) linked the characters with primate arboreality, assuming implicitly (but with no evidence) that the ancestral mammalian condition was of a small ground-living shrew-like nature, and that the ancestral primates differentiated as a result of changes in sense organs (Smith 1912) and limbs (Wood Jones 1916), which were adaptations for arboreality. The absence of many of these characters in other equally arboreal mammalian groups was not explained by these authors, but Le Gros Clark (1970) later proposed that the ancestral mammalian condition included many of these arboreal adaptations, which have been subsequently lost in most other mammalian groups. This seems likely to be the case, but it must be questioned whether any of these adaptations was related in any way to arboreality. It seems that the non-cursorial adaptations of the generalized mammal were confused with similar characters in the primates i.e. retention of the clavicle), so that interpretations became hopelessly confused. Either both the primates and the generalized early mammals were arboreal, or the primates must be seen as generalized (Cartmill 1974); it does not seem possible to argue that, with apparently the same adaptations, the primates were arboreal and other mammals were not.

These difficulties have led to the belief, expressed for instance by Simpson (1945), that primates lack any clear-cut diagnostic features. This view has been current until very recently (MacPhee *et al.* 1983). It has resulted in various prevarications, with some doubt that primates are a monophyletic group (Mivart 1873) and, more recently, with the attempt by Le Gros Clark (1959) to express characters as evolutionary trends. This has been popular with primatologists for the last 25 years, although a recent, but not altogether successful, attempt has been made to define primates cladistically (Szalay and Delson 1979). These authors identified a number of characters present in all primates, such as the presence of a bony bulla composed in part by the petrosal, loss of medial internal carotid artery, rounded promontorium, and a complex of derived postcranial characters related to arboreal existence, but there are problems with a number of them (see below). In 1986, Martin published a comprehensive list of characters diagnostic of the primates, and this will be used here as a basis for defining the ancestral primate morphotype in terms of primate autapomorphies, with additions from the work of Szalay (1975; Szalay and Delson 1979) and MacPhee and Cartmill (1981; MacPhee *et al.* 1983).

The list of characters brought together by Martin (1986) serves to define the Primates, but it does so by severely limiting our concept of the order. First of all, it is essentially a horizontal definition, combining those characters common to a single time period, the present. There is no possibility (at this stage) of including a vertical element of ancestral–descendant character change. Secondly, it excludes from consideration any uncertain groups such as the tree-shrews among living animals, and the plesiadapoids among the fossils, Thirdly, it is of little help in exploring the relationships of primates with other mammals. The defining characters are autapomorphies of the order, and they establish that it is monophyletic even in the absence of adequate outgroup comparison: the comparison is made with mammals in general, so that for the characters listed below, primates are uniquely different from all or most other groups of mammals.

These limitations are imposed because there is an element of circularity in seeking the definition of any group of animals. It must first be decided what is to be included in the group to be defined, before the definition can be attempted. To overcome this problem, it is more reasonable to arrange species in hierarchical relationships to one another, and to recognize higher taxonomic units based on the resulting groups. The characters uniquely shared by these groups can then be listed, and comparisons can be made with fossils, which are much less complete than living species.

It is generally agreed that primates encompass lemurs and lorises, often grouped in the suborder Strepsirhini; the Tarsiiformes are grouped with higher primates in the suborder Haplorhini, the New World monkeys are grouped with the infraorder Platyrrhini, and the Old World monkeys and apes are grouped with the infraorder Catarrhini. In terms of the living species only, there has never been any serious doubt that they are other than a monophyletic group, although, as stated earlier, there have been problems in defining this group. In the definition that follows, which is enlarged from Martin (1986), the characters used are those that unite living primates (for more detailed references for each character, see Martin 1986). The characters are divided into a number of complexes, each containing several intercorrelated characters: such characters cannot be treated independently, but in so far as they may be identifiable singly on fossil specimens, and can be taken as indicative of the whole complex when so identified, they will be mentioned briefly here.

Locomotor complex

1. Presence of grasping foot with divergent hallux: modifications of the tarsus and metatarsus exemplify this adaptation.

2. Presence of nail on hallux.

3. Elongation of distal part of calcaneus.

4. Presence of tarsi-fulcrumation: with the divergent hallux, the point of thrust during arboreal locomotion lies between the hallux and the second digit near the proximal ends of the metatarsals, so that the increased length of the tarsals increases the leverage with the fulcrum at their distal end.

5. Hindlimb dominance during locomotion (behavioural adaptation).

6. Centre of gravity of the body displaced towards the hindlimb.

7. Associated with these characters of the hindlimb is the frequent occurrence of grasping hands, with opposable thumbs, serving similar functions to the grasping feet.

8. Nails usually present on all or most digits, but sometimes secondarily lost.

9. All digits with tactile pads regardless of whether nails or claws are present, but since these are also present on some other groups of mammals their significance is not certain.

Orbital complex

10. Presence of some degree of forward rotation of the orbits, always associated with the postorbital bar (which can occur without the forward rotation of the orbits in some other mammals) and enlarged orbit size.

11. Narrow interorbital distance compared with other mammals, resulting from the forward rotation of the orbits.

12. Exposure of the ethmoid bone on the medial wall of the orbits—this also is a direct consequence of forward rotation, whereby the orbits can be said to move into the sphere of action of the ethmoid, which remains static.

13. Presence of stereoscopic vision with 40–50% of retinal fibres with ipsilateral projection (as opposed to <10% in other mammals): the entire left half of the total visual field from both eyes passes to the right half of the brain, and the entire right half passes to the other side of the brain; this provides a different, presumably more efficient, basis for stereoscopic vision (the high ipsilateral component arises because of the forward rotation of the orbits and the resulting high degree of overlap in the fields of view of the two eyes).

Brain complex

14. Presence of generally increased brain size: overall size is a weak character, but fetal brain size is large in relation to fetal body weight, and as this is true throughout gestation, neonate brain size is also greater relative to body size than in other mammals.

15. Presence of Sylvian sulcus.

16. Presence of triradiate calcarine sulcus.

Precocial complex

17. Precocial descent of testes: always occurs early in life, often soon after birth.

18. Females have loss of urogenital sinus.

19. Gestation length is long, relative to body weight, and is related to the small litters of precocial young.

20. Fetal growth is slow relative to maternal body weight.

21. Sexual maturity is late.

22. Life span is typically long.

Auditory complex

23. Presence of bony auditory bulla (McKenna 1966; Archibald 1977) with the ventral floor composed entirely of the petrosal bone. (In some non-primate taxa, such as erinaceoids, part of the bony wall of the bulla develops in ontogeny as an outgrowth of the petrosal.)

24. Lateral extension of the ectotympanic into an auditory meatus: since it is absent in the earliest and most primitive members of each primate lineage, it is inferred that it is not homologous (Szalay 1977).

25. Loss of medial internal carotid artery (Archibald 1977; Szalay 1975); this has been included in the phylogenetic analysis (see below, Table 6.1), although it has since been brought to my notice that it is no longer considered to be a valid character (Wible 1986).

26. Moderately large promontory and smaller stapedial branch of the lateral internal carotid artery: this is probably the primitive eutherian character state—see discussion below (character 34).

Dentition

27. A number of dental characters are generally held to be primate synapomorphies: lower molars have an expanded talonid region, with development of an M3 hypoconulid, and upper molars have an anteriorly placed protocone, with an enlarged talon region and development of a postprotocingulum (=*Nannopithex*-fold). These characters are by no means universally present in primates, and they are variably developed in primitive condylarths such as *Protungulatum* and Scandentia.

28. Loss of one incisor and one premolar from the presumed mammalian ancestral state:

$$\text{dental formula is } \frac{2\ 1\ 3\ 3}{2\ 1\ 3\ 3}$$

The characters making up this definition are present in the living species composing the Primates, unless exceptions are noted. They are mostly absent from the tree-shrews, long considered to be primitive primates (Le Gros Clark 1959). In the locomotor complex, tree-shrews and some other mammalian groups have tactile pads on the extremities of their digits, but they have none of the other characters making up this complex. Tree-shrews have relatively large eyes with postorbital bars, long thought to be evidence of primate relationship, but they lack the other primate specializations of the orbital complex, especially the primate form of stereoscopic vision. They lack a petrosal bulla and all of the characters of the brain or precocial complexes. On the other hand, they do resemble primates in numerous superficial characters, probably related to their shared arboreal way of life (Martin 1986), but these have been shown in the introduction to be almost certainly ancestral retentions, so that in terms of the definition outlined here, they cannot be considered indicative of relationship.

Fossil evidence for primate origins

The primate fossil record is unfortunately far from complete (Fig. 6.1). Even at its most complete, less than half of the expected number of taxa is known in the fossil record, assuming comparable diversity in the past to the present (183 living species—Fig. 6.1), and the great majority

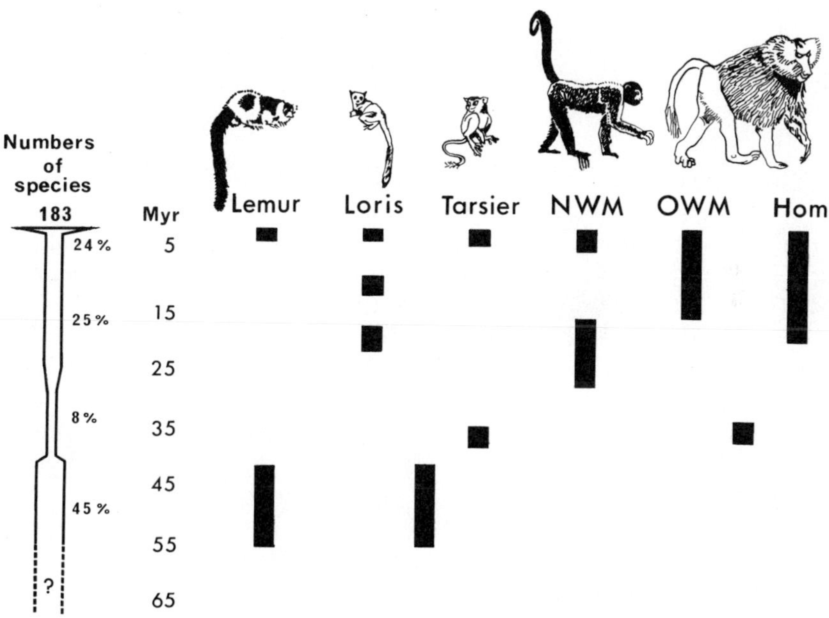

Fig. 6.1. Primate fossil record. The width of the column on the left of the figure shows the proportions of species present at different times compared with the total primate species living today of 183 species. The time scale is in millions of years. Plesiadapiformes are not included. Abbreviations: Lemur, Lemuroidea; Loris, Lorisoidea; Tarsier, Tarsiiformes; NWM, New World monkeys (Platyrrhini); OWM, Old World monkeys (Catarrhini); Hom, Hominoidea.

(~75%) of the fossil taxa are represented by extremely fragmentary remains, which tell us little about morphological trends. This limitation has come to be expected in the vertebrate fossil record, but in the early stages it is enhanced by geographical limitations: almost no pre-Oligocene fossils are known from Africa or South America, and Oligocene fossil primates are limited to restricted areas of North Africa and the Americas. The fossils, therefore, provide only temporally intermittent and anatomically incomplete evidence on problems of primate origins and relationships.

The first primate group to be considered in relation to the definition outlined above is Adapiformes, the rather diverse group from Eocene deposits, recognized as a distinct infraorder of the Strepsirhini by Szalay and Delson (1979). These authors make Adapiformes the sister-group to the extant Lemuriformes, but the phylogenetic analysis presented later in this paper does not support this relationship. Adapiformes are represented most completely by *Adapis* and *Leptadapis* in Eurasia, and *Northarctus* and *Smilodectes* in the Americas. There is some evidence that these two geographical groups may not share a near common ancestor, but there is no doubt of their primate status, since they have many of the characters listed above as diagnostic of primates. All the most-parsimonious trees in the phylogenetic analysis make them the sister-group to the living euprimates (primates of modern aspect), but with no particular relationship to any one part (see below, and Fig. 6.3).

The same conclusion can be reached for the second primate group of Eocene age to be considered here, Omomyoidea. This group includes three distinct families (or subfamilies, depending on the classification—Szalay and Delson 1979): Anaptomorphidae, Omomyidae, and Microchoeridae. Despite their lack of some diagnostic haplorhine characters, they are often considered to be related to the extant tarsiers, and are grouped with them in the suborder Haplorhini (Szalay and Delson 1979). The omomyoids are even more heterogeneous than the adapiforms, with major differences in morphology, and in the phylogenetic analysis (see below) they consistently appear as a paraphyletic group. They do, however, share all of the primate defining characters that can be recognized in such relatively incomplete fossils, and there is considerable support here for their link with tarsiers and anthropoid primates in Haplorhini.

The third group to be considered is again a loose conglomeration of diverse taxa, referred to various taxonomic levels by different authors: Plesiadapiformes (Szalay and Delson 1979; Rose and Fleagle 1981), Plesiadapoidea (Kay and Cartmill 1977), Paromomyiformes (Szalay 1975), and so on. Their primate status is usually unquestioned, although Cartmill (1972) and Martin (1972) have been arguing for some years now that there is no evidence for considering them primates, whatever name they go by.

Plesiadapiformes are known generally from relatively incomplete fossil remains, with only the teeth and jaws represented in any abundance. They come mainly from Palaeocene deposits, but they range back in time into the Upper Cretaceous and on into the Late Eocene. They were very abundant, forming a major constituent of Middle and Late Palaeocene faunas, and being twice as abundant as extant prosimians today. All workers seem to agree that they were too specialized in their dentition, or showed too great a dental reduction, to be ancestral to later primates, but they have been claimed to share a number of primate specializations (Gingerich 1977) in the occlusal pattern of the teeth, brain size and pattern (Jerison 1973), and the presence of the petrosal in the formation of the floor of the auditory bulla. The first of these is of dubious value, for although there are many and minutely detailed similarities in molar morphology shared by plesiadapiforms and Eocene primates, many of these are also shared by Late Cretaceous and Early Palaeocene rodents and protungulates, and would appear likely on this basis to be of limited phylogenetic significance. It was for this reason that, in the definition above, no characters of the teeth could be provided.

Brain morphology is not well known in plesiadapiforms (Radinsky 1977), but the best estimate of brain volume of the specimens available for study (Kay and Cartmill 1977) indicates a size relatively smaller than any primate (Radinsky 1977). The presence of a petrosal bulla has been suggested by Szalay and Delson (1979) to be diagnostically primate, but the petrosal forms part of the bulla in other mammals, i.e. in erinaceoids. Some of these character states are also not present in all plesiadapiforms, being absent in the Microsyopidae (Kay and Cartmill 1977; Rose and Fleagle 1981). This is one of five families currently recognized in the Plesiadapiformes, so that it seems better at this stage to omit them until there is better evidence for justifying their inclusion in the Primates.

In the phylogenetic analyses plesiadapiforms were consistently placed outside the tree-shrew primate grouping (Fig. 6.3; see also Chapter 3). In other words, the analysis indicate that tree-shrews are more closely related to living primates than were plesiadapiforms. This result, however, is based on the interpretation of just two characters described below (Table 6.1). Additional characters of the postcranial skeleton have also been used to support primate affinities for plesiadapiforms (Szalay *et al.* 1975; Szalay and Dagosto 1980), but many of these are present in tree-shrews as well (Le Gros Clark 1929). These have more recently been shown to represent primitive retentions in the larger grouping Archonta, which includes Dermoptera and

Chiroptera, as well as Scandentia, Primates, and Plesiadapiformes (Wible and Covert 1987). Even within this larger grouping, the tree-shrews and primates appear more closely related to each other than either is to the plesiadapiforms (Wible and Covert 1987), and it is apparent that there is little support for either to be considered as primate.

While the exclusion of the Plesiadapiformes limits the size and time range of the order, so that there are no recognizable Palaeocene primates, it does not exclude any one group from consideration as possible ancestral primates. Rather, it widens the choice and provides a more firmly established basis for the recognition of possible primate ancestors. The remainder of this paper will consider the broad divisions within the order.

Strepsirhine/haplorhine transition

The Strepsirhini, as usually envisaged, includes the lemurs and lorises among living primates and the fossil adapiforms. The strepsirhine nose complex, with naked rhinarium, slit nostrils, and unfused nasal prominences, is seen in many groups of mammals, in addition to the lemurs and lorises. As a result, it must be recognized as a primitive retention in the primates, and thus of no use in defining the suborder. On the other hand, the lemurs and lorises can readily be identified as a monophyletic clade on the basis of the shared presence of a tooth comb composed of canines as well as incisors, and of a grooming claw on the second digit of the foot. Both of these characters may be lost, e.g. by the divergent *Daubentonia*, or modified, as in subfossil lemurs of Madagascar such as *Hadropithecus*, and they are absent in the Eocene adapiforms (although there is some evidence for the presence of a grooming claw in the unidentified primate from Messel, assumed to be an adapid by Koenigswald 1979). Most definitions of the Strepsirhini, e.g. that of Szalay and Delson (1979: 99), go on to list many primitive retentions as diagnostic of the suborder, such as the postorbital bar, the grasping hallux, etc., but these are clearly inadmissible as evidence of monophyly, and so the case for the suborder rests on just the tooth comb and the grooming claw.

The lack of strepsirhine defining characters has resulted in some ambiguity in the phylogenetic analyses. Of the three most-parsimonious trees (see below, Fig. 6.3), only one grouped lemurs and lorises on the basis of the shared presence of the tooth comb (character 43, Table 6.1). This character is absent from adapiforms, and the assumed shared presence of the grooming claw (character 41, Table 6.1) based on the Messel specimen has been over-ridden by greater change in other characters. It was found more parsimonious to assume that the presence of a grooming claw, even if true, was a parallel development in adapiforms, because the tree that links adapiforms to lemurs and lorises, but is similar in all other respects to the computer-generated trees, is three branch lengths greater than the most-parsimonious tree, involving four reversals to adapids (characters 3, 4, 14, and 28, Table 6.1) and the one adapid/lemur/loris synapomorphy (character 41, Table 6.1). The most-parsimonious solution, on the contrary, makes this apparent synapomorphy of the grooming claw an ancestral primate character lost in later primates (node 6, Fig. 6.3), and eliminates the four reversals. The reduction of the promontory branch of the lateral carotid in both adapiforms and lemurs appears to be ancestral for the primates, and is present also in tree-shrews (node 3, Fig. 6.3). On the strength of these character assignations, and also on the uncertainty of the identification of the grooming claw to an adapid primate (Koenigswald 1979), it is concluded that adapiforms do not belong to the lemur/loris clade, and the characters that link them are parallelisms (characters 36 and 41, Table 6.1) or primitive retentions (characters 10, 11, 12, 34p, and 34s, Table 6.1). In Fig. 6.3, the characters linking adapiforms with other primates are listed, and the same 15 characters support this node whatever the subsequent branching pattern within the euprimate clade. A further six characters support node 5 of Fig. 6.3, from which diverge the lemur/loris and haplorhine clades. Although it is slightly more parsimonious to have separate derivations of lemurs and lorises (nodes 5 and 5A, Fig. 6.3), the evidence for this rests on nothing more than the greater orbital frontation in lorises (characters 10 and 11).

By distinguishing between adapiforms and living strepsirhines, the suborder Strepsirhini, as presently defined, becomes paraphyletic. This seems an acceptable state of affairs at this stage of our knowledge and it is implicit in the classification of Szalay and Delson (1979), who divide the suborder into two infraorders, Adapiformes and Lemuriformes. There is some evidence also that Adapiformes is a paraphyletic grouping. The apparent ancestral eutherian character states in the postcranial skeleton of *Adapis* and *Leptadapis* have been shown convincingly to be reversals from the ancestral primate condition, as seen in North American adapiforms, back to a condition resembling (but not homologous to) the ancestral eutherian states (Dagosto 1983). It is possible that some, at least, of the European adapids do not share a unique common ancestor with the American notharctines.

Within Lemuriformes are Lemuroidea, containing the living and subfossil lemurs of Madagascar, and Lorisoidea, containing lorises and bushbabies. The former are distinguished by upper molars with compressed cusps, broadening and lowering of the protocone distally, mesial elongation of the anterior premolars, and some distension of the lower molar metaconid (Schwartz and Tattersall 1987). The other lemuriform clade, Lorisoidea, is characterized by the development of an enlarged ascending pharyngeal artery, an anterior carotid foramen, prehypocone cristae on upper M1, lower molar protoconids and metaconids that are melded at their bases, a paracristid that rings the base of these cusps, and on the calcaneus a narrow and elongated trochlear facet (Schwartz and Tattersall 1987). The stapedial branch of the internal carotid artery has been lost (except in *Phaner*), which in the phylogenetic analyses presented here is seen as parallel to the condition in haplorhines, and not homologous. Within Lorisoidea, the lorisids and galagids are united by the shared presence of fused ectotympanics, and the extensive mastoid pneumatization distinguishes the latter from the former. The presence of pneumatized mastoids and fused ectotympanics in the Miocene *Progalago* from East Africa identifies this fossil with the galagids (Le Gros Clark and Thomas 1952). Similarly, the adaptations of the humerus for slow powerful locomotion, seen in the later Miocene *Nycticeboides* from Pakistan (Jacobs 1981), links this fossil with lorisids, which are distinguished from galagids by these characters. These two Miocene fossils provide the earliest evidence of the lorisiform lineage (Fig. 6.1).

The other main primate branch, suborder Haplorhini, was first advocated by Pocock (1918), and later by Wood Jones (1929). It was based on the haplorhine adaptations of the nose seen in higher primates and tarsiers, combined with fusion of the upper lip and absence of the moist naked rhinarium, together with reduction of the nasal capsule and lack of an olfactory recess (Hill 1955; Cave 1973). Also present are a number of highly correlated changes that are clearly part of a single functional complex of amnion formation and haemochorial placentation (Luckett 1974, 1980); presence of a fovea centralis in the eye and absence of tapetum lucidum (Wolin and Massoupust 1978; Martin 1973; Aiello 1986); perbullar path of the internal carotid artery and loss of the subtympanic recess (Cartmill *et al.* 1981).

A number of other characters have been put forward as haplorhine synapomorphies. The significance of many of these relates particularly to the phylogenetic position of tarsiers and their supposed fossil relatives, which will be discussed in the next section.

The position of tarsiers

For much of this century, the relationships of tarsiers have been the subject of controversy. What might be called the 'traditional' view, groups tarsiers with other prosimians (Gregory 1925; Simpson 1945). Various taxonomic levels have been proposed for the classification of

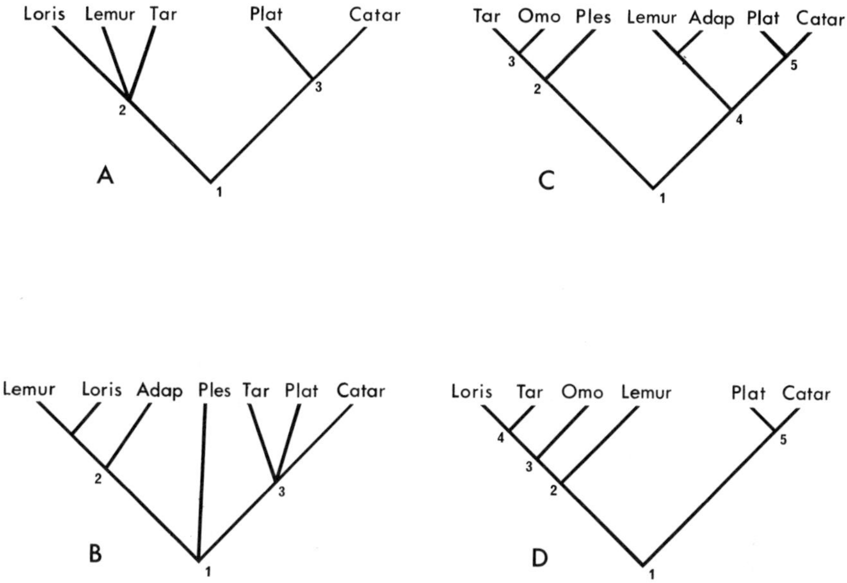

Fig. 6.2. Four sets of relationships of the Primates. (A) The traditional view accepted by Simpson (1945), with lower primates (Prosimii) separated from the higher primates (Anthropoidea) as separate grades; (B) the more generally accepted notion of Strepsirhini, including Lemuriformes and sometimes Plesiadapiformes, and Haplorhini, including Tarsiiformes and Anthropoidea (Pocock 1918; Szalay and Delson 1979); (C) the phylogeny proposed by Gingerich (1975, 1976, 1980), grouping tarsiiforms, omomyoids, and plesiadapiforms in his taxon Plesitarsiiformes, and lemuriforms, adapiforms, and anthropoids in Simiolemuriformes; (D) the recently proposed (Schwartz and Tattersall 1987) grouping of tarsiiforms with lorisoids in an otherwise similar division to Simpson's classification of Prosimii and Anthropoidea.

this grouping, e.g. that of Simpson, who proposed two infraorders of suborder Prosimii: Lemuriformes for lemurs and lorises, and Tarsiiformes for tarsiers (Fig. 6.2A). This division is essentially grade-based but has long been maintained, despite the proposal by Pocock (1918) that the lack of a naked rhinarium and the presence of a haemochorial placenta in living tarsiers linked them with the higher primates. This view (Fig. 6.2B) is widespread today (Szalay and Delson 1979). A third view (Fig. 6.2C) was espoused by Gingerich (1975, 1978, 1980, 1981), who relegated the tarsiiforms to a peripheral position in primate phylogeny as the sister-group of the combined group Lemuriformes plus Simiiformes (=Anthropoidea). This was based on his view that the adapiforms were ancestral to higher primates, and since adapiforms were also considered ancestral to living lemurs, and these could not be separated from living lorises, they all had to be grouped with the higher primates. Gingerich (1984) has since rescinded this view, but it has been included here as a matter of historical interest. Finally, a fourth view has been presented recently by Schwartz and Tattersall (1987), which groups tarsiers with lorises in a clade distinct from lemurs (Fig. 6.2D). Although no higher-order classification is provided by Schwartz and Tattersall (1987), their view essentially requires a return to Simpson's (1945) suborder distinction of Prosimii and Anthropoidea.

All four sets of relationship have in common the recognition of a broad distinction between lemurs and lorises on the one hand, and monkeys and apes on the other. Simpson groups tarsiers broadly with the former, Szalay and Delson group them broadly with the latter, and Gingerich makes them the outgroup to all other primates. Schwartz and Tattersall are the most radical in grouping the tarsiers with one particular group, the lorises. Nobody has yet grouped the tarsiers with the lemurs, or with one or other of the monkey families, but this is probably not for want of trying, and with such diversity of interpretation, it is necessary to look at the characters.

The character states most pertinent to the question of tarsier affinities are listed below, with the numbers following on sequentially from the previous list for ease of reference in the figures.

29. Haplorhine nose. The strepsirhine nose has a median, ovoid part of the nostril separated from the slit-like aboral part by the atrioturbinal, and this is seen typically, for example, in didelphids, tupaiids, and lemurids (Hill 1953; Hofer 1976, 1980). This condition is also seen in lorisids, but correlated features of the rhinarium are different, with reduction of the pars labialis, so that evidently the strepsirhine condition overall is of considerable complexity (Hill 1953). In comparison to this, the haplorhine condition consists of aboral rounding of the nostrils with no slit (Hofer 1980). In three tarsier specimens examined by Hofer, two had aboral slits and one did not, so that the haplorhine condition as strictly defined is only variably present in tarsiers. This variation was not observed by Hill (1955), who equated tarsiers with the haplorhine condition, as primates with nostrils 'possessing entire margins, due to the complete fusion below them of the medial and lateral nasal processes of the embryo, with consequent loss of the slit-like character observed in the nares of Strepsirhini'. The presence of the aboral slit in some tarsiers, illustrated by Hofer (1980, Fig. 6.4A), demonstrates the imperfect development of the haplorhine nose in tarsiers and calls into question their allocation to Haplorhini.

30. Fusion of the upper lip and loss of vibrissae and naked rhinarium. These character states are generally associated with the haplorhine nose, and are certainly shared by tarsiers and higher primates, but probably not by omomyoids, judging by the size of the

infraorbital foramen in these fossils (Cartmill and Kay 1978). They appear to be good synapomorphies, for the alternative characters of split lip and naked rhinarium are common to insectivores, tupaiids, lemuriforms, and lorisiforms, and would seem to represent the ancestral primate condition. The homology of upper lip fusion has been questioned by Hofer (1980), and it has also been pointed out (Szalay 1977) that upper lip fusion and loss of naked rhinarium are common mammalian adaptations, so that some doubt must also be cast on these characters (Schwartz and Tattersall 1987).

31. Reduction of the nose. Hill (1955) points out that the changes in the nose make the lips and chin the most anterior part of the skull in tarsiers and higher primates, so that they have a face but not a muzzle. This reduction of the nose has been given as a synapomorphy of tarsiers and higher primates by Cave (1973), who describes their nasal fossa as showing 'drastic reduction both anteriorly and posteriorly and marked increase in height'. This character state is also present in omomyoids. Schwartz and Tattersall (1987) point out that the mechanism of reduction may be different in tarsiers, the result of orbital enlargement which encroaches on the nasal region and olfactory recess (Starck 1975; Aiello 1986), so that these character states may not be homologous with those present in higher primates (Starck 1975).

32. Apical interorbital septum. This has been put forward as a tarsier/higher primate synapomorphy (Cartmill 1972; Aiello 1986). The ancestral interorbital area is considered to have had a series of complex ethmoturbinals, a large nasoturbinal, and an olfactory recess enclosing posterior ethmoturbinals, all functioning to augment the sense of smell. The simplification and loss of most of these in omomyoids, tarsiers, and higher primates, which resulted from the enlargement of the orbits and the changed emphasis on vision as the principal sensory mechanism, is considered to be a good haplorhine synapomorphy (Cave 1973). In most of these, the orbits are separated by a simple bony septum, but so great is the orbit enlargement in tarsiers that the bony septum may be missing altogether (Cave 1973), so that caution is needed in homologizing it with that of higher primates because of the all-pervading influence of the massive orbits.

33. Other character states of the orbit, such as the presence of the alisphenoid in the postorbital septum (Aiello 1986), must be viewed with caution because of the likelihood of convergent change resulting from the extreme enlargement of the orbits in tarsiers. The tarsier has similarities with lorisoids in ethmoid exposure (Schwartz and Tattersall 1987), arising from the enlargement of the orbits. Both are similar in this respect to anthropoids, although it is not certain that these

character states are homologous; the broad palatine–frontal contact in tarsiers is similar also to that in lemurs; and the broad infralacrimal maxillomaxillary contact in tarsiers and anthropoids is present also in omomyoids (Cartmill 1978). Given the variable nature of these discontinuous character states, great caution is needed in drawing phylogenetic conclusions from them.

34. The enlargement of the promontory branch of the internal carotid artery, and loss of the stapedial branch, have been described as a diagnostic character state of the haplorhines (Szalay and Delson 1979; Rosenberger and Szalay 1980; Aiello 1986). The presence of moderate-sized promontory and stapedial branches in insectivores and microchiropteran bats (MacPhee *et al.* 1983) is taken here as indicative of the primitive eutherian character state from which two character states are derived in primates: the enlargement of the promontory and reduction or loss of the stapedial in lorises, omomyoids, tarsiers, and higher primates; and reduction of the promontory and retention of a primitively large stapedial in lemurs and adapiforms. This differs from the interpretation of Luckett and Szalay (1978), who presumed that the promontory was primitively small in primates because this is the condition in adapiforms and some plesiadapiforms, and this is the conclusion indicated here by the phylogenetic analyses (see below, Fig. 6.3).

35. Retinal morphology. A number of characters of the retina have been proposed as synapomorphies of tarsiers and higher primates (Aiello 1986; Schwartz and Tattersall 1987), but these have been shown to be either incorrectly identified or inconsistently developed in tarsiers, and so not indicative of relationship (Castenholz 1965; Schwartz and Tattersall 1987).

36. Posteromedial position of the carotid foramen. This character state is shared by omomyoids, tarsiers, and higher primates (Szalay 1975; Rosenberger and Szalay 1980), but Cartmill *et al.* (1981) have shown that this condition is probably primitive for primates. Lemuriforms and adapiforms share a posterolateral position of the carotid foramen, and a similar state is also present in plesiadapiforms.

37. Perbullar course of the carotid artery. This appears to be a higher primate synapomorphy shared also by tarsiers (Cartmill *et al.* 1981; Aiello 1986).

38. Loss of the subtympanic recess. This appears to be a higher primate synapomorphy shared also by tarsiers (Cartmill *et al.* 1981; Aiello 1986).

39. Anterior accessory capsule of the middle-ear. This character of

higher primates is shared by tarsiers (Cartmill *et al.* 1981). It has also been claimed to be present in lorisiforms (Cartmill 1975; Schwartz and Tattersall 1987), but more recent ontogenetic evidence has shown that the lorisiform condition is not homologous with that of tarsiers and anthropoids (MacPhee and Cartmill 1986).

40. Haemochorial placenta. A large complex of characters associated with haemochorial placentation provides apparently strong evidence of synapomorphy in tarsiers and higher primates (Luckett 1974, 1976, 1980). Conversely, it has been argued by Schwartz (1978; Schwartz and Tattersall 1987) that, since the initial phase of placental and amnion formation is different in tarsiers and anthropoids, the subsequent similarities, such as mesodermal body stalk, rudimentary allantoic vesicle, a small free yolk-sac, etc., are probably convergences. Similar convergence is seen in other mammals that also have haemochorial placentation. If this criticism is correct, the tarsier would have a unique complex of characters linking it with neither the lemurs and lorises, nor the higher primates. On the other hand, Luckett (1975; Luckett and Szalay 1978) places the monodiscoidal haemochorial placentation of tarsiers as directly transitional to the bidiscoidal placentation of anthropoids.

41. Grooming claw. It has been seen earlier that the presence of a grooming claw on the second pedal digit is one of two characters establishing the relationship between lemurs and lorises (Hill 1955). Tarsiers also have grooming claws, but on the second and third digits, instead of the second alone. Schwartz and Tattersall (1987) argue that the sharing of the second digit grooming claw indicates the relationship of tarsiers within the strepsirhine clade, and the extra grooming claw is a unique development in tarsiers. It has been shown, however, that while the grooming claws are degenerate in both tarsiers and lemur/lorises, the former have a thin deep layer of claw, whereas lemurs and lorises do not (Le Gros Clark 1936; Thorndyke 1968). This makes it unlikely that the grooming claw in tarsiers is homologous with the single one on lemurs and lorises.

42. Dental evidence. A number of characters of the dentition are shown by Schwartz (1984; Schwartz and Tattersall 1987) to be similar in tarsiiforms and lorisiforms. The presence of a prehypocone crista on the first and second upper molars is one such character; and the tall protoconid and metaconid, melded at their bases, is a second. These may be homologous, and are certainly uniquely shared by tarsiers and lorises, so they may indicate a relationship (see node 2, Fig. 6.2).

43. Dental tooth comb. Lemuriforms and lorisiforms share a highly derived condition of the incisors and canines, which project forwards as

a tooth comb. This is not seen in any omomyoid, although anterior tooth enlargement occurs, and microwear studies indicate functional use for grooming. This character state also differs from the functionally similar tooth comb in tupaiids, in which it is composed of the incisors only (Rose *et al.* 1981). The lack of a tooth comb in tarsiers and anthropoids must be considered a shared primitive character.

Before discussing the relationship of tarsiers, it is necessary first to consider fossil tarsiiforms, because two separate issues sometimes become confused. The primary question is whether there is a tarsiiform clade combining recent and fossil taxa, and if so, which fossil taxa should be considered. Secondly, the more general question of the relationship of any of them with haplorhines or strepsirhines must be considered.

Fossil evidence for tarsiiform origins

Many fossil taxa have been identified as having tarsiiform affinities. These have been included in a single superfamily, Omomyoidea (or Family Omomyidae by Szalay and Delson 1979), despite many differences between them; justification is seldom made for linking omomyoids with tarsiers. This has compounded the problems over the relationships of the living tarsiers, and it is necessary to investigate further the character states said to be characteristic of omomyoids before commenting on tarsier relationships.

The omomyoids are traditionally considered to be a monophyletic group related to tarsiers. They lived during the Early Eocene to Early Oligocene, less common than adapiforms but with greater species diversity. Like the adapiforms, they are clearly primate in the sense defined here (see above), and are divided into a North American family, Omomyidae, a European family, Microchoeridae, and a third family common to both continents, Anaptomorphidae. They have primitive teeth similar to those of living tarsiers and Paleocene Microsyopidae, but some may be more derived than living tarsiers (Schwartz 1984), and many are distinguished by enlargement of their anterior teeth or tooth loss (Rose and Fleagle 1981). In a number of characters, they are apparently uniquely specialized, although there is a problem with parallel enlargement of anterior teeth in plesiadapiforms (Schwartz 1984). Their skulls and postcrania are not well known, but in most respects these seem to retain ancestral primate character states.

Some characters that are of importance in omomyoids are listed below, in addition to the ones already listed.

44. Enlarged anterior teeth. Some of the omomyoids have enlarged

anterior teeth, and in this respect they are similar to plesiadapiforms. This similarity is claimed to be based on dentally homologous teeth (Schwartz 1978), and it has been used as evidence of the relationship between plesiadapiforms and omomyoids by Gingerich (1975, 1976, 1981; but see Gingerich 1984). They were described as forming a monophyletic group distinguished from other primates (which were also considered to be derived in having vertically placed and spatulate upper incisors: Gingerich 1976). The earliest omomyoids lack anterior tooth enlargement, however, so that this cannot be an omomyoid synapomorphy, and even less an omomyoid–plesiadapiform synapomorphy. In fact, the homologies on which this conclusion is based have been shown to be incorrect (Cartmill and Kay 1978; Luckett and Maier 1982), and there seems to be no basis in the upper teeth for grouping plesiadapiforms and omomyoids together. In the case of the lower teeth, the central incisors of omomyoids are large, whereas in lemuriforms and higher primates they are smaller than the lateral incisors. It is difficult to predict the ancestral primate condition because of the variety of adaptations seen in other mammals, as well as in primates, but it is likely that the adapid/anthropoid condition represents the ancestral state.

45. Length of face. The facial skeleton is reduced and is hafted low on the neurocranium in omomyoids (Rosenberger and Szalay 1980). This is similar to the condition in tarsiers and higher primates, but it would seem to be a fairly superficial point of similarity. In tarsiers and omomyoids, a number of characters have been described for the basicranium which would appear likely to be correlates of basicranial flexion, such as reduction in length of the medial pterygoid plate, and encroachment on the pterygoid fossa by the bulla (Rosenberger 1985). These character states are not present in higher primates, and this raises the question of whether the overall similarity between tarsiers and anthropoids is significant.

46. Proportions of brain. The olfactory lobes are reduced and the occipital lobes are expanded in omomyoids, and they lack an olfactory recess. This is in line with the changed emphasis in sensory behaviour from smell to eyesight, with corresponding changes already noted in the size of the nose and disposition of the orbits, similarities shared with tarsiers and higher primates.

47. Mandibular symphysis. This is never fused in omomyoids, which is probably the ancestral primate condition. Higher primates are derived in having fused symphyses, and this condition is also present in some subfossil lemurs and in adapiforms, apparently independently acquired in several lineages (Gingerich 1980). This would appear to be

related to dietary change, with fusion of the symphysis being a product of counteracting the stresses associated with a vegetarian diet.

48. Canine occlusion. Omomyoids have primitively limited canine occlusion, in contrast to adapiforms and higher primates which have interlocking canines, some degree of canine dimorphism, and canine/premolar honing (Gingerich 1980).

49. Molar forms. Some omomyoids have primitively tritubercular molars, whereas adapiforms and higher primates have more expanded, quadrate molars (Gingerich 1980). This also appears to be an adaptation to increased herbivory, related to increased body size and fusion of the symphysis, and perhaps also to the canine interlocking described in the previous character.

50. Loss of paraconid. Omomyoids show differences in this character. Its presence on lower molars is almost certainly an ancestral primate character, and its loss in lemuriforms, lorisiforms, and anthropoids appear to be the result of parallel change. Similarly, its reduction and loss in some microchoerid omomyoids also seems to be an independent acquisition.

51. Pyrimidal processes. These arise medially to the last molar in tarsiers and omomyoids, compared to their position in most other primates which is posterior to the maxillary tuberosities. This also appears to be a tarsier–omomyoid synapomorphy (Rosenberger 1985).

52. Choanal regions. In many primates the choanae are shallow and gently arched, whereas in tarsiers and microchoerids the choanal roofs are peaked (Rosenberger 1985). This is an unusual and apparently highly derived character state, and there is some evidence for its presence in anaptomorphids as well.

53. Temporomandibular joint. In tarsiers and microchoerids, this is a trough-like gutter running the length of the zygomatic root, a morphology that is unique among primates (Szalay 1976; Rosenberger 1985). It is most strongly developed in tarsiers and in the microchoerid *Necrolemur*, but is less so in other omomyoids, although still distinct from the biconcave mandibular fossa that is probably primitive for primates.

54. Entoglenoid crest. The development of this crest is related to the guttering of the temporomandibular joint, being most strongly developed in tarsiers and microchoerids where the joint is most deeply guttered (Rosenberger 1985). In omomyoids it is present, but less well defined. It is absent in other primates (except plesiadapiforms, where it would not seem likely to be homologous).

Phylogenetic analysis

The position of tarsiers can now be discussed on the basis of these characters, with a view to deciding between the competing hypotheses presented in Figs. 6.2 and 6.3. They have been analysed by PAUP (version 2.4.1) phylogenetic analysis, for eight primate taxa compared with three possible outgroups: plesiadapiforms, tupaioids, and a reconstructed eutherian outgroup that has been assumed to be primitive for all the characters used in the analysis, since they were all selected as probable primate synapomorphies (Table 6.1). The branch-and-bound algorithm was used to find the most-parsimonious trees, and three were found of equal length: branch length 99 and CI = 0.636. There was a considerable similarity in the three trees, and they are illustrated by a single phylogeny in Fig. 6.3, which shows what I consider the most likely set of relationships together with the equally parsimonious variations from this.

All the most-parsimonious trees put tupaioids as the nearest outgroup to primates, more closely related to them than the plesiadapiforms. This has been discussed earlier, as has also the position of adapiforms outside the lemuriform clade, which is also consistently indicated by the most-parsimonious trees. The monophyly of the haplorhine clade is supported by all most-parsimonious trees, but two of the trees link tarsiers more closely with anthropoids, with successive outgroups of microchoerids, and omomyids, and one tree links these two fossil groups with tarsiers in a tarsiiform clade.

A combined recent and fossil tarsiiform clade is supported by three characters of the ear region and temporomandibular joint (characters 51, 53, and 54, Fig. 6.3, node 8). A further four characters link tarsiers and microchoerids within the tarsiiform clade (node 9, Fig. 6.3: extreme orbital frontation (character 24), and further derived characters of the ear region (characters 52–4)). Three reversals are required at node 8 (characters 34s, 39 and 50) because of characters shared by tarsiers and anthropoids but not by omomyoids, but no reversals are required at node 9. Characters 30 and 37–38 are assumed to have been independently acquired in tarsiers and anthropoids.

The alternative trees showing a closer relationship between tarsiers and anthropoids (node 6b, Fig. 6.3) are supported by five characters. These are loss of the stapedial artery (34s), perbullar course of the carotid (37), loss of subtympanic recess (38), anterior accessory cuspule (39), and the haplorhine features of fusion of the upper lip and loss of naked rhinarium (30). None of these is present in omomyoids, but the characters shared by omomyoids and tarsiers, listed above for nodes 8 and 9 of Fig. 6.3, must be assumed to be convergent if this tree is correct, and likewise for shared tarsier–lorisid characters (characters

Table 6.1. Data matrix for phylogenetic analysis. The characters are referred to by the number under which they are described in the text, and they are grouped into functionally related character complexes. The taxa used are as follows: out: inferred eutherian outgroup; Tup: Tupaiaoidea; Ples: Plesiadapiformes; Adap: Adapiformes; Micro: Microchoeridae; Omo: Omomyidae; Lemur: Lemuroidea; Loris: Lorisoidea; Tars: Tarsiiformes; Plat: Platyrrhini; Catar: Catarrhini

	Primate taxa										
	out	Tup	Ples	Adap	Micro	Omo	Lemur	Loris	Tars	Plat	Catar
Locomotor character complex											
1 present=1	0	0	0	1	1	1	1	1	1	1	1
2 present=1	0	0	0	1	1	1	1	1	1	1	1
3 present=1	0	0	0	0	1	1	1	1	1	1	1
4 present=1	0	0	0	0	1	1	1	1	1	1	1
5 present=1	0	0	0	9	1	1	1	1	1	1	1
6 present=1	0	0	0	9	1	1	1	1	1	1	1
7 present=1	0	0	0	1	1	1	1	1	1	1	1
8 present=1	0	0	0	1	1	1	1	1	1	1	1
9 present=1	0	1	9	9	9	9	1	1	1	1	1
41 present=1	0	0	0	1	0	0	1	1	2	0	0
two=2											
Auditory character complex											
23 present=1	0	0	1	1	1	1	1	1	1	1	1
24 present=1	0	0	1	0	1	0	0	0	1	0	1
25 loss=1	0	1	1	1	1	1	1	1	1	1	1
34p reduced=1	0	0	0	1	2	2	1	2	2	2	2
increased=2											
34s reduced=1	0	0	1	0	1	1	0	2	2	2	2
absent=2											
36 postlat.=1	0	0	1	1	0	0	1	0	0	0	0

37 perbull.=1	0	0	0	0	0	0	0	0	1	1	1
38 loss=1	0	0	0	0	0	0	0	0	1	1	1
39 present=1	0	0	0	0	0	0	0	1	1	1	1
51 medial=1	0	0	0	0	1	1	0	0	1	1	1
52 peaked=1	0	0	0	0	1	0	0	0	1	0	0
53 moderate=1	0	0	0	0	2	1	0	0	2	0	0
v. deep=2											
54 crest=1	0	0	1	0	2	1	0	0	2	0	0
Orbital character complex											
10 some rot.=1	0	0	0	1	2	1	1	2	2	2	2
extreme=2											
11 narrow=1	0	0	0	1	2	1	1	2	2	2	2
extreme=2											
12 exposed=1	0	0	0	1	1	1	0	1	1	1	1
13 present=1	0	0	9	9	9	9	1	1	1	1	1
32 present=1	0	0	0	0	1	1	0	0	1	1	1
33p present=1	0	1	0	1	1	1	1	0	1	0	1
33m present=1	0	0	0	0	1	1	0	0	1	1	1
35 loss=1	0	0	9	9	9	9	0	0	1	1	1
45 reduced=1	0	0	0	0	1	1	0	0	1	1	1
Nasal character complex											
29 haplor.=1	0	0	9	9	9	9	0	0	0	1	1
30 present=1	0	0	0	0	0	0	0	0	1	1	1
31 reduced=1	0	0	0	0	1	0	0	0	1	1	1
Brain character complex											
14 increase=1	0	0	0	0	1	1	1	1	1	1	1
15 present=1	0	0	9	1	1	1	1	1	1	1	1
16 present=1	0	0	9	1	1	1	1	1	1	1	1
46 red.lobes=1	0	0	0	0	1	1	0	0	1	1	1
46a rec.lack=1	0	0	0	0	1	1	0	0	1	1	1

Table 6.1. Cont.

	Primate taxa										
	out	Tup	Ples	Adap	Micro	Omo	Lemur	Loris	Tars	Plat	Catar
Precocial character complex											
17 to											
22 present=1	0	0	9	9	9	9	1	1	1	1	1
40 present=1	0	0	9	9	9	9	0	0	1	2	2
Dental character complex											
28 3 1 4 3=0											
2 1 4 3=1	0	2	0	1	2	2	2	2	2	2	3
2 1 3 3=2											
2 1 2 3=3											
42 present=1	0	0	0	0	0	0	0	1	1	0	0
43 toothcomb=1	0	1	0	0	0	0	2	2	0	0	0
44a I2<I1=1	0	0	0	1	0	0	0	0	0	1	1
44c enlarged=1	0	0	1	0	1	0	0	0	1	0	0
47 fusion=1	0	0	0	1	0	0	0	0	0	1	1
48 interlock=1	0	0	0	1	0	0	0	0	0	1	1
49 quadrate=1	0	0	0	1	1	0	1	0	0	1	1
50 loss=1	0	0	0	0	1	0	1	1	0	1	1

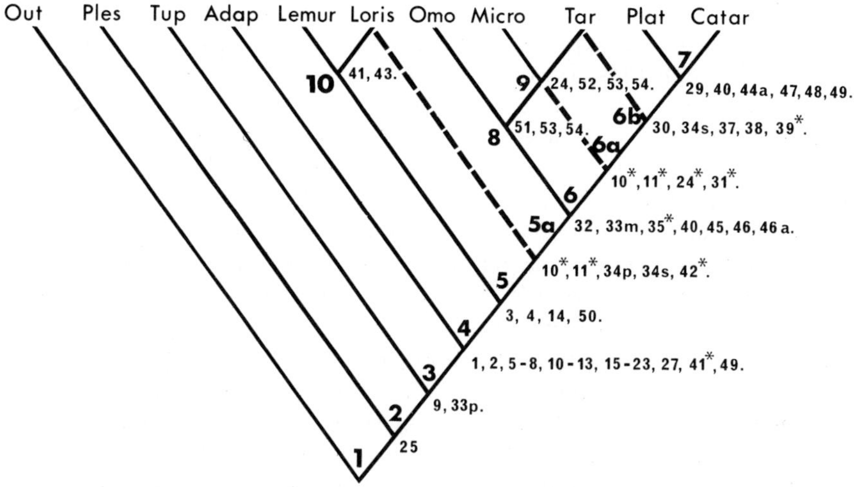

Fig. 6.3. Primate phylogeny computed by the PAUP package using 54 characters listed in the text and Table 6.1. Three equally parsimonious trees were generated, one having anthropoid and lemuriform clades but not a tarsiiform clade; the second having tarsiiform and anthropoid clades but not a lemuriform one; and the third having only an anthropoid clade. These alternatives are shown with dashed lines. The nodes are identified in bold lettering, and the character numbers, which refer to the characters listed in the text, are shown to the right of their respective nodes. Abbreviations of taxa are the same as in Table 6.1.

39 and 42). Similarly, characters 49 and 50 must have been independently lost in omomyoids and tarsiers after nodes 6 and 6b. Node 6a is weakly supported, mainly on the basis of the greater frontation of the orbits, but this is not considered to be a very strong character.

Both of these possibilities, a tarsiiform clade with tarsiers linked with omomyids, and a haplorhine clade with tarsiers more closely related to anthropoids than are omomyoids, have some support from this analysis, and both require several homoplasies. On the grounds that the hypothesis with the tarsiiform clade requires fewer reversals and convergences, it is the one that is preferred here.

In order to test tarsier relationships further, and to compare the phylogeny here with other phylogenies that are current in the literature, the four alternative trees shown in Fig. 6.2 were compared with the best-fit trees described above using the user-tree option of PAUP.

A. Tarsiers as prosimians This is the traditional view seen in Simpson's 1945 classification (Fig. 6.2A). It groups tarsiers (and omomyoids) with lemurs (and adapiforms) and lorises at node 2, and the higher primates—ceboids, cercopithecoids, and hominoids—at node 3. A variation of this is the phylogenetic scheme of Schwartz and Tattersall (1987), linking tarsiers with the prosimian clade and lorisids in particular (Fig. 6.2D). This latter scheme was tested by phylogenetic analysis, because it is more rigorously defined cladistically. It was found to have a branch length of 115, with a CI of 0.548 (compared with values for the maximum parsimony trees of 99 and 0.636). The shared presence of an anterior accessory cuspule of the middle-ear (character 39), and several other characters, including the dental evidence described as a single character (42) above, supports a sister-group relationship between tarsiers and lorises. However, almost all of the characters listed for nodes 6–6b of Fig. 6.3 would have to be considered primitive or convergent, given this relationship. These are the haplorhine characters shared by tarsiers and/or omomyoids with anthropoids: characters 31, 32, 33m, 33p, and 35 would be primitively retained from the primate common ancestor, and characters 30, 31, 37, 38, 39, 40, 45, 46, 46a, and 51 would be considered convergent. The last 10 characters (the four characters described earlier as being convergent because adapiforms are grouped with lemuriforms, and another three characters when plesiadapiforms are made the sister-group to primates) show that for the gain of two synapomorphies, 17 homoplasies have to be accommodated within this scheme. This is clearly not supported by the data set used in this analysis.

B. Tarsiers as haplorhines This is currently the most strongly supported view, although in fact it pre-dates Simpson's classification (Pocock 1918; Szalay and Delson 1979). It groups tarsiiforms, which include omomyoids and tarsioids, with anthropoids in the suborder Haplorhini (Fig. 6.2B), and there is some support for this also from the molecular evidence (Beard and Goodman 1976; Goodman 1981). Lemuriforms and adapiforms are grouped in the suborder Strepsirhini. Plesiadapiformes is classified by Szalay and Delson as a third suborder of equal rank to the other two, but in the analysis it was put as the sister-group to the combination of Haplorhini plus Strepsirhini. This is the most parsimonious of the three user-trees analysed, although its branch length of 111 and CI of 0.568 are well short of the best-fit trees described above. This is mainly due to two of the differences already described: the position of plesiadapiforms relative to tupaioids, and the assumed monophyly of the strepsirhines. In addition, however, the two omomyoid families were analysed in a sister-group relationship, following Szalay and Delson, and this added a further four reversals (characters 34s, 37, 38, and 39), since omomyoids and microchoerids are similar only in shared primitive characters. This is the only new result to come out of this user-tree analysis, that whichever option is most likely in the relationships of the tarsiiforms and anthropoids, the least likely one is the grouping of fossil omomyoids and microchoerids in a single taxonomic group. Either tarsiers are the sister-group to anthropoids, with the two fossil groups the progressively more remote sister-groups, or microchoerids and tarsiids are sister-groups and, together with omomyoids, form the sister-group to the anthropoids.

C. Tarsiers as primitive primates This groups tarsiers with plesiadapids as the sister-group to other primates. Gingerich (1981) has provided evidence that a branch of the adapiforms was ancestral to the higher primates, so that adapiforms and lemuriforms must necessarily be more closely related to anthropoids (Fig. 6.2C). This is the least parsimonious of the comparative analyses, with a branch length of 125 and a CI of 0.504. It is an interesting analysis because of the very large number of primate synapomorphies at the node leading to tarsiers and plesiadapiforms on the one hand, and lemuriforms and anthropoids on the other. The former share three synapomorphies (characters 24a, 44c, and 54), and the latter share four (characters 44a, 47, 48 and 49). The two groups share 17 synapomorphies, but these are matched by the 14 reversals made necessary to the plesiadapiforms, and the 10 convergences between anthropoids and some or all of the tarsiiforms (Fig. 6.3). There are also many other homoplasies in the other groups, more than twice as many as were found in the other comparative analyses, so that this phylogeny is not well supported by the evidence.

Anthropoidea

Because of space limitations and the well-established monophyly of the anthropoid primates, this group need only be mentioned briefly. It contains the Platyrrhini (the New World monkeys), Catarrhini (the Old World monkeys and apes), Cercopithecoidea, and Hominoidea (Fig. 6.4). The monophyly of the catarrhines is well documented (Andrews 1985), as is that of its constituent groups (Strasser and Delson 1987; Andrews and Martin 1987). Similarly, the New World monkeys are probably monophyletic (Ciochon and Chiarelli 1980), although some doubts persist about this. The combined group of platyrrhines plus catarrhines, known collectively as the higher primates, shares numerous character states, i.e. partial or complete postorbital closure, which separates the orbits from the temporal fossa, a large inferior orbital fissure (Harrison 1982), large maxillary and sphenoidal sinuses (Cave 1967), and a number of characters of the teeth, including increased molarization of the premolars, particularly the bicuspid P_4, strong development of the hypocone, and relatively

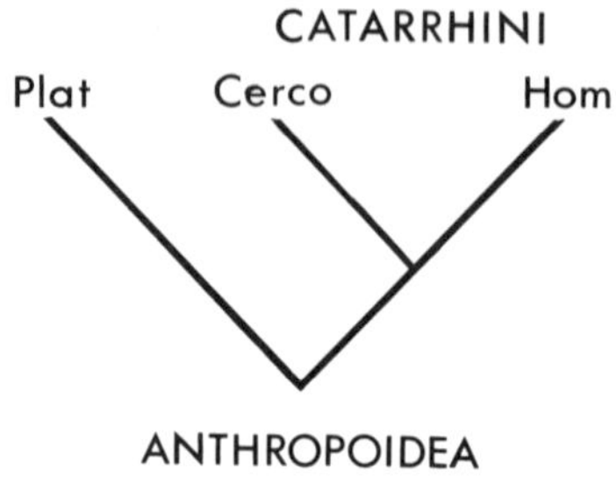

Fig. 6.4. Anthropoid clade with the New World monkeys (Platyrrhini), the sister-group to the Old World monkeys and apes (Catarrhini).

large canines compared with incisors. None of these characters has been used in the phylogenetic analysis because I was interested in seeing whether the existing data set was sufficient to distinguish the anthropoid clade from the other primates. It clearly has been sufficient, for Anthropoidea is distinguished as a monophyletic clade in all the most-parsimonious analyses.

Conclusions

The living species of the Order Primates have been shown to share 28 character states, in whole or in part. Many of these character states are present singly in other mammals, but these are not held to invalidate the definition of the order. Tree-shrews (Scandentia) share three of the primate characters, and fossil plesiadapiforms share two different ones. It is a matter of personal preference whether these are seen as potential synapomorphies, justifying the inclusion of these taxa in Primates, or whether the lack of the other 25 or 26 primate-defining characters is more significant. If one or other were to be included in the Primates, the defining characters of the order would be correspondingly reduced to the two or three shared by each with the living primates. If both were included, there would be no defining characters left at all.

The Eocene Adapiformes share 20 of the character states defining the Order Primates (together with two additional characters, all listed in Table 6.1 and Fig. 6.3). Their inclusion in the order is based on these, but the definition of the order would have to be modified to accommodate them. This is the same choice as before, only the primate status of the adapiforms is more strongly supported.

The addition of a further 26 characters, making 54 in all, and their analysis by a maximum parsimony method, indicate problems in the identity of Adapiformes. It appears to be a paraphyletic grouping, with the European genera *Adapis* and *Leptadapis* divergent from the other

European and North American adapiforms. There is no evidence that any adapiform group shared a unique common ancestor with the lemuriforms or lorisiforms, and Strepsirhini is seen as a paraphyletic grouping in which adapiforms, however defined, share only primitive characters with lemuriforms and lorisiforms.

The Eocene Omomyoidea share the same 20 characters with the living primates that are present in adapiforms, with the addition of three additional ones within the original 28. A further nine characters are also listed in Fig. 6.3 (nodes 5 and 6). This also appears to represent a paraphyletic grouping. The Family Microchoeridae is linked either with the living tarsier in a putative tarsiiform clade, or as outgroup to the tarsier–anthropoid clade, but never with Omomyoidea. In both cases, omomyoids and tarsiers could justifiably be linked with anthropoids in the suborder Haplorhini. Primitive anaptomorphids like *Teilhardina* were not included in the analysis, because the material available is too fragmentary, but they would appear to represent a primitive primate clade.

Suborder Strepsirhini, as already mentioned, appears to be paraphyletic, but Haplorhini can be recognized as a monophyletic clade including some, at least, of the omomyoids, living tarsiers, and anthropoids. Anthropoidea also is monophyletic.

Acknowledgements

I am very grateful to Jeremy Hooker, Bob Martin, Jeff Schwartz, Bernard Wood, Chris Stringer, and an anonymous reviewer for helpful comments on this paper, and to Chris Humphries for help in explaining and operating the PAUP analysis.

References

Aiello, L. (1986). The relationships of the Tarsiiformes: a review of the case for the Haplorhini. In *Major Topics in Primate and Human Evolution* (eds. B. Wood, L. Martin, and P. Andrews), pp. 47–65. Cambridge University Press, Cambridge.

Andrews, P. (1985). Family groups, systematics, and evolution among catarrhine primates. In *Ancestors: the Hard Evidence* (ed. E. Delson), pp. 14–22. Alan Liss, New York.

Andrews, P. and Martin, L. (1987). Cladistic relationships of the Hominoidea. *J. Hum. Evol.* **16**, 101–18.

Archibald, J. D. (1977). Ectotympanic bone and internal carotid circulation of eutherians in reference to anthropoid origins. *J. Hum. Evol.* **6**, 609–22.

Beard, J. M. and Goodman, M. (1976). The hemoglobins of *Tarsius bancanus*.

In *Molecular Anthropology* (eds. M. Goodman and R. E. Tashian), pp. 239–55. Plenum Press, New York.

Cartmill, M. (1972). Arboreal adaptations and the origin of the Order Primates. In *The Functional and Evolutionary Biology of the Primates* (ed. R. Tuttle), pp. 97–122. Alding–Atherton, New York.

Cartmill, M. (1974). Pads and claws in arboreal locomotion. In *Primate Locomotion* (ed. F. Jenkins), pp. 45–83. Academic Press, London.

Cartmill, M. (1975). Strepsirhine basicranial structures and the affinities of the Cheirogaleidae. In *Phylogeny of the Primates* (eds. W. P. Luckett and F. S. Szalay), pp. 313–54. Plenum Press, New York.

Cartmill, M. (1978). The orbital mosaic in prosimians and the use of variable traits in systematics. *Folia Primatol.* **30**, 89–114.

Cartmill, M. and Kay, R. F. (1978). Craniodental morphology, tarsier affinities and primate suborders. In *Recent Advances in Primatology* (eds. D. J. Chivers and K. A. Joysey), pp. 205–14. Academic Press, London.

Cartmill, M., MacPhee, R. D. E., and Simons, E. L. (1981). Anatomy of the temporal bone in early anthropoids, with remarks on the problem of anthropoid origins. *Am. J. Phys. Anthrop.* **56**, 3–21.

Castenholz, A. (1965). Uber die Struktur der Netzhautmitte bei Primaten. *Z. Zellforsch.* **65**, 646–61.

Cave, A. J. E. (1967). Observations on the platyrrhine nasal fossa. *Am. J. Phys. Anthropol.* **26**, 277–88.

Cave, A. J. E. (1973). The primate nasal fossa. *Biol. J. Linn. Soc.* **5**, 377–87.

Ciochon, R. L. and Chiarelli, A. B. (1980). *Evolutionary Biology of the New World Monkeys and Continental Drift.* Plenum Press, New York.

Dagosto, M. (1983). Postcranium of *Adapis parisiensis* and *Leptadapis magnus* (Adapiformes, Primates). *Folia Primatol.* **41**, 49–101.

Gingerich, P. D. (1975). A new genus of Adapidae (Mammalia, Primates) from the late Eocene of southern France, and its significance for the origin of higher primates. *Contrib. Mus. Paleontol. Univ. Mich.* **24**, 163–70.

Gingerich, P. D. (1976). Cranial anatomy and evolution of early Tertiary Plesiadapidae (Mammalia, Primates). *Univ. Mich. Pap. Paleontol.* **15**, 1–141.

Gingerich, P. D. (1977). Radiation of Eocene Adapidae in Europe. *Géobios, Mém. Spéc.* **1**, 165–85.

Gingerich, P. D. (1978). Phylogeny, reconstruction and the phylogenetic position of *Tarsius*. In *Recent Advances in Primatology* (eds. D. J. Chivers and K. A. Joysey), pp. 249–55. Academic Press, London.

Gingerich, P. D. (1980). Eocene Adapidae, paleobiogeography, and the origin of the South American Platyrrhini. In *Evolutionary Biology of the New World Monkeys and Continental Drift* (eds. R. L. Ciochon and A. B. Chiarelli), pp. 123–38. Plenum Press, New York.

Gingerich, P. D. (1981). Early Cenozoic Omomyidae and the evolutionary history of tarsiiform primates. *J. Hum. Evol.* **10**, 345–74.

Gingerich, P. D. (1984). Paleobiology of tarsiiform primates. In *Biology of Tarsiers* (ed. C. Niemitz), pp. 33–44. Gustav Fischer, Stuttgart.

Goodman, M. (1981). Decoding the pattern of protein evolution. *Prog. Mol. Biol.* **38**, 105–64.

Gregory, W. K. (1925). I. On the relationship of the Eocene lemur *Notharctus* to the Adapidae and to other primates. II. On the classification and phylogeny of the Lemuroidea. *Bull. Geol. Soc. Am.* **26**, 419–46.
Harrison, T. (1982). *Small-bodied apes from the Miocene of East Africa*, unpublished Ph.D. Dissertation. University of London.
Hill, W. C. D. (1953). *Primates, Comparative Anatomy and Taxonomy*: Volume 1, *Strepsirhini*. University of Edinburgh Press, Edinburgh.
Hill, W. C. O. (1955). *Primates, Comparative Anatomy and Taxonomy*: Volume 2, *Haplorhini*. University of Edinburgh Press, Edinburgh.
Hofer, H. (1976). Preliminary study of the comparative anatomy of the external nose of South American monkeys. *Folia Primatol.* **25**, 193–214.
Hofer, H. (1980). The external anatomy of the oro-nasal region of primates. *Z. Morphol. Anthropol.* **71**, 233–49.
Jacobs, L. L. (1981). Miocene lorisid primates from the Pakistan Siwaliks. *Nature (London)* **289**, 585–7.
Jerison, H. (1973). *Evolution of the Brain and Intelligence*. Academic Press, New York.
Kay, R. F. and Cartmill, M. (1977). Cranial morphology and adaptations of *Palaechthon nacimienti* and other Paromomyidae (Plesiadapoidea, ?Primates) with a description of a new genus and species. *J. Hum. Evol.* **6**, 19–53.
Koenigswald, W. von (1979). Ein Lemurenrest aus dem eozänen Ölschiefer der Grube Messel bei Darmstadt. *Paläontol. Z.* **53**, 63–76.
Le Gros Clark, W. E. (1929). On the anatomy of the pen-tailed tree-shrew (*Ptilocercus lowii*). *Proc. Zool. Soc. London* **1929**, 1179–309.
Le Gros Clark, W. E. (1936). The problem of the claw in primates. *Proc. Zool. Soc. London* **1936**, 1–24.
Le Gros Clark, W. E. (1959). *The Antecedents of Man*. Edinburgh University Press, Edinburgh.
Le Gros Clark, W. E. (1970). *History of the Primates*. British Museum (Natural History), London.
Le Gros Clark, W. E. and Thomas, D. P. (1952). The Miocene lemuroids of East Africa. *Fossil Mammals of Africa*, Vol. 5. British Museum (Natural History), London.
Luckett, W. P. (1974). Comparative development and evolution of the placenta in primates. *Contrib. Primates* **3**, 142–234.
Luckett, W. P. (1975). Ontogeny of the fetal membranes and placenta: their bearing on primate phylogeny. In *Phylogeny of the Primates* (eds. W. P. Luckett and F. S. Szalay), pp. 157–82. Plenum Press, New York.
Luckett, W. P. (1976). Cladistic relationships among primate higher categories: evidence of the fetal membranes and placenta. *Folia Primatol.* **25**, 245–76.
Luckett, W. P. (1980). Monophyletic or diphyletic origins of Anthropoidea and Histricognathi: evidence of the fetal membranes. In *Evolutionary Biology of the New World Monkeys and Continental Drift* (eds. R. L. Ciochon and A. B. Chiarelli), pp. 347–68. Plenum Press, New York.
Luckett, W. P. and Maier, W. (1982). Development of deciduous and permanent dentition in *Tarsius* and its phylogenetic significance. *Folia Primatol.* **37**, 1–36.

Luckett, W. P. and Szalay, F. S. (1978). Clades versus grades in primate phylogeny. In *Recent Advances in Primatology* (eds. D. J. Chivers and K. A. Joysey), pp. 227–35. Academic Press, London.

MacPhee, R. D. E. and Cartmill, M. (1981). Further evidence for a tarsier–anthropoid clade within Haplorhini. *Am. J. Phys. Anthropol.* **54**, 248.

MacPhee, R. D. E. and Cartmill, M. (1986). Basicranial structures and primate systematics. In *Comparative Primate Biology*, Vol. I, *Systematics, Evolution and Anatomy* (ed. D. Swindler), pp. 219–75. Alan Liss, New York.

MacPhee, R. D. E., Cartmill, M., and Gingerich, P. D. (1983). New Paleogene primate basicrania and the definition of the Order Primates. *Nature (London)* **301**, 509–11.

Martin, R. D. (1972). Adaptive radiation and behaviour of the Malagasy lemurs. *Phil. Trans. R. Soc. London* **264**, 295–352.

Martin, R. D. (1973). Comparative anatomy and primate systematics. *Symp. Zool. Soc. London* **33**, 301–37.

Martin, R. D. (1986). Primates: a definition. In *Major Topics in Primate and Human Evolution* (eds. B. Wood, L. Martin, and P. Andrews), pp. 1–31. Cambridge University Press, Cambridge.

McKenna, M. (1966). Paleontology and the origin of the primates. *Folia Primatol.* **4**, 1–25.

Mivart, St. G. (1873). On *Lepilemur* and *Cheirogaleus*, and on the zoological rank of the Lemuroidea. *Proc. Zool. Soc. London* **1873**, 484–510.

Pocock, R. I. (1918). On the external characters of the lemurs and of *Tarsius*. *Proc. Zool. Soc. London* **1918**, 19–53.

Radinsky, L. (1977). Early primate brains: fact and fiction. *J. Hum. Evol.* **6**, 79–86.

Rose, K. D. and Fleagle, J. G. (1981). The fossil history of nonhuman primates in the Americas. In *Ecology and Behaviour of Neotropical Primates* (eds. A. F. Coimbra-Filhol and R. A. Mittermeier), pp. 111–67. Academia Brasiliera de Ciencias, Brasil.

Rose, K. D., Walker, A. C., and Jacobs, L. L. (1981). Function of the mandibular tooth comb in living and extinct mammals. *Nature (London)* **289**, 583–5.

Rosenberger, A. L. (1985). In favor of the *Necrolemur–Tarsier* hypothesis. *Folia Primatol.* **45**, 179–94.

Rosenberger, A. L. and Szalay, F. S. (1980). On the tarsiiform origins of Anthropoidea. In *Evolutionary Biology of the New World Monkeys and Continental Drift* (eds. R. L. Ciochon and A. B. Chiarelli), pp. 139–57. Plenum Press, New York.

Schwartz, J. (1978). If *Tarsius* is not a prosimian, is it a haplorhine? In *Recent Advances in Primatology* (eds. D. J. Chivers and K. A. Joysey), pp. 195–204. Academic Press, London.

Schwartz, J. (1984). What is a Tarsier? In *Living Fossils* (ed. S. M. Stanley), pp. 38–49. Springer Verlag, New York.

Schwartz, J. and Tattersall, I. (1987). Tarsiers, adapids and the integrity of Strepsirhini. *J. Hum. Evol.* **16**, 23–40.

Simpson, G. G. (1945). The principles of classification and a classification of mammals. *Bull. Am. Mus. Nat. Hist.* **85**, 1–350.

Smith, G. E. (1912). The origin of man. *Smiths. Inst. Ann. Rep.* **1912**, 553–72.
Starck, D. (1975). The development of the chondrocranium in primates. In *Phylogeny of the Primates* (eds. W. P. Luckett and F. S. Szalay), pp. 127–55. Academic Press, New York.
Strasser, E. and Delson, E. (1987). Cladistic analysis of cercopithecid relationships. *J. Hum. Evol.* **16**, 81–99.
Szalay, F. S. (1975). Phylogeny, adaptation and dispersal of the tarsiiform primates. In *Phylogeny of the Primates* (eds. W. P. Luckett and F. S. Szalay), pp. 357–404. Plenum Press, New York.
Szalay, F. S. (1976). Systematics of the Omomyidae (Tarsiiformes, Primates): taxonomy, phylogeny and adaptations. *Bull. Am. Mus. Nat. Hist.* **156**, 157–450.
Szalay, F. S. (1977). Constructing primate phylogenies: a search for testable hypotheses with maximum empirical content. *J. Hum. Evol.* **6**, 3–18.
Szalay, F. S. and Dagosto, M. (1980). Locomotor adaptations as reflected on the humerus of Paleogene primates. *Folia primatol.* **34**, 1–45.
Szalay, F. S. and Delson, E. (1979). *Evolutionary History of the Primates.* Academic Press, New York.
Szalay, F. S., Tattersall, I., and Decker, R. L. (1975). Phylogenetic relationships of *Plesiadapis*: postcranial evidence. In *Approaches to Primate Paleobiology* (ed. F. S. Szalay), pp. 136–66. Karger, Basel.
Thorndyke, E. E. (1968). A microscopic study of the marmoset claw and nail. *Am. J. Phys. Anthropol.* **28**, 247–62.
Wible, J. R. (1986). Transformation in the extracranial course of the internal carotid artery in mammalian phylogeny. *J. Vertebr. Paleontol.* **6**, 313–25.
Wible, J. R. and Covert, H. H. (1987). Primates: a cladistic diagnosis and relationships. *J. Hum. Evol.* **16**, 1–21.
Wolin, L. and Massoupust, L. (1978). Morphology of the primate retina. In *The Primate Brain* (eds. C. Noback and W. Montagna), pp. 1–27. Appleton–Century–Crofts, New York.
Wood Jones, F. (1916). *Arboreal Man.* Edward Arnold, London.
Wood Jones, F. (1929). *Man's Place among the Mammals.* Edward Arnold, London.

7. Rodent phylogeny: new data and old problems

J.-J. JAEGER
Laboratoire de Paléontologie des Vertébrés et de Paléontologie Humaine, Université Pierre et Marie Curie, Paris, France

Abstract

The interrelationships of the major rodent groups are poorly resolved because of variability in only a few characters related to feeding, the questionable polarity of many soft-part characters, the rapid differentiation of most modern families in the Palaeocene/Eocene, geographically limited fossil occurrences, and scarce molecular data. The Hystricomorpha appears to be a valid clade, with *Cocomys*, Ctenodactylidae and Anomaluridae as primitive members. The Hystricognatha include Caviomorpha, Hystricidae, Bathyergidae, and Thryonomyidae. There are several possible cladograms of primitive rodents, myomorphs, and other sciuravid derivatives. In these, Eurymylidae and Ischyromyoidea are the primitive sister-groups, with a possible clade consisting of (Muroidea+Dipodoidea+Geomyoidea+Cylindrodontidae+Sciuravidae). Certain morphological and molecular data suggest that the nearest living sister-group of Rodentia is Lagomorpha.

Introduction

The Order Rodentia is, among the mammals, the most diversified and one of the most successful. Fossil rodents are also very common and, mainly because of their generally small size and high density, easy to collect. For the palaeontologist, rodents are a group of fast-evolving animals in which anatomical changes take place very quickly, i.e. in

The Phylogeny and Classification of the Tetrapods, Volume 2: Mammals (ed. M. J. Benton), Systematics Association Special Volume No. 35B, pp. 177–99. Clarendon Press, Oxford, 1988.

several successive levels within one sedimentary basin (Fahlsbuch 1970). Many lineages have been described from all over the world, which contribute to biochronological studies and associate rodent palaeontology with the study of Tertiary continental basins.

Despite this, very little is known about the interrelationships of the main families of rodents. Several reasons seem to explain this fact:

1. Morphological characters that show rapid changes are mainly concerned with the intake of food, including zygomasseteric structure, cheek–tooth morphology, and palate anatomy. It has been suspected for a long time that there was a bias introduced by palaeontologists, as these parts of the skeleton are the most easily preserved elements in mammalian taphonomy. This is only partially true because some postcranial remains are also very common, but do not show such important evolutionary changes. In a general way, it appears that most other osteological structures, particularly of the postcranial skeleton, are very conservative among rodents, and even primitive. Adaptive changes through time mostly affect the intake of food, and also probably the digestive system as a whole, and these changes concern only a very small number of characters in comparison with the whole organism. Consequently, there is a tremendous amount of parallel evolution, which makes the study of phylogenetic relationships at the familial level very difficult.

2. The small number of morphological characters involved in significant evolutionary changes means that a great deal of soft anatomical data, whose polarities are usually unknown or speculative, have to be introduced into a cladistic analysis.

3. The very rapid differentiation of rodents at superfamilial and familial level between the Late Palaeocene and the Late Eocene makes the study of interrelationships particularly difficult, and confers on the group a very original evolutionary pattern (see Hartenberger 1985 for a review).

4. As in many other mammalian groups, most of the fossil record originates from Europe and North America. Despite some recent contributions from Asia, and particularly China, the fossil record from the southern area of the northern continents and from the southern continents is clearly very incomplete.

5. Molecular data are very scarce in comparison to other mammalian groups. One of the main reasons seems to be the difficulty in obtaining fresh material for analysis.

Since Waterhouse (1839), the systematics and phylogeny of rodents has been a matter of debate. At that time, only anatomical characters were used, some of which still play a dominant role in the present debate. This has been followed by a great number of new observations in many different fields, and today one can observe a period of renewed interest. However, in the last ten years, despite the increasing strength of some molecular methods, unfortunately mostly applied to other vertebrate groups, the main progress in rodent phylogeny has come from ontogenetic data and from palaeontological discoveries in areas from which very few data, or no data at all, were previously known, such as China, Pakistan, or Africa.

The Paris meeting organized by W. P. Luckett and J. L. Hartenberger in 1984, and published in 1985, represents a major advance in the field of rodent phylogeny. In this paper, only new developments, or new interpretations that have occurred since this Paris meeting, will be emphasized.

Hystricomorpha revisited

Major progress in rodent phylogeny recently seems to have focused on Hystricomorpha. This infraorder has been recognized since Brandt (1855) and, despite some opposing views (e.g. Wood 1974, 1975, 1984, 1985), evidence has been accumulating for several years to indicate that the unique character of the *M. masseter medialis* may be inherited from a common ancestor and did not appear several times independently. The Hystricomorpha are characterized by their large infraorbital opening (hystricomorphy), associated with a tendency for the development of a complete hypolophid, and the anterior displacement of the entoconid on the lower molars.

The proposed phylogeny (Fig. 7.1) is derived from the attempts of Hussain *et al.* (1978), Jaeger *et al.* (1985*b*), and Flynn *et al.* (1985, 1986), and deserves several specific comments.

Cocomys

According to Flynn *et al.* (1986), the genus *Cocomys* Dawson *et al.* represents the sister-group of all hystricomorphs. The character that unites *Cocomys* to the later hystricomorphs is, in the author's opinion, the large hypocone of the upper molars, which develops from a distolingual cingulum, rather than the occurrence of a mesoconid on the lower molars, as suggested by Flynn *et al.* (1986).

The first major dichotomy separates ctenodactylids from all other hystricomorphs. Ctenodactylids are characterized, among several other derived morphological characters (Wood 1977; Flynn *et al.* 1986), by a strong reduction of the posterior part of P4, compensated for by an

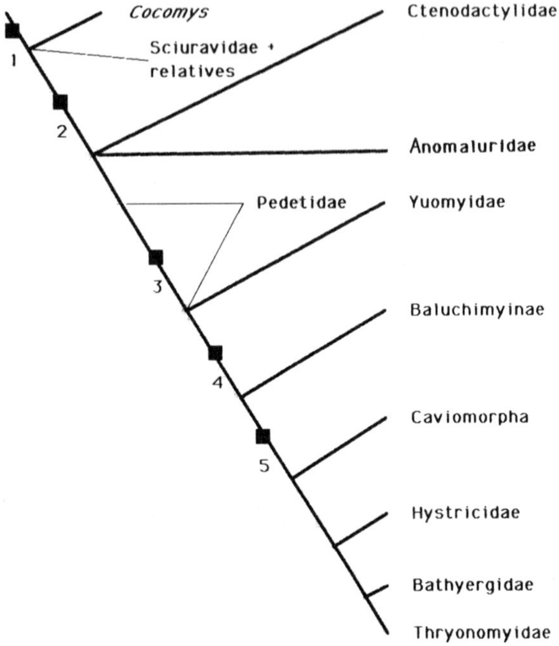

Fig. 7.1. Hystricomorph interrelationships. 1: Strong hypocone on molars; hypocone on P^4; large hypoconulid. 2: Hystricomorphy; hypolophid development. 3: Molariform and enlarged P_4. 4: Metaconule inflated; loss of paraconule; hypolophid in anterior position. 5: Hystricognathy; foetal membrane characters (Luckett 1985), reproductive biology characters and others (George 1985; Woods and Hermanson 1985; Luckett and Hartenberger 1985).

increase of tooth size from M2 to M3. This dichotomy has been attributed to vicariant evolution of two groups of early hystricomorphs on either side of the Early Eocene Tethys Sea (ctenodactylids in the North, chapattimyids in the South) by Jaeger *et al.* (1985*b*), a working hypothesis that has since been accepted by others (Flynn *et al.* 1986). Further morphological evidence of this relationship can be found in the foetal membranes (Luckett 1980, 1985), and in reproductive and musculoskeletal features (George 1985). However, there is still no biochemical evidence to confirm that ctenodactylids are the sister-group of hystricomorphs, pending the results on pancreatic ribonuclease undertaken by Beintema (1985).

Baluchimyinae

A major contribution to hystricomorph evolution and interrelationships has been the discovery and detailed description by Flynn *et al.* (1986) of a new group of chapattimyids from Pakistan, the

Baluchimyinae. This group arose from local evolution of the Eocene chapattimyids in the Indo-Pakistan province. All material comes from a single locality to which a Lower Miocene age (between 18.3 and 21 Ma) has been attributed on the basis of geological considerations and general faunal evolution.

These age arguments are weak, and other interpretations, also based on general faunal evolution, can be made. For instance, it is well known that a major faunal change occurred in Africa around 18 Ma ago as a consequence of its collision with Eurasia. The faunistic consequence of that collision was the exchange of many terrestrial vertebrate groups between these two provinces. Among them, the migration to Africa of ctenodactylids, cricetids, sciurids, and lagomorphs is well recorded in the East African Lower Miocene deposits. Most of these Eurasiatic taxa are supposed to have originated in Asia, but are as yet unrecorded from the North Asiatic realm; this is the case, for example, for the Afrocricetodontinae Lavocat 1973, which cannot be related to any other known tribe of cricetids. Therefore, it is not possible to eliminate southern Asia from the dispersal route between Asia and Africa, at least after the welding of Indo-Pakistan to mainland Asia, which occurred during or slightly before the Lower Eocene (Patriat and Achache 1984; Besse *et al.* 1984). Recent investigations on the large mammals from the Bugti fauna, supposedly from overlying strata of the same formation, show that it represents a mixture of heterochronic elements, containing some typical Oligocene taxa associated with characteristic Lower Miocene taxa (Pickford 1986; Pickford and Rogers 1987). These data indicate, in our opinion, that this new rodent locality is older than 18 Ma, and that it is most probably of Late Oligocene age, or even older.

Despite this minor controversy about the precise age of the Baluchimyinae, their discovery has many consequences for understanding the evolution of the whole group of hystricomorph rodents.

First, the description of the genus *Fallomus* as a chapattimyid derivative, and as a possible relative of the Baluchimyinae (Flynn *et al.* 1986), suggests the possible origin of the pedetids from these primitive rodents, through the succession of morphotypes illustrated by the genera *Fallomus*, *Diatomys* Li, and modern *Pedetes*. The Pedetidae would therefore appear as survivors of an old and primitive southern Asiatic group of hystricomorph and sciurognath rodents (chapattimyids), which would also explain their numerous plesiomorphic features, and the difficulty in relating them to other living groups of rodents (Luckett 1985; Luckett and Hartenberger 1985). However, one should not minimize the conflicting results obtained by George (1985) and Luckett (1985), who relate the pedetids to the ctenodactylids, on the basis of several reproductive and soft anatomical characters. It is also likely that the polarity of some of these characters has been misinterpreted.

The living Pedetidae are sciurognathous, with the primitive condition of the rodent lower jaw. Presumably, their ancestor, and any possibly related Baluchimyinae, were also sciurognathous. However, this depends on the true relationships between *Fallomus* and the Baluchimyinae. If one includes this genus in the Baluchimyinae, this would then indicate that all Baluchimyinae were probably sciurognathous. If one excludes *Fallomus* from this subfamily, as proposed by Flynn *et al.* (1986), then the possibility that some of these Baluchimyinae (or all) were hystricognathous, or incipiently hystricognathous, cannot be excluded.

The Baluchimyinae derive from the local evolution of Eocene southern Asiatic chapattimyids, and therefore share several derived characters with them (see Flynn *et al.* 1986). They also share several derived characters with the North African genus *Protophiomys*, which has been described (Jaeger *et al.* 1985*b*) as an early representative of the African phiomyids (thryonomyoids). Like the Baluchimyinae in Pakistan, these early African phiomyids derive from Lower or Middle Eocene Asiatic chapattimyids, most probably from taxa whose molars did not develop a fifth molar crest (mesolophid). It is therefore highly probable that the chapattimyids are paraphyletic, as is the case for most stem groups, and that they represent a composite association of Baluchimyinae and hystricognath ancestors (at least thryonomyoid ancestors). This scenario was clearly suggested by Flynn *et al.* (1986), but these authors also proposed (their fig. 25) in two other possible phylogenetic relationships for hystricognaths that we do not support. On the basis of the available North African fossil record (Jaeger *et al.* 1985*b*), the Eocene genus *Protophiomys* shares many derived characters with both Chapattimyidae and Baluchimyinae. It differs from them by the more central position of the metaconule (if this structure really represents the metaconule). According to Flynn *et al.* (1986), the North African taxa should be attributed to the Chapattimyidae because they are represented only by isolated molars, which do not indicate whether they share the two most important autapomorphies of the Phiomyidae, the hystricognathy of the lower jaw and the retention of DP_4.

However, *Protophiomys* shares several other characters with Phiomyidae (characters 15, 13, 12, 10, 9, 6, and 4 from table 18 in Flynn *et al.* 1986). It differs from Oligocene phiomyids (thryonomyoids) in having a very large metaconule, which is transformed in later phiomyids into a mesoloph. It is this large metaconule that unites them with the Baluchimyinae. They should be considered as early phiomyids, 'faute de mieux', since they originated in Africa, show continuous variation to a typical phiomyid molar structure, and share several derived characters with the Phiomyidae. In fact, since we consider the phiomyids and other hystricognaths as the sister-group of the Baluchimyinae, both the Chapattimyidae, and the Ctenodactyloidea (as defined by Flynn *et al.* 1986) become paraphyletic groups, and these African Eocene rodents cannot therefore be described as chapattimyids.

Hystricognath ancestors

In the opinion of the author there are no palaeontological data demonstrating the differentiation of a hystricognathous lower jaw from a sciurognathous lower jaw, but as this last morphotype is the only one reported among almost all Early Eocene rodents, it is highly probable that the sciurognathous structure is primitive. From an accumulation of data from numerous areas of biology and palaeontology (Luckett and Hartenberger 1985), it appears now that the living hystrocomorphous and hystricognathous rodents are monophyletic. As noted earlier, the oldest fossils, belonging most probably to the Phiomyidae (Thryonomyoidea), are those from the Eocene of Algeria (Jaeger *et al.* 1985*b*), which document the molar structure of possible early hystricognaths. Their occlusal surface is basically tetralophodont, and the mesoloph(-id) shows a primitive stage of development. Recently, Jaeger (1988) has suggested that the fifth crest of the upper molars of the Caviomorpha was a mesoloph that developed in a different way from that of the early African thryonomyoids. This suggests that the dispersal of the Hystricognatha occurred at a time when their molars lacked a mesoloph(-id) but after they had developed a hystricognathous lower jaw structure. We can therefore reconstruct the ancestral morphotype of these earliest hystricognaths, and compare this reconstruction with the supposed North American ancestors of Caviomorpha described by Wood (1973, 1974, 1975, 1985) under the name of Franimorpha.

In the light of these new data, it is clear that none of the Franimorpha can be included in the definition of Hystricognatha, as most of them do not even show hystricomorphy. The hystricognathy of some taxa like *Prolapsus* is either denied by some authors (Dawson 1977; Korth 1984), or recognized (Flynn *et al.* 1986), but most authors agree in excluding the Franimorpha from the Hystricognatha, at least at the present stage of knowledge. These fossils are interpreted then as the result of convergent evolution of isolated hystricognath characters.

Also the enamel microstructure (Koenigswald 1985) excludes the Franimorpha from the ancestry of Hystricognatha. This is multiserial in Chappatimyidae and their derivatives, including the Baluchimyinae (Flynn *et al.* 1986) and the earliest African Thryonomyidae, as in all living hystricognaths. One can therefore reasonably expect that the ancestors of Caviomorpha also had multiserial enamel. *Prolapsus* has pauciserial enamel (Wood 1973), which excludes it from the ancestry of the Caviomorpha (Koenigswald 1985, p. 415). Parallel evolution is otherwise well documented among rodents, as for instance the perforation of the pterygoid fossa that occurs in extant geomyoids, *Spalax* and *Aplodontia* (Patterson and Wood 1982), and the development of a sciuromorphous infraorbital foramen structure among the geomyoids (Fahlbusch 1985). Parallel evolution seems to occur very frequently in rodent evolution, and may be characteristic. However, it is more often suspected than demonstrated.

Hystricognath interrelationships

In comparison to the general views on hystricognath monophyly, there is no agreement about the interrelationships between hystricognath families, and conflicting data are available concerning even the monophyly of the classical subgroups, as for instance for the Caviomorpha.

Caviomorpha

A comparison of the caviomorph relationships based on myology (Woods and Hermanson 1985), with those based on soft anatomical data and chromosomes (George 1985), and on different molecular methods (Sarich 1985), indicates that no agreement has been achieved. Data are conflicting but converge in indicating a completely different pattern of hystricognath interrelationships from the classical ones, which were based on the fossil record. All these attempts represent stimulating working hypotheses but cannot yet be used to modify the taxonomy at the familial level. However, they seem to agree in the recognition of the paraphyly of the Caviomorpha. We therefore propose here a very conservative pattern of relationships with an early dichotomy between the South American hystricognaths and the others.

The dispersal routes of the earliest South American hystricognaths still remain mysterious. In order to explain their sudden origin in South America, two main conflicting biogeographic scenarios have been proposed:

1. The hystricognaths originated and diversified initially in Africa, followed by one or several dispersals to South America over the South Atlantic before the Early Oligocene (Lavocat 1971, 1973, 1980; Hoffstetter 1972; and others). The main weakness of this theory is identification of the dispersal route over an ocean that had already reached a great width, although Tarling (1982) has given some

geophysical support to the probable existence of land connections through the South Atlantic up to the Oligocene.

2. The hystricognaths originated in Asia, and there were two dispersal events, one to Africa and the other, through northern Asia and North America, to South America. If the Asiatic stem group is now known (the Baluchimyinae), there is no record of any possible baluchimyine from the well-documented North American Eocene fossil record.

Other scenarios are also possible with numerous different combinations, but they cannot be taken into account at present. However, even if we do not know the dispersal routes, it is interesting to notice the major importance of dispersal in the history of these hystricognath rodents.

Hystricidae

We give up our earlier hypothesis of parallel evolution of the hystricognathy of the Hystricidae (Jaeger *et al.* 1985*b*) because it is not parsimonious (Luckett and Hartenberger 1985). The origin, age, and place of differentiation of the earliest Hystricidae also remains unknown.

Bathyergidae

There is a major controversy, about the origin and evolution of the Bathyergidae, between Wood (1984, 1985) and Lavocat (1973). The latter author has based his interpretation on fossil evidence, which indicates that there was a trend among the Lower Miocene East African bathyergids to reduce secondarily the infraorbital foramen. Most probably, this reduction is the consequence of an adaptation to fossorial life, as suggested by several authors. For Lavocat (1973), the bathyergids are therefore closely related to the thryonomyoids, and this interpretation is also corroborated by albumin immunological data (Sarich 1985) and recent ontogenetic data. Schrenk and Maier (1986) have shown that, during the ontogenetic development of the masseteric muscle in *Bathyergus*, a branch of the Masseter medialis anterior stretches through the infraorbital foramen. During the later phases of development, this branch of the masseteric muscle regresses and disappears (Maier and Schrenk 1987).

Wood (1985) suggested that the bathyergids derived from the Asiatic Oligocene tsaganomyids, a phylogenetically still isolated group of imprecise affinities (Luckett and Hartenberger 1985). He suggested that both bathyergids and tsaganomyids were derived from the protrogomorphous cylindrodontids, a North American and Asiatic Eocene group with sciuravid affinities. This is based on the fact that Tsaganomyidae are fully hystricognathous, although with a small, protrogomorphous infraorbital foramen, and that both groups share a number of structural features not found in any other rodents, even among other burrowing rodents. In the light of the monophyly of the Hystricognatha, and their derivation from hystricomorphous ancestors, this interpretation is certainly not the most parsimonious, and cannot be supported in view of the possible paraphyly of the Hystricognatha.

In fact, Wood's principal objective was to demonstrate the paraphyly of the Hystricomorpha in order to support the Franimorpha as possible ancestors of the Caviomorpha. His philosophy is understandable, but his interpretations are not parsimonious. He may be right on some points, because the parsimonious scenario defended here has some weaknesses when confronted with the whole data set. For example, an alternative interpretation based on the fossil record would imply paraphyly of the Baluchimyinae, as suggested with caution by Flynn *et al.* (1986). If, for instance, new discoveries show that some Baluchimyinae were hystricognathous and some were sciurognathous, an Asiatic origin of the Hystricidae as supported by Jaeger *et al.* (1985*b*), Wood (1985), and Flynn *et al.* (1986) would be a better explanation than the scenario necessitated by the present parsimonious tree, which requires an African origin, a migration to Asia, an extinction in Africa, followed by a new migration from Asia to Africa during the Miocene (Jaeger *et al.* 1985*b*). The same is true of the possible origin of the Bathyergidae from the baluchimyine *Downsimys* (Flynn *et al.* 1986). All these alternative hypotheses must be considered as highly speculative, as is the multiple origin of hystricognathous and hystricomorphous rodents in the present state of knowledge.

Anomaluridae

Palaeontological data do not corroborate the hypothesis of close relationships of pedetids with anomalurids that has been proposed on the basis of middle-ear features (Lavocat and Parent 1985) and the pattern of carotid arterial branches and anastomoses (Bugge 1985). The phylogenetic position of the Anomaluridae still remains obscure. Recent reinvestigation of their cranial anatomy (Chami-Khazraji 1986) does not provide any more synapomorphies with other hystricomorphs. Instead, new autopomorphic features were found, such as the extreme reduction of the pterygoid fossa (Chami-Khazraji 1986). The generally proposed close relationship between Anomaluridae and Theridomyidae (Chaline and Mein 1979; Luckett and Hartenberger 1985) seems to rely only on shared derived molar morphology and on the fact that they are both hystricomorphous and sciurognathous

(Jaeger *et al.* 1985*b*). The earliest anomalurids combine hystricomorphy, hypolophid development, and molarized (complex) P_4. These characters could characterize the sister-group of the Ctenodactylidae among the Hystricomorpha. A possible position between the Ctenodactylidae and the Hystricognatha remains therefore a possible working hypothesis, which is strengthened by their South Tethysian distribution, but weakened by the occurrence of uniserial enamel among the extant Anomaluridae (Koenigswald 1985); the Ctenodactylidae and Hystricognatha have multiserial enamel. However, the enamel microstructure of the oldest known anomalurids (Jaeger *et al.* 1985*b*) from the Late Eocene of North West Africa has not yet been investigated. The earliest theridomyid, *Protadelomys* Hartenberger from the Middle Eocene of Western Europe, shows unmolarized P_4, a character that led Hartenberger (1969) to consider them as more probably derived from some local ischyromyoids.

Decisive information could arise from the discovery of Early Eocene rodents from the land masses situated between Africa and Europe (Dercourt *et al.* 1985), where possible common ancestors of these two groups could have evolved. The number of characters at present is too restricted to draw any valuable conclusion. The origin of theridomyids from ischyromyoids remains nevertheless a major problem, especially because it would challenge the monophyly of the Hystricomorpha. Therefore, we propose to leave this problem in its present state of uncertainty, and to consider tentatively the Theridomyidae as a sister-group of the Anomaluridae.

Sciuravids and derivatives

Sciuravids are a North American stem group, which appears very early in the Early Eocene record (Korth 1984; Flanagan 1986). They differ from the sympatric Ischyromyoidea by the strong development of the hypocone and the metaconule on the upper molars. They have been considered as derivatives of the Microparamyidae (Wilson 1986), but Lavocat and Parent (1985) do not share this opinion, on the basis of the more primitive middle-ear anatomy. This observation gives more weight to the suggestion of Korth (1984) that they are derivatives of *Cocomys*. Many North American authors consider them as ancestral to both the Cylindrodontidae (Wood 1973) and the Geomyoidea (Wood 1974, 1980; Dawson 1977). Cylindrodontids show a rather unusual mixture of primitive and derived characters: they are protrogomorphous, have a large P_4, and pauciserial and uniserial enamel. They have no hypocone on the upper molars, but the polarity of that character is unknown. If they arose from some unknown sciuravids,

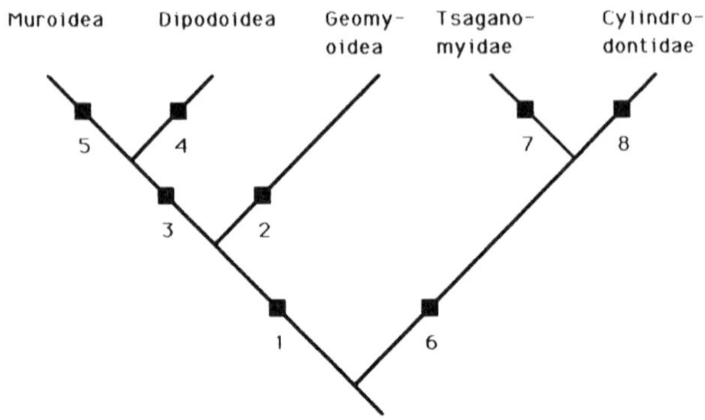

Fig. 7.2. Interrelationships of sciuravid derivatives. 1: 'Cricetid' molar structure; cranial foramina (Wahlert 1978, 1985); myological and skeletal characters (Hill 1937); foetal membranes (Luckett 1985); uniserial incisor enamel structure. 2: Development of a sciuromorphous infraorbital foramen. 3: Enlargement of infraorbital foramen (hystricomorphy); myology (Klingener 1964); cephalic arterial pattern (Bugge 1971); foetal membranes (Luckett 1985). 4: Infraorbital foramen with independent lower vascular groove. 5: Distinct anterocone (-id) on M1; loss of P4. 6: Hystricognathy. 7: Highly developed hystricognathy. 8: Stapedial artery absent; pauciserial to uniserial incisor enamel; hypocone reduction.

one must admit the secondary loss of the hypocone. The same is true if they arose from the Yuomyidae, as suggested by Flynn *et al.* (1986). Wahlert (1974), however, has suggested, on the basis of the cranial foramina, that cylindrodontids may be more closely related to ischyromyoids than to sciuravids.

There are now many arguments that unite the Myodonta (Muroidea plus Dipodoidea) with the Geomyoidea as the Myomorpha (Fig. 7.2) (Flynn *et al.* 1985), but this association raises two sets of objections. Firstly, albumin immunological studies (Sarich 1985) do not support it, and secondly, these relationships are associated with a complex evolution of the infraorbital foramen. In particular, the earliest Myodonta had a large hystricomorphous-like infraorbital foramen (Vianey-Liaud 1985), supposed to be derived from a protrogomorphous structure before it evolved into a typical myomorphous structure. Therefore, some authors like Hartenberger (1985) and Vianey-Liaud (1985) prefer to derive Myodonta directly from early Hystricomorpha (Fig. 7.3C), also because they probably originated in Asia, from which no unchallenged sciuravids have yet been recorded.

I support the common origin of all Myomorpha from a single sciuravid ancestor, but there are three equally possible phylogenetic patterns (Fig. 7.3).

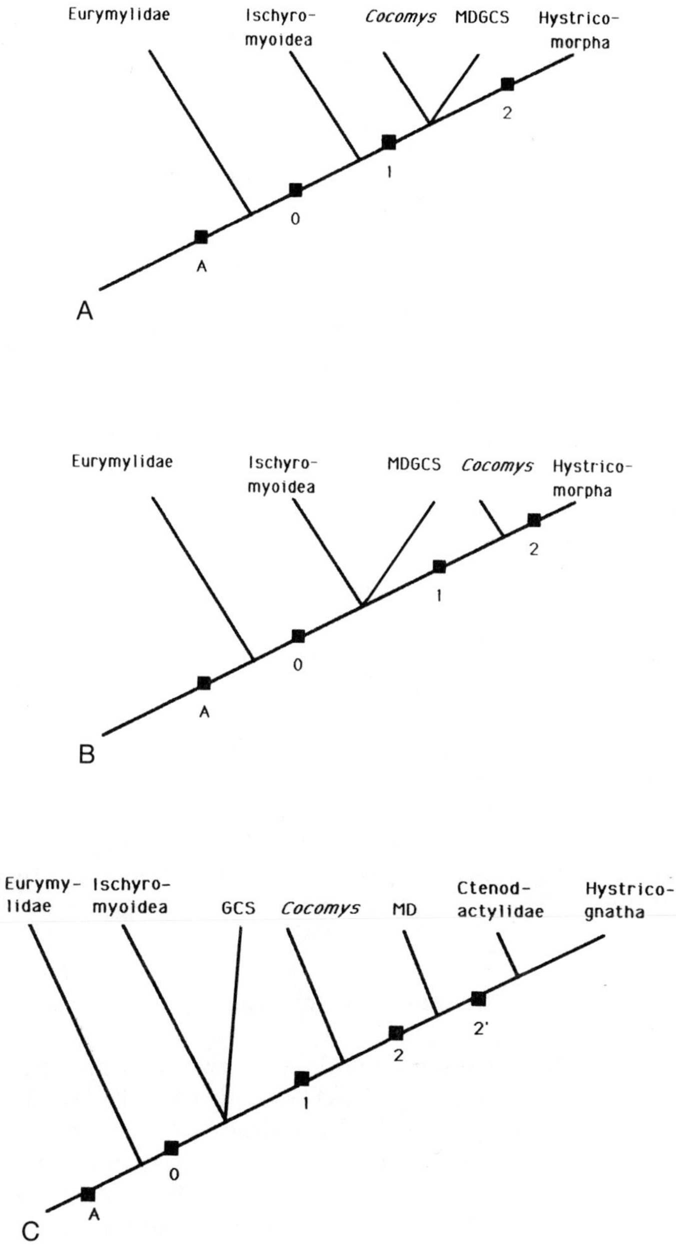

Fig. 7.3. Possible interrelationships between primitive rodents and myomorphs, and other possible sciuravid derivatives. Abbreviations: S, Sciuravidae; C, Cylindrodontidae; G, Geomyoidea; D, Dipodoidea; M, Muroidea. A: Dental formula=1/1 0/0 2/2 3/3; hypsodont incisors; gnawing and chewing separate. O: Loss of P_3. 1: Strong hypocone on molars; hypocone on P^4; large hypoconulid. 2: Hystricomorphy; hypolophid development.

Muroids

Muroid phylogeny is far from being elucidated, and only a few recent works have focused on the interrelationships of the numerous muroid families. For Flynn *et al.* (1985), the monophyly of the muroids is demonstrated by several derived characters, such as the lack of P_4, the appearance of an anterocone (-id) on the first upper (lower) molar, and the confluence of the masseteric and neurovascular foramina of the anterior zygoma. For Vianey-Liaud (1985), the oldest muroid (cricetid) occurs in the Upper Eocene of China and in the latest Eocene (Suevian) of Southern Germany. For Flynn *et al.* (1985), the oldest true muroid, referred to the genus *Eumys*, is Early Oligocene in age, and occurs in Chadronian deposits of North America. The interrelationships of the numerous muroid families, and specifically of the cricetids and the murids, are unclear, as shown by the work of Carleton and Musser (1984) based on the study of the extant muroids, and that of Flynn *et al.* (1985) based on fossil forms. One recent result unites some African and Asiatic muroid families, the Dendromuridae, Cricetomyidae, and Muridae, and makes them a sister-group of the Gerbillidae (including the fossil Myocricetodontidae from South Asia and Africa) (Jaeger *et al.* 1985*a*, 1986; Flynn *et al.* 1985). This unexpected relationship between murids and gerbillids has also been demonstrated by Brownell (1983, pers. comm.) on the basis of DNA hybridization studies. However, a character analysis of fossil and extant murids and gerbillids has led to a controversy between molecular data and palaeontological data on ages of divergence. Molecular data (Sarich 1985) indicate much older divergence ages between *Mus* and *Rattus* within the Muridae, and therefore between Muridae and Gerbillidae, than observed in the fossil record (Jaeger *et al.* 1986).

The sister-group of Hystricomorpha and *Cocomys*

As a consequence of the numerous new discoveries from the Palaeocene and Lower Eocene of China (Dawson *et al.* 1984; Li and Ting 1985), many authors now consider the Eurymylidae as the sister-group of all rodents. Also, good agreement exists for the monophyly of Rodentia, and for a basic dichotomy among early rodents separating Ischyromyoidea and their derivatives from *Cocomys* and its derivatives. However, there is a controversy over the polarity of characters used to infer these phylogenetic relationships. Dawson and Krishtalka (1984), Korth (1984), and Hartenberger (1985) consider that the Cocomyidae (not only *Cocomys* but also some genera that are considered by Flynn *et al.* 1986, and here, as Ctenodactylidae) are the most ancient and earliest differentiated group of rodents. The main argument is that *Cocomys* is morphologically very close to the eurymylid *Heomys* Li, with which it shares derived and primitive characters. Among these shared derived characters (for a detailed list, see Li and Ting 1985) are a strong hypocone on the upper molars and a strong hypoconulid on the lower molars. As the most primitive ischyromyoids lack such a hypocone, several authors, following Korth (1984), have claimed that the hypocone was inherited from the eurymylids and secondarily lost (Fig. 7.4A). As several derivatives of these ischyromyoids have a strong hypocone, one must admit a reversion for this character, which seems neither parsimonious nor demonstrated by the fossil record.

Alternatively, Butler (1985) interprets some *Cocomys* dental characters as synapomorphies with Ctenodactylidae. He argues that the development in *Cocomys* of a large hypocone caused the anterior shift of the entoconid and the development of the hypolophid. The Ischyromyoidea, and particularly the Reithroparamyinae (included by Wood in the Franimorpha), are considered to illustrate the most primitive rodent molar structure, closest to that of the primitive eutherian condition. In this interpretation, the absence of a hypocone among the earliest ischyromyoids would characterize the primitive rodent condition, and the resemblance between *Heomys* and *Cocomys*

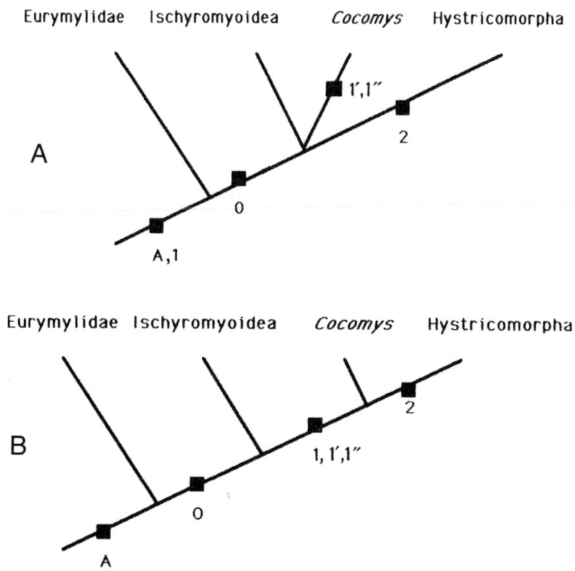

Fig. 7.4. Possible interrelationships of the earliest rodents. A: Dental formula=1/1 0/0 2/2 3/3; hypsodont incisors; gnawing and chewing separate. O: Loss of P_3. 1: Strong hypocone on molars. 1′: Hypocone on P^4. 1″: Large hypoconulid. 2: Hystricomorphy; hypolophid development.

would be the result of convergent evolution (Fig. 7.4B). As this interpretation is based on functional aspects, we consider it to have strong weight. Following this interpretation, *Heomys* cannot be considered as the direct rodent ancestor any longer, an interpretation which is sustained by the high crown elevation of the lingual side of its upper molars, an autapomorphic character that excludes it from this position. The rodent ancestors must therefore be sought among other Asiatic Palaeocone eurymylids.

Gliridae

This group of rodents has an unusually good fossil record. The earliest glirid characters appear in *Eogliravus* from the Middle Eocene of western Europe, which is related to the Microparamyinae (Hartenberger 1971). A continuous fossil record of Gliridae follows in western Europe up to the extant genera. The infraorbital foramen of these early glirids shows a protrogomorphous structure that evolves later into a typical myomorphous structure, called pseudomyomorphous by Vianey-Liaud (1974). This phylogenetic relationship has been supported by Lavocat and Parent (1985) on the basis of the middle-ear structure, and by Bugge (1985) on the basis of the carotid pattern. The albumin immunological data (Sarich 1985) that cluster the glirid genus *Eliomys* with the Myodonta (Muroidea plus Dipodoidea) seems therefore questionable, as do some basicranial features (Wahlert 1978), that cluster them with Sciuridae and Aplodontidae.

Sciuromorpha

For many years (Wilson 1949) the sciurids and castorids, on the basis of their sciuromorphous infraorbital foramen structure, have been united with the protrogomorphous aplodontids, which are supposed to have retained their primitive infraorbital structure. They are considered as ischyromyoid derivatives. The sciuromorphous geomyoids are excluded from that group because it has been demonstrated on the basis of a rather continuous fossil record that their infraorbital structure was independently derived from an ancestral protrogomorphous structure (Fahlbusch 1985).

Eocene sciuromorphous fossils are unknown, since most of the described possible Eocene ancestors still show a protrogomorphous structure. On the basis of the fossil evidence, Vianey-Liaud (1985) considers Sciuridae and Aplodontidae as sister-groups, a relationship that is also sustained by Wahlert (1985) on the basis of cranial foramina, and by Lavocat and Parent (1985) on the basis of the middle-ear anatomy. Albumin immunological data (Sarich 1985) do not conflict with that interpretation.

Castorids are even more isolated owing to the lack of detailed studies, despite a relatively complete North American record. Since the earliest rodents that are generally considered to be true members of the Sciuridae were protrogomorphous, and since the earliest castorids antedated these sciurids, some authors, such as Wood (written communication), suppose that sciuromorphy must have arisen independently in these families. They cannot be closely assigned to any other living group of rodents, and need more investigation, especially in the field of molecular phylogeny. They are therefore assigned 'faute de mieux' to the Sciuromorpha.

These groups are considered by most authors, such as Korth (1984), to derive from Ischyromyoidea. However, other authors do not eliminate an alternative possibility, as for instance Vianey-Liaud (1985), who suggests a possible derivation from Sciuravidae, at least for the Sciuridae and the Aplodontidae.

Conclusions on rodent interrelationships

For more than 70 years, the North American Ischyromyoidea, which were the first-discovered and well-studied Eocene rodents, have been considered as the stem group of the rodents. The present analysis, which follows many new discoveries of Eocene rodents in other continents, illustrates an alternative view in which it is actually difficult to find any later relative of the Ischyromyoidea. As a consequence of the discovery of other possible ancestors, such as the Chinese *Cocomys*, the polarity of many characters has changed. One can therefore suspect that future discoveries may greatly modify the present interpretations, particularly if more molecular data on extant taxa become available.

However, most recent discoveries indicate that all the main dichotomies between rodent higher taxa were achieved during the Eocene. The bush-like evolution of rodents, predicted by some authors many years ago, now becomes a demonstrated characteristic of rodent evolution. Another characteristic of rodent evolution is the high survivorship among the many evolving clades (Pedetidae, Anomaluridae, Aplodontidae, Gliridae, etc.), which should make them more attractive for molecular phylogenetic analysis than many other mammalian groups.

The Lagomorpha as the sister-group of the rodents

The concept of Glires was first proposed by Linnaeus in 1758, and most recent work supports this relationship (see Chapter 3) through the Asiatic fossil eurymyloids. Following Li and Ting (1985), the Mimotonidae would be related to the Lagomorpha, and the Eurymylidae to the Rodentia. However, since the supposed divergence is older than 55–60 Ma, this long separate evolution makes a character analysis between the living representatives difficult; controversial but stimulating. Luckett (1985) advanced a related argument that both groups share homology of incisors and similar fetal membrane development (see Hartenberger (1985) for a review of several other hypotheses). A character analysis of skull anatomy (Novacek 1985) supports this interpretation, as does a phylogenetic analysis using parsimony (PAUP) on a total of 182 non-dental osteological characters of 26 orders of mammals (Shoshani 1986). Despite this, some morphologists disagree with these results, such as Szalay (1985) on the basis of some postcranial attributes, and Martinez (1985) who considers that the number of synapomorphies shared between Rodentia and Lagomorpha is too small to be significant. In the light of such a debate, the molecular data are of special interest.

Albumin immunological studies (Sarich 1985) and α- and β-haemoglobin sequences (Shoshani *et al.* 1985) fail to recognize sister-group relationships between these two groups. However, following Miyamoto and Goodman (quoted in Shoshani 1986), the sequence analysis of a combination of seven protein molecules in one file gives a branching pattern in which Lagomorpha appear as the sister-group of Rodentia. This result, if confirmed, would indicate that a certain minimum number of sequences is necessary to obtain useful results when using sequence analysis with the maximum parsimony method. The relationship between Lagomorpha and Rodentia are therefore strongly supported by different sets of data, but cannot be considered as definitively established.

Acknowledgements

I thank A. E. Wood, J. H. Wahlert, and J.-L. Hartenberger for their helpful comments on the manuscript.

References

Beintema, J. J. (1985). Amino acid sequence data and evolutionary relationships among hystricognaths and other Rodents. In *Evolutionary Relationships among Rodents, a Multidisciplinary Analysis* (eds. W. P. Luckett and J. L. Hartenberger), pp. 549–65. Plenum Press, New York.

Besse, J., Courtillot, V., Pozzi, J. P., Westphal, M., and Zhou, Y. X. (1984). Palaeomagnetic estimates of crustal shortening in the Himalayan thrusts and Zangbo suture. *Nature (London)* **311**, 621–6.

Brandt, J. F. (1855). Beiträge zur Kentniss der Säugethiere Russlands. *Mem. Acad. Imp., St. Petersburg* **9**(6), 1–375.

Brownell, E. (1983). DNA/DNA hybridization studies of muroid rodents: symmetry and rates of molecular evolution. *Evolution* **37**, 1034–51.

Bugge, J. (1971). The cephalic arterial system in mole-rats (Rhizomyidae), jumping mice and jerboas (Dipodoidea) and dormice (Gliroidea) with special reference to the systematic classification of rodents. *Acta Anat.* **76,** 313–6.

Bugge, J. (1985). Systematic value of the carotid arterial pattern in rodents. In *Evolutionary Relationships among Rodents, a Multidisciplinary Analysis* (eds. W. P. Luckett and J. L. Hartenberger), pp. 355–79. Plenum Press, New York.

Butler, P. M. (1985). Homologies of molar cusps and crests, and their bearing on assessments of rodent phylogeny. In *Evolutionary Relationships among Rodents, a Multidisciplinary Analysis* (eds. W. P. Luckett and J. L. Hartenberger), pp. 381–401. Plenum Press, New York.

Carleton, M. D. and Musser, G. G. (1984). Muroid rodents. In *Orders and Families of Recent Mammals of the World* (eds. S. Anderson and J. K. Jones), pp. 289–379. Wiley, New York.

Chaline, J. and Mein, P. (1979). *Les Rongeurs et l'Evolution*, pp. 1–235. Douin Editeurs, Paris.

Chami-Khazraji, S. (1986). Relations phylogénétiques des Anomaluridae (Mammalia, Rodentia): apport de l'ostéologie cranienne. *Mémoires des Sciences de la Terre*, Vol. 86-23, pp. 1–159. Université Pierre et Marie Curie, Paris.

Dawson, M. R. (1977). Late Eocene rodent radiations: North America, Europe and Asia. *Géobios, Mém. Spéc.* **1**, 195–209.

Dawson, M. R. and Krishtalka, L. (1984). Fossil history of the families of recent mammals. In *Orders and Families of Recent Mammals of the World* (eds. S. Anderson and J. K. Jones), pp. 11–57. Wiley, New York.

Dawson, M. R., Li Chuan-Kuei, and Qi Tao (1984). Eocene ctenodactyloid rodents (Mammalia) of Eastern and Central Asia. In *Papers in Vertebrate Paleontology Honoring Robert Warren Wilson* (ed. R. M. Mengel), *Carnegie Mus. Nat. Hist., Spec. Publ.* **9**, 138–50.

Dercourt, J., Zonenschein, L. P., Ricou, L. E., Kazmin, G., Le Pichon, X., Kiepper, A. L., Grandjacquet, C., Sborshchikov, L. M., Bouillin, J., Sorokhlin, O., Geyssant, J., Lepvrier, C., Biju-Duval, B., Sibuet, J. C., Savostin, L. A., Westphal, M., and Lauer, J. P. (1985). Présentation de neuf cartes paléogéographiques au 1/20,000,000 s'étendant de l'Atlantique au Pamir pour la période du Lias à l'Actuel. *Bull. Soc. Géol. France*, **6**, pp. 637–52 et supplément.

Fahlbusch, V. (1970). Populationsverschiebungen bei tertiären Nagetieren, eine Studie an oligozänen und miozänen Eomyidae Europas. *Abh. Bayer. Akad. Wiss. Math.- Naturw. Kl.*, N. F. **145**, 1–136.

Fahlsbusch, V. (1985). Origin and evolutionary relationships among geomyoids. In *Evolutionary Relationships among Rodents, a Multidisciplinary Analysis* (eds. W. P. Luckett and J. L. Hartenberger), pp. 617–29. Plenum Press, New York.

Flanagan, K. M. (1986). Early Eocene rodents from the San Jose Formation, San Juan Basin, New Mexico. In *Vertebrates, Phylogeny and Philosophy* (eds. K. M. Flanagan and J. A. Lillegraven), *Contrib. Geol.*, pp. 197–220. The University of Wyoming.

Flynn, L. J., Jacobs, L. L., and Lindsay, E. H. (1985). Problems in muroid phylogeny: relationship to other rodents and origin of major groups. In *Evolutionary Relationships among Rodents, a Multidisciplinary Analysis* (eds. W. P. Luckett and J.L. Hartenberger), pp. 589–616. Plenum Press, New York.

Flynn, L. J., Jacobs, L. L., and Cheema, I. U. (1986). Baluchimyinae, a new ctenodactyloid Rodent subfamily from the Miocene of Baluchistan. *Am. Mus. Novit.* **2841**, 1–58.

George, W. (1985). Reproductive and chromosomal characters of ctenodactylids as a key to their evolutionary relationships. In *Evolutionary Relationships among Rodents, a Multidisciplinary Analysis* (eds. W. P. Luckett and J. L. Hartenberger), pp. 453–74. Plenum Press, New York.

Hartenberger, J. L. (1969) Les Pseudosciuridae (Mammalia, Rodentia) de l'Eocène moyen de Bouxwiller, Egerkingen et Lissieu. *Palaeovertebrata* **3**, 27–61.

Hartenberger, J. L. (1971). Contribution à l'étude des genres *Gliravus* et *Microparamys* (Rodentia) de l'Eocène d'Europe. *Palaeovertebrata* **4**, 97–135.

Hartenberger, J. L. (1985). The order Rodentia: major questions on their evolutionary origin, relationships and suprafamilial systematics. In *Evolutionary Relationships among Rodents, a Multidisciplinary Analysis* (eds. W. P. Luckett and J. L. Hartenberger), pp. 1–33. Plenum Press, New York.

Hill, J. E. (1937). Morphology of the pocket gopher mammalian genus *Thomomys*. *Univ. Calif. Publ. Zool.* **42**, 81–171.

Hoffstetter, R. (1972). Origine et dispersion des rongeurs hystricognaths. *C. R. Acad. Sci. Paris, Ser. D* **274**, 2867–70.

Hussain, S. T., de Bruijn, H., and Leinders, J. M. (1978). Middle Eocene rodents from the Kala Chitta Range (Punjab Pakistan). *Proc. Ned. Akad. Wettensch., Ser. B* **81**, 101–12.

Jaeger, J.-J. (1988). L'évolution de la pentalophodontie chez les rongeurs (Mammalia, Rodentia) caviomorphes. *Géobios, Mém. Spéc.*, in press.

Jaeger, J.-J., Tong, H., Buffetaut, E., and Ingavat, R. (1985*a*). The first fossil rodents from the Miocene of Northern Thailand and their bearing on the problem of the origin of the Muridae. *Rev. Paleobiol.* **4**, 1–7.

Jaeger, J.-J., Denys, C., and Coiffait, B. (1985*b*). New Phiomorpha and Anomaluridae from the Late Eocene of North West Africa. In *Evolutionary Relationships among Rodents, a Multidisciplinary Analysis* (eds. W. P. Luckett and J. L. Hartenberger), pp. 403–22. Plenum Press, New York.

Jaeger, J.-J., Tong, H., and Denys, C. (1986). Age de la divergence *Mus-Rattus*: comparaison des données paléontologiques et moléculaires. *C. R. Acad. Sci., Paris* **302**, 917–22.

Klingener, D. J. (1964). The comparative myology of four dipodoid rodents (genus *Zapus, Napeozapus, Sicista* and *Jaculus*). *Misc. Publ. Mus. Zool. Univ. Mich.* **124**, 1–100.

Koenigswald, W. von (1985). Evolutionary trends in the enamel of rodents incisors. In *Evolutionary Relationships among Rodents, a Multidisciplinary Analysis* (eds. W. P. Luckett and J. L. Hartenberger), pp. 403–22. Plenum Press, New York.

Korth, W. W. (1984). Earliest Tertiary evolution and radiation of rodents in North America. *Bull. Carnegie Mus.* **24**, 1–71.

Lavocat, R. (1971). Affinités systématiques des Caviomorphes et des Phiomorphes et origine Africaine des Caviomorphes. *An. Acad. Bras. Cienc.* **43**, 515–22.

Lavocat, R. (1973). *Les Rongeurs du Miocène d'Afrique Orientale. I. Miocène inférieur*, Vol. I, pp. 1–284. École Pratique des Hautes Études, Institut de Montpellier.

Lavocat, R. (1980). The implications of rodent paleontology and biogeography to the geographical sources and origin of the platyrrhine primates. In *Evolutionary Biology of the New World Monkeys and Continental Drift* (eds. R. L. Ciochon and A. B. Chiarelli), pp. 93–102. Plenum Press, New York.

Lavocat, R. and Parent, J. P. (1985). Phylogenetic analysis of middle ear features in fossil and living rodents. In *Evolutionary Relationships among Rodents, a Multidisciplinary Analysis* (eds. W. P. Luckett and J. L. Hartenberger), pp. 333–54. Plenum Press, New York.

Li Chuan Kuei and Ting Su-Yen (1985). Possible phylogenetic relationship of Asiatic eurymylids and rodents, with comments on mimotonids. In *Evolutionary Relationships among Rodents, a Multidisciplinary Analysis* (eds. W. P. Luckett and J. L. Hartenberger), pp. 35–58. Plenum Press, New York.

Linnaeus, C. (1758). *Systema naturae per regna tri naturae, secundum classis, ordines, genera, species cum characteribus, differentiis, sysnonymis, locis*, 10th edn., Vol. 1, 824 pp. Laurentii Salvii, Stockholm.

Luckett, W. P. (1980). Monophyletic or diphyletic origin of Anthropoidea and Hystricognathi: evidence of the fetal membranes. In *Evolutionary Biology of the New World Monkeys and Continental Drift* (eds. R. L. Ciochon and B. Chiarelli), pp. 347–68. Plenum Press, New York.

Luckett, W. P. (1985). Superordinal and intraordinal affinities of Rodents: developmental evidence from the dentition and placentation. In *Evolutionary Relationships among Rodents, a Multidisciplinary Analysis* (eds. W. P. Luckett and J. L. Hartenberger), pp. 227–76. Plenum Press, New York.

Luckett, W. P. and Hartenberger, J. L. (1985). Evolutionary relationships among Rodents: comments and conclusions. In *Evolutionary Relationships among Rodents, a Multidisciplinary Analysis* (eds. W. P. Luckett and J. L. Hartenberger), pp. 685–712. Plenum Press, New York.

Maier, W. and Schrenk, F. (1987). The hystricomorphy of the Bathyergidae as determined from ontogenetic evidence. *Z. Säugetierkunde*, **52**, 156–64.

Martinez, N. L. (1985). Reconstruction of ancestral cranioskeletal features in the order Lagomorpha. In *Evolutionary Relationships among Rodents, a Multidisciplinary Analysis* (eds. W. P. Luckett and J. L. Hartenberger), pp. 151–89. Plenum Press, New York.

Novacek, M. J. (1985). Cranial evidence for rodent affinities. In *Evolutionary Relationships among Rodents, a Multidisciplinary Analysis* (eds. W. P. Luckett and J. L. Hartenberger), pp. 59–81. Plenum Press, New York.

Patriat, P. and Achache, J. (1984). India–Eurasia collision chronology has implications for crustal shortening and driving mechanism of plates. *Nature (London)* **311**, 615–21.

Patterson, B. and Wood, A. E. (1982). Rodents from the Deseadan Oligocene of Bolivia and the relationships of the Caviomorpha. *Bull. Mus. Comp. Zool.* **149**, 371–543.

Pickford, M. (1986). Première découverte d'un Hyracoïde paléogène en Eurasie. *C. R. Acad. Sci. Paris* **303**, 1251–4.

Pickford, M. and Rogers, D. (1987). Révision des Suiformes de Bugti, Pakistan. *C. R. Acad. Sci. Paris*, **305**, 643–6.

Sarich, V. M. (1985). Rodent macromolecular systematics. In *Evolutionary Relationships among Rodents, a Multidisciplinary Analysis* (eds. W. P. Luckett and J. L. Hartenberger), pp. 423–52. Plenum Press, New York.

Schrenk, F. and Maier, W. (1986). Ist *Bathyergus* hystricomorph? In *Kurzfassungen der Vorträge und Posterdemonstrationen, Deutsche Gesellschaft für Säugetierkunde, 60. Hauptversammlung, Säugetierkunde*, p. 72. Verlag Paul Parey.

Shoshani, J. (1986). Mammalian phylogeny: comparison of morphological and molecular results. *Mol. Biol. Evol.* **3**, 222–42.

Shoshani, J., Goodman, M., Czelusniak, J., and Braunitzer, G. (1985). A phylogeny of Rodentia and other Eutherian orders: parsimony analysis utilizing amino-acid sequences of alpha and beta hemoglobin chains. In *Evolutionary Relationships among Rodents, a Multidisciplinary Analysis* (eds. W. P. Luckett and J. L. Hartenberger), pp. 191–210. Plenum Press, New York.

Szalay, F. S. (1985). Rodent and lagomorph morphotype adaptations, origins, and relationships: some postcranial attributes analyzed. In *Evolutionary Relationships among Rodents, a Multidisciplinary Analysis* (eds. W. P. Luckett and J. L. Hartenberger), pp. 83–132. Plenum Press, New York.

Tarling, D. H. (1982). Land bridges and plate tectonics. In *Phylogénie et Paléobiogéographie, Géobios, Mém. Spéc.* **6**, 361–74.

Vianey-Liaud, M. (1974). Les Rongeurs de l'Oligocène inférieur d'Escamps. *Palaeovertebrata* **6**, 197–241.

Vianey-Liaud, M. (1985). Possible evolutionary relationships among Eocene and Lower Oligocene rodents of Asia, Europe and North America. In *Evolutionary Relationships among Rodents, a Multidisciplinary Analysis* (eds. W. P. Luckett and J. L. Hartenberger), pp. 277–309. Plenum Press, New York.

Wahlert, J. H. (1974). The cranial foramina of protrogomorphous rodents: an anatomical and phylogenetic study. *Bull. Mus. Comp. Zool.* **146**, 363–410.
Wahlert, J. H. (1978). Cranial foramina and relationships of the Eomyoidea (Rodentia, Geomorpha). Skull and upper teeth of *Kansasimys*. *Am. Mus. Novit.* **2645**, 1–16.
Wahlert, J. H. (1985). Cranial foramina of Rodents. In *Evolutionary Relationships among Rodents, a Multidisciplinary Analysis* (eds. W. P. Luckett and J. L. Hartenberger), pp. 311–32. Plenum Press, New York.
Waterhouse, G. E. (1839). Observations on the Rodentia, with a view to point out the groups, as indicated by the structure of the crania, in this order of Mammals. *Mag. Nat. Hist.* N.S. **3**, 90–6.
Wilson, R. W. (1949). Early Tertiary rodents of North America. *Carnegie Inst. Wash. Publ.* **584**, 67–164.
Wilson, R. W. (1986). The Paleogene record of rodents: fact and interpretation. In *Vertebrates, Phylogeny and Philosophy* (eds. K. M. Flanagan and J. A. Lillegraven), *Contrib. Geol.*, pp. 163–76. The University of Wyoming.
Wood, A. E. (1973). Eocene rodents, Pruett Formation, Southwest Texas; their pertinence to the origin of the South American Caviomorpha. *Texas Mem. Mus., Pearce-Sellards Ser.* **20**, 1–40.
Wood, A. E. (1974). The evolution of Old World and New World hystricomorphs. *Symp. Zool. Soc. London* **34**, 21–60.
Wood, A. E. (1975). The problem of the hystricognathous rodents. In *Studies on Cenozoic Paleontology and Stratigraphy in Honor of Claude W. Hibbard* (ed. G. E. Smith), *Univ. Mich. Pap. Paleontol.* **12**, 75–80.
Wood, A. E. (1977). Evolution of the rodent family Ctenodactylidae. *J. Palaeontol. Soc. India* **20**, 120–37.
Wood, A. E. (1980). The origin of the caviomorph rodents from a source in Middle America: a clue to the area of origin of the platyrrhine primates. In *Evolutionary Biology of the New World Monkeys and Continental Drift* (eds. R. L. Ciochon and B. Chiarelli), pp. 79–91. Plenum Press, New York.
Wood, A. E. (1984). Hystricognathy in the North American Oligocene rodent *Cylindrodon* and the origin of the Caviomorpha. In *Papers in Vertebrate Paleontology Honoring Robert Warren Wilson* (ed. R. M. Mengel), *Carnegie Mus. Nat. Hist. Spec. Publ.* **9**, 151–60.
Wood, A. E. (1985). The relationships, origin and dispersal of the hystricognathous Rodents. In *Evolutionary Relationships among Rodents, a Multidisciplinary Analysis* (eds. W. P. Luckett and J. L. Hartenberger), pp. 475–513. Plenum Press, New York.
Woods, C. A. and Hermanson, J. W. (1985). Myology of hystricognath rodents: an analysis of form, function, and phylogeny. In *Evolutionary Relationships among Rodents, a Multidisciplinary Analysis* (eds. W. P. Luckett and J. L. Hartenberger), pp. 515–48. Plenum Press, New York.

8. The phylogeny of the ungulates

DONALD R. PROTHERO
Department of Geology, Occidental College, Los Angeles, California, USA

EARL M. MANNING
Museum of Geoscience, Louisiana State University, Baton Rouge, Louisiana, USA

MARTIN FISCHER
Institut für Biologie III, Universität Tübingen, Tübingen, West Germany

Abstract

Cladistic analysis of the major living and extinct ungulates, including the living hoofed mammals, was performed on dental, cranial, and postcranial characters. Both dental and non-dental characters clearly show that hyraxes and perissodactyls are a monophyletic group, followed by progressively more distantly related outgroups in the following order: tethytheres (proboscideans, sirenians, and desmostylians), arsinoitheres, uintatheres, phenacodonts and meniscotheres, whales and mesonychids, hyopsodonts and periptychids, arctocyonids, and artiodactyls. Molecular phylogenies of living ungulates typically cluster artiodactyls, perissodactyls, and whales, but differ from the morphologically-based phylogeny in that they separate hyraxes and tethytheres as 'paenungulates'. We find several derived characters that support a closer affinity of hyraxes with other perissodactyls than with tethytheres. Indeed, Owen included hyraxes in his original definition of the Perissodactyla. Contrary to common assumption, perissodactyls are closer to whales and elephants than they are to artiodactyls. There are characters of the skull, dentition, and postcranial skeleton that support the monophyly of ungulates and conflict with the molecular

The Phylogeny and Classification of the Tetrapods, Volume 2: Mammals (ed. M. J. Benton), Systematics Association Special Volume No. 35B, pp. 201–34. Clarendon Press, Oxford, 1988.

phylogenies that place 'paenungulates' at more primitive nodes within the Eutheria. The phylogenetic pattern supported here clearly shows that the 'Condylarthra' is a paraphyletic waste-basket taxon that has long outlived its usefulness, and should be abandoned.

Introduction

The ungulates, including the living hoofed mammals, are the largest, most diverse, and ecologically most dominant group of mammals. According to Simpson (1945), ungulates make up over a third of the named genera (1049 out of 2864) and families (97 out of 257) of mammals. In this respect, they outnumber even the rodents. Outside Australia and Antarctica, ungulates have been the dominant large herbivores since the beginning of the Cenozoic. The largest known animal (the blue whale), the largest land mammal that ever lived (the extinct hyracodontid rhinocerotoid *Paraceratherium*), and the largest living land mammal (the African elephant) are all ungulates. There were very primitive housecat-sized Late Cretaceous ungulates, but even then they were the largest placental mammals around.

Although they dominate the large herbivore niche in most ecosystems, ungulates are much more ecologically diverse. The mesonychids and arctocyonids were probably bear-like omnivores, and whales feed on fish, squid, or plankton. Whales, desmostylians, and sirenians have secondarily become aquatic. Some ungulates became huge and graviportal, such as elephants, arsinoitheres, uintatheres, pyrotheres, toxodonts, rhinoceroses, titanotheres, and hippopotami. Others became highly cursorial, especially among the litopterns, perissodactyls, and artiodactyls. There is even an arboreal ungulate (*Dendrohyrax*). A few ungulate groups, particularly the early ones, retain unspecialized patterns of feeding and locomotion.

Living ungulates are best known from the abundant 'even-toed' artiodactyls (pigs, hippos, peccaries, llamas, camels, chevrotains, deer, giraffes, pronghorns, antelope, cattle, sheep, goats, musk oxen), the 'odd-toed' perissodactyls (horses, tapirs, rhinos), and elephants. Although not usually considered 'hoofed mammals', whales (including dolphins and porpoises), sirenians (manatees and dugongs), and hyraxes are also grouped in the Ungulata. In the fossil record, ungulates are even more diverse; 75% of the families and 87% of the genera of ungulates are extinct (Simpson 1945). Any discussion of ungulate phylogeny based solely on living groups neglects their great diversity, and may be misled by limited character information (*contra* Patterson 1981).

In spite of the great number of fossil and living ungulates, and their abundance and excellent preservation in the fossil record, there have been remarkably few comprehensive statements about their phylogeny. Some of the earliest efforts were reviewed by Gregory (1910), who was one of the first to sketch a phylogenetic diagram of the major groups of ungulates. Mammal classifications, from Linnaeus (1758) to Simpson (1945) to McKenna (1975), have suggested phylogenetic hypotheses, but none was fully supported by character evidence. The most explicit recent hypotheses (Sloan 1970, 1987; Van Valen 1978*b*) show only ancestor–descendant sequences of taxa, with no supporting evidence. Novacek (1982, 1986), and Novacek and Wyss (1986) have presented cladograms of the major eutherian groups, but have dealt only with the living ungulate groups. Our own phylogenetic work began in the 1970s, culminating in an unpublished cladogram using mainly dental characters by Manning in 1975, which was widely circulated. Part of this work was published by McKenna and Manning (1977). As an independent test of the 1975 cladogram, Prothero and Fischer collected non-dental evidence for the relationships of the major ungulate groups. Finally, we examined the evidence presented by molecular phylogenies of recent years. In this paper, we compare the three hypotheses generated from independent data sources by different authors. In this way, we hope that the resulting hypothesis will be well tested and robust.

Phylogenetic analysis

Materials and methods

There are many reasons for the confusion and lack of explicit hypotheses of ungulate relationships. Part of the problem is that ungulate groups are often strangely derived (e.g. uintatheres, elephants, arsinoitheres, whales, and sirenians) or almost completely primitive (most of the so-called 'condylarths'). In very specialized groups, it is often difficult to screen out unusual characters and find underlying common patterns. In groups with few specializations, it can be difficult to find anything that would be clearly synapomorphic.

A more significant problem with previous phylogenetic schemes is methodological. Too often, previous researchers have presented diagrams or classifications with no explicit character analysis in support. When evidence is presented, the great majority of the resemblances are clearly symplesiomorphic. For many Early Tertiary lineages, the phylogeny consists of ancestor–descendant sequences of genera allied only on dental features. This prevents any comparison or analysis of non-dental features, which are known for most ungulate genera. 'Tooth phylogenies' often degenerate to a 'join up the dots' approach that has been widely criticized (McKenna *et al.* 1977). Since major ungulate groups have migrated among Africa, Eurasia, and the Americas throughout the Early Cenozoic, a 'tooth phylogeny' of a single basin or continent (e.g. Sloan 1970, 1987; Van Valen 1978*b*) is not likely to produce good results (Lazarus and Prothero 1984). Such phylogenies not only emphasize similarity based on shared primitive features, but also neglect the non-dental evidence and the taxa from other areas. Certainly, dental evidence is important, and for many genera teeth are all that we have. The same patterns of relationship should also be evident from the rest of the anatomy.

Most of the dental features discussed here were examined in the excellent collection of fossil and living ungulates in the American Museum of Natural History. A number of very primitive ungulates were very briefly described and named by Van Valen (1978*b*); some of this material has also been described by Archibald (1982; Archibald *et al.* 1983*a*, 1983*b*). Unfortunately, the most important specimens, the skulls, have not been described. When the material is more completely described and illustrated, it will be possible to test some of our hypotheses. Non-dental comparisons were based on the specimens and sources in Table 8.1. The matrix of non-dental characters was analysed using the MacClade software package for the Apple MacIntosh, developed by Wayne Maddison at the Museum of Comparative Zoology, Harvard University.

Outgroup

Cifelli (1983*b*) has shown that *Kennalestes* and other Late Cretaceous eutherians make the best outgroups for determining character polarity with the Eutheria. Where appropriate, primitive carnivores (e.g. *Oodectes*), rodents (e.g. *Paramys*), primates (e.g. *Purgatorius*), and certain 'insectivores' (e.g. *Didelphodus*) were examined to determine whether a character state is widely distributed outside the ungulates. Many authors have suggested that *Protungulatum* (Sloan and Van Valen 1965; Van Valen 1978*b*) is primitive in all known features with respect to the other ungulates. We disagree, and prefer to use comparisons outside the ungulates.

Simpson's (1945) concept of the 'Ferungulata' has been used widely to suggest a close relationship between carnivores and ungulates. The influence of this concept can be seen in mammalogy literature, notably in Walker *et al.* (1975), where carnivores are placed next to the ungulate orders. Unfortunately, 'Ferungulata' was based entirely on the erroneous placement of certain ungulates. Cope (1884) greatly expanded his taxon 'Creodonta' (which previously included only carnivores, hyaenodonts, and oxyaenids) to include the arctocyonid (*sensu lato*) *Triisodon* and the mesonychids. Even though these primitive ungulates were later distinguished from all other creodonts by Matthew (1909) as 'Acreodi', they were still considered to be related to carnivores. Thus Simpson (1945) connected carnivores and ungulates because of the connection between 'arctocyonid carnivores' and ungulates. Simpson (1953) even used this mistaken connection as a defence for horizontal classification. According to Simpson (1953, p. 345), 'there is one genus, *Protogonodon*, in which on balance of resemblance in small details some species would be classified as carnivores and others as ungulates'. When arctocyonids were later removed from the Carnivora (Patterson and McGrew 1962), all basis for the 'Ferungulata' disappeared (see Chapter 3).

Anatomical characters

The majority of Palaeogene fossil ungulates have been identified exclusively from their teeth. The cladogram in Fig. 8.1 is an attempt to reduce the complexity of dental variation to a meaningful pattern of derived character states. Some primitive ungulate genera were not included due to lack of space, but enough are shown to suggest a framework of relationships. Non-dental characters are summarized in Table 8.1. Much of this evidence goes back to the classic work of Gregory (1910, 1920), although Gregory did not always distinguish primitive from derived features.

Ungulate skulls bear the hallmarks of their relationships in many ways, in addition to their teeth. In the preorbital region, most artiodactyls show an unusually large facial portion of the lacrimal, which often contains a preorbital fossa or vacuity. Artiodactyls also have an unusually large orbitosphenoid bone, which separates the frontal from the alisphenoid (Russell *et al.* 1983). Perissodactyls (except some hyraxes, tapirs, and certain rhinocerotids) have developed a broad contact between the nasals and lacrimals, separating the frontals from the maxillaries.

In the ear region, there are many striking and unique characters, which were analysed in an excellent study by Cifelli (1982). Their distribution (Cifelli 1982, tables 1 and 2) is placed in the context of the present hypothesis in Table 8.1. Wible (1986, 1987) has analysed a number of characters of the cranial circulation. According to Wible (1987), in all ungulates the superior ramus of the stapedial artery no longer lies between the petrosal and squamosal; it either passes through the tegmen tympani of the petrosal (in *Arctocyon*, *Arctocyonides*, and *Pleuraspidotherium*, according to Russell 1964) or is absent altogether (all other ungulates). In addition, cetaceans, perissodactyls, hyracoids, and tethytheres are distinguished from artiodactyls by the loss of the

Table 8.1. Distribution of non-dental features in selected ungulate taxa

Character	Taxa												
	Art	Arc	Hyp	Prp	Mes	Cet	Phe	Men	Uin	Ars	Pro	Hyr	Prs
Cranial characters													
Posterior petrosal epitympanic sinus	–	–	–	?	–	–	+	+	?	?	?	–	–
Lose promontory sulcus	–	–	–	?	?	?	–	–	?	–	–	+	+
Ventral, slit-like aquaeductus cochleae	–	–	–	?	?	?	+	?	?	?	?	+	+
Inflated tegmen tympani	–	–	+	?	?	?	+	+	?	?	?	+	+
Tympanic aperture facial canal anterior to fenestra ovalis	–	+	+	?	?	?	+	+	?	?	?	+	+
Large, bridged tympanohyal	–	–	–	?	–	–	–	–	?	?	?	+	+
Long tympanic process (also Tu)	–	–	–	?	–	–	–	–	?	?	?	+	+
Lose stapedial sulcus	–	–	+	?	–	+	–	–	?	+	+	+	+
Lose proximal stapedial artery	–	–	–	?	+	+	+	+	?	?	+	+	+
Superior ramus of stapedial in petrosal or lost	+	+	+	?	+	+	+	+	?	?	+	+	+
Bulla (if present) made of ectotympanic only (also Ro, La)	+	?	?	?	?	+	?	?	?	?	+	–	+
Weak occipital exposure of mastoid (also Ch, Ed)	+	–	–	?	+	+	+	+	+	+	+	+	+
Lose mastoid foramen	+	+	+	?	+	+	+	+	+	+	+	+	+
Postglenoid foramen small or absent (also Tu)	–	–	–	?	+	+	+	+	+	+	+	+	+
Large posterior lacerate foramen coalesces with cochlear fissure (also De, Ch)	–	–	–	?	+	+	+	?	+	?	+	–	+
Ethmoid foramen above posterior palate	–	?	?	?	?	–	?	?	?	?	–	+	+

Eustachian sac	–	?	?	?	?	–	?	?	?	?	–	+	+
Extrabullar anterior internal carotid artery (also some Ro, La)	–	–	–	?	–	–	–	–	?	?	?	+	+
Lose carotid sulcus	–	?	?	?	?	+	–	–	–	–	–	+	+
Tuber maxillaris	–	–	–	?	–	–	–	–	?	?	+	+	+
Postcranial characters													
Lose clavicle	–	–	–	–	+	+	+	+	+	+	+	+	+
Reduced acromion, coracoid process on scapula	–	–	–	–	+	+	+	–	+	+	+	+	+
Reduced medial epicondyle, entepicondylar foramen	–	–	–	–	–	–	+	+	+	+	+	+	+
Ventral ulna straight or concave (some No)	–	–	–	–	+	+	+	+	+	+	+	+	+
Weak deltoid crest on humerus	–	–	–	–	+	+	+	+	+	+	+	+	+
19 or more thoracic vertebrae	–	–	–	–	–	–	–	–	+	+	+	+	+
Reduced astragalus/cuboid contact (also Tu, Pa)	–	–	–	–	–	?	+	+	+	+	+	+	+
Large, distally shifted third trochanter on femur (also Tu, Pa)	–	+	+	+	+	?	+	+	+	+	+	+	+

(–) Indicates primitive condition; +=derived condition; ?=condition unknown or not yet examined. Petrosal characters based on Cifelli (1982); vertebral characters from Gregory (1910); skull characters based on Novacek (1986) and Gregory (1920), plus personal observations. Sources for specific taxa include Rose (1982, 1985), Russell *et al.* (1983) for artiodactyls; for arctocyonids: Russell (1964) and *Arctocyon ferox*, American Museum of Natural History (AMNH) 16542, 16543; hyopsodonts: Gazin (1965); periptychids: Matthew (1937) and *Periptychus rhabdodon*, AMNH 17075; mesonychids: Scott (1888) and *Pachyaena ossifraga*, AMNH 4262, *Mesonyx obtusidens*, AMNH 12643; whales: Kellogg (1936), Kumar and Sahni (1986); phenacodonts: Matthew (1937), Radinsky (1966), and *Phenacodus wortmani*, AMNH 4378, *P. primaevus*, AMNH 4369; meniscotheres: Gazin (1968); uintatheres: Flerov (1957); arsinoitheres: Andrews (1906); Proboscidea: Andrews (1906); hyraxes: recent material of *Procavia capensis*; perissodactyls: MacFadden (1976), Radinsky (1966). Abbreviated taxa listed below in the following order: artiodactyls, arctocyonids, hyopsodonts, periptychids, mesonychids, Cetacea, phenacodonts, meniscotheres, uintatheres, arsinoitheres, proboscideans, hyraxes, and perissodactyls (excluding hyraxes). Abbreviations for other taxa showing some of these features in parallel are: Di=didelphids, De=dermopterans, Ch=bats, Ro=rodents, La=lagomorphs, Ed=edentates, Tu=aardvarks, Pa=pantodonts, No=some notoungulates.

proximal stapedial artery. Perissodactyls and hyraxes share an extrabullar path of the internal carotid artery, a feature found only in some Glires within the Eutheria. All living ungulates, except hyraxes and some perissodactyls, have an auditory bulla composed of the ectotympanic bone, although this feature occurs in some rodents and rabbits (Novacek 1986). Many fossil ungulates do not have an auditory bulla, although in some cases this may be due to lack of preservation. Cranial foramina, particularly in the orbital and basicranial regions, were reviewed by Novacek (1986), and many have relevant distributions. The most striking is the loss of the mastoid foramen on the occiput between the mastoid and the supraoccipital. The mastoid foramen secondarily reappears in some artiodactyls and a few palaeotheres. Outside the ungulates, this feature is found only in didelphids (Novacek 1986).

Several characters are also available in the postcranial skeletons of members of most ungulate groups. Surprisingly, very few of these have been used as indicators of phylogenetic relationship. Some have an erratic distribution that shows no phylogenetic pattern. For example, the astragalar foramen is lost in nearly every ungulate group, apparently independently. Many features, however, have been described, discussed, and then ignored. We suspect that this may be due to an inherent functional bias. When organisms have very similar functional anatomy, there is reason to suspect that similarities are due to convergence. Yet many important characters have been dismissed or ignored, even when they occur widely in groups with completely different ecologies. We feel that each character should be examined initially without *a priori* weighting or biases, and rejected when its distribution within or between groups clearly leads to unparsimonious hypotheses.

It is very striking that living ungulates are among the few mammals that lack a clavicle. Among fossil ungulates, only primitive artiodactyls, arctocyonids, hyopsodonts, and periptychids retain this feature; advanced artiodactyls lose it. Cifelli (1983*b*) doubted the value of this character, but we are impressed with its consistent distribution within the ungulates. Once the clavicle is lost, it never reappears in a group. Its loss seems to be independent of functional considerations as well. Even though cursorial ungulates lack a clavicle, so do many (but not all) graviportal, aquatic, and arboreal ungulates. According to the primate literature, a clavicle is essential to arboreality, yet the tree hyrax does not have one. Although organisms are influenced by functional considerations, their anatomy is not infinitely flexible. Once a feature is lost, it never reappears, despite functional demand for it.

The ungulate scapula is also distinctive. Mesonychids and all higher ungulates (except whales and meniscotheres) have greatly reduced the acromion and coracoid processes of the scapula in concert with the loss of the clavicle. Some artiodactyls have independently reduced these processes. In the humerus, most higher ungulates have reduced the deltopectoral crest. Primitive ungulates have a large knob-like medial epicondyle with a large entepicondylar foramen. This is highly reduced or lost in all higher ungulates. The ventral border of the ulna is curved and posteriorly convex in most mammals, including primitive ungulates. The olecranon flexes toward the limb axis, accentuating this curvature. In mesonychids and higher ungulates, the ulna has a posterior border that is straight or posteriorly concave, with the olecranon flexed away from the limb axis. Although this feature can be related to limb efficiency, it also occurs in animals that seem to have no functional similarity.

The carpal region has long featured in discussions of ungulate relationships. In most mammals, the carpal elements overlap one another in a 'brickwork' fashion, known as the 'alternating' or 'diplarthral' carpus. Certain ungulates have shifted the carpal elements so that they are in line or 'serial'. Cope (1897) considered the serial (or 'taxeopod') carpus to be primitive for ungulates, based on the complete skeleton of *Phenacodus primaevus*, which had the most complete postcranial material of any fossil 'condylarth' known at the time. As a result, Osborn (1898) placed the periptychids (with their alternating carpus) with the pantodonts in the waste-basket taxon 'Taligrada', as part of another waste-basket taxon 'Amblypoda' (which then included pantodonts and uintatheres). Many later authors kept this arrangement, largely on Osborn's authority.

Ironically, in the same year that Cope argued for a primitive serial carpus, Matthew (1897) showed that the alternating carpus is primitive for ungulates, based on the skeleton of *Tetraclaenodon (='Euprotogonia') puercensis*. Matthew's and Simpson's classic monographs of 1937 finally cleared up the confusion over alternating versus serial carpus. Although the serial carpus and the corresponding loss of the lunar–unciform contact is still used (Novacek and Wyss 1986, character 62) as a derived character supporting some phylogenetic arrangements, there are reasons to doubt its usefulness. As Gregory (1910, p. 452) showed, it occurs in some rodents and insectivores, *Hyaenodon*, and certain ungulates (some phenacodontids, meniscotheres, elephants, sirenians, hyraxes). It might be used to unite these ungulate taxa at node 38 in Fig. 8.1, except that it does not occur in primitive members of some of these groups (such as the phenacodont *Tetraclaenodon*), and even varies within genera. For example, *Phenacodus primaevus* is serial, but *P. copei* is alternating (Radinsky 1966). Similarly, a free centrale occurs in a number of primitive forms, but the fusion of this element with the scaphoid appears to have happened in several groups independently

(Gregory 1910). By the Oligocene, the carpals and tarsals of all main ungulate groups are easily distinguished.

As early as 1816, Blainville used the relative proportions of their podial elements to group ungulates. Artiodactyls are united by their 'paraxonic' foot, with the main axis between digits 3 and 4. These two digits have become symmetrical by the enlargement of the fourth metapodial and digit, and give artiodactyls their diagnostic 'even-toed' foot. Digits 2 and 5 are also equal in proportion to each other, and if they are reduced, they are reduced equally. The first digit is reduced in the earliest artiodactyls to a non-functional nubbin, since it has no symmetrical counterpart in the paraxonic foot. Agriochoeres, however, have apparently redeveloped the first metacarpal.

In most other ungulates, all five digits are present and approximately equal in length. Phenacodonts, hyraxes, and perissodactyls have strengthened the central digit (as well as metapodial 3) to various degrees. This produces a derived form of the mesaxonic foot. The reduction of the side digits in phenacodonts is not as extreme as it is in hyraxes and perissodactyls. In the last two groups, the first metapodials are extremely reduced (Radinsky 1963), and metatarsal 5 is lost. In more derived perissodactyls, metacarpal 5 is also lost and the third metapodial becomes enlarged at the expense of the lateral metapodials.

The features that give the ungulates their name are hooves, or unguals, the laterally broadened terminal phalanges that stabilize the feet on flat ground when the number of digits has been reduced. Unfortunately, associated phalanges are seldom preserved in most Early Tertiary fossil taxa. Several Early Tertiary ungulates had claws. Van Valen (1978*b*, p. 292) reported that *Protungulatum* had claws, although these have not yet been described. Claws occur in *Arctocyon* (Russell 1964), *Chriacus* (Rose 1987), *Hyopsodus* (Gazin 1968), and *Tetraclaenodon* (Radinsky 1966). More hoof-like phalanges occur in the most primitive members of nearly all the remaining groups, including the early artiodactyl *Diacodexis* (Rose 1982, 1985), the periptychids *Eoconodon* and *Periptychus* (Matthew 1937), *Phenacodus* (Radinsky 1966), *Mesonyx* (Scott 1888), *Meniscotherium* (Gazin 1965), and in primitive uintatheres, arsinoitheres, proboscideans, hyraxes, and perissodactyls. Thus, hooves developed several times independently within the ungulates, and a few groups (such as chalicotheres, agriochoeres, and homalodotheres) secondarily reverted to claws (Coombs 1983).

One feature of the axial skeleton seems phylogenetically significant. Primitively, most eutherian mammals have 12–15 thoracic vertebrae (Gregory 1910, p. 275). Hyopsodonts, uintatheres, elephants, sirenians (except manatees), hyraxes, and perissodactyls have all increased this number to 19–21 (Table 8.1).

Most advanced ungulate femora (except those of artiodactyls) are easily recognized by a well-developed, distally-shifted third trochanter, to which the superficial gluteus muscles attach. Among ungulates, only artiodactyls, hyraxes, and some gravigrade forms lack an enlarged third trochanter. The most primitive artiodactyl, *Diacodexis*, had a faint ridge-like third trochanter, but all higher artiodactyls have lost it completely (Rose 1982, 1985).

Like the carpus, the tarsus can be difficult to interpret. Some features, such as the loss of the astragalar foramen mentioned above, show no phylogenetic pattern. As Szalay (1977) has shown, however, the ungulate calcaneum and astragalus are readily distinguished from those of other groups. The astragalus of *Protungulatum* is derived in having a relatively short, robust neck. The reduced contact between the astragalus and cuboid, indicating the beginnings of a 'serial tarsus', occurs in the phenacodonts, meniscotheres, and higher non-gravigrade ungulates. Uintatheres, arsinoitheres, elephants, hyraxes, and perissodactyls are also united by having a flat, rather than convex or knob-like, navicular facet on the head of the astragalus. Perissodactyls have further modified this by developing a concave navicular facet, and corresponding convex astragalar facet on the navicular. Artiodactyls, of course, are unique in having the famous 'double pulley' astragalus, with well-developed trochlea at both ends (Schaeffer 1947). Ungulates never developed a fully plantigrade tarsus (such as those of humans or bears), nor has any ungulate become fully bipedal (although the modern gerenuk, *Litocranius*, comes close).

Results

Dental and non-dental anatomical evidence produces a remarkably consistent phylogenetic pattern (Fig. 8.1). Considering that dental, cranial, petrosal, and postcranial features are largely functionally independent, this consistency is even more striking. Some of the dental and/or postcranial anatomy might be dismissed as being due to functional or ecological convergence, but this does not apply to most of the petrosal and cranial features. Because of the congruence of a wide variety of anatomical evidence, we conclude that the hypothesis is robust.

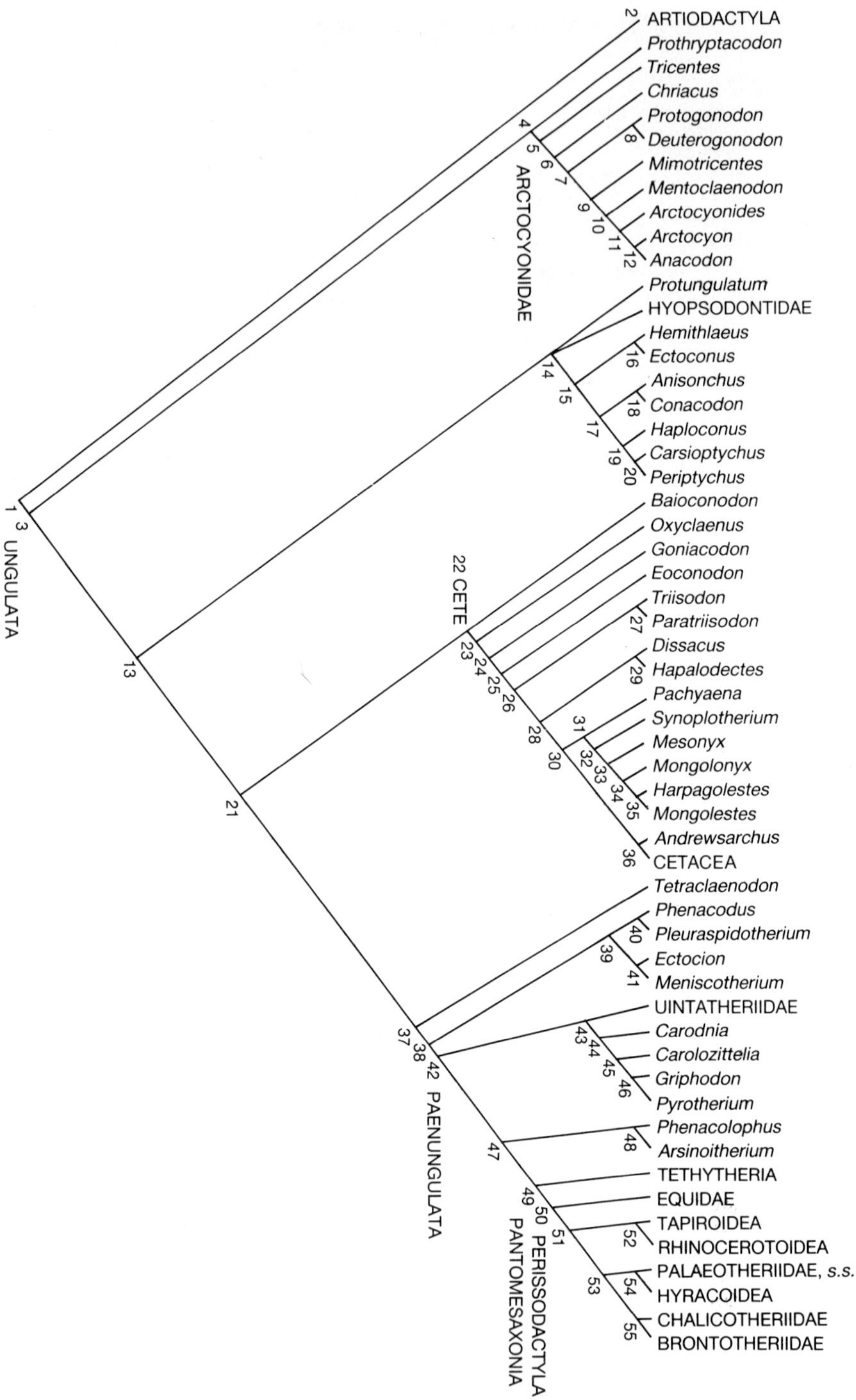

ARTIODACTYLA
Prothryptacodon
Tricentes
Chriacus
Protogonodon
Deuterogonodon
Mimotricentes
Mentoclaenodon
Arctocyonides
Arctocyon
Anacodon
ARCTOCYONIDAE
Protungulatum
HYOPSODONTIDAE
Hemithlaeus
Ectoconus
Anisonchus
Conacodon
Haploconus
Carsioptychus
Periptychus
UNGULATA
Baioconodon
Oxyclaenus
Goniacodon
Eoconodon
Triisodon
Paratriisodon
Dissacus
Hapalodectes
Pachyaena
Synoplotherium
Mesonyx
Mongolonyx
Harpagolestes
Mongolestes
Andrewsarchus
CETACEA
22 CETE
Tetraclaenodon
Phenacodus
Pleuraspidotherium
Ectocion
Meniscotherium
UINTATHERIIDAE
Carodnia
Carolozittelia
Griphodon
Pyrotherium
Phenacolophus
Arsinoitherium
TETHYTHERIA
EQUIDAE
TAPIROIDEA
RHINOCEROTOIDEA
PALAEOTHERIIDAE, s.s.
HYRACOIDEA
CHALICOTHERIIDAE
BRONTOTHERIIDAE
PAENUNGULATA
PERISSODACTYLA
PANTOMESAXONIA

Fig. 8.1. Cladogram of selected ungulate higher taxa. Characters at numbered nodes as follows: 1. UNGULATA: superior ramus of stapedial artery shifted to petrosal or lost; mastoid foramen lost; bulla (if present) composed of ectoctympanic; relatively bunodont teeth with low cusp relief; lower molar trigonids shortened anteroposteriorly; large, posteriorly projecting hypoconulid on M_3; astragalus with shorter, more robust head. 2. ARTIODACTYLA: lower molar trigonids very narrow, with paraconid and metaconid closely appressed; alisphenoid canal lost; pars facialis of lacrimal enlarged; enlarged orbitosphenoid, separating frontal from alisphenoid; paraxonic foot, with double-pulley astragalus, enlarged third and fourth metapodials, and symmetrically-reduced second and fifth metapodials. 3. Add hypocone to upper molars; third trochanter of femur enlarged and distally shifted; tympanic aperture of canalis facialis anterior to fenestra ovalis. 4. Complete lingual cingulum on upper molars. 5. Anterolingual cingulum on M^3 enlarged. 6. Size larger. 7. Parastyles and metastyles reduced. 8. Posterocingular cusp on M^3. 9. Upper molars square. 10. Enamel crenulate. 11. Labial cingulum rounded. 12. Canines enlarged and saber-like, serrated posteriorly; mandibular condyle low; $P_{3,4}$ serrated posteriorly. 13. Inflated tegmen tympani. 14. $P^{3,4}$ protocone and paracone bulbous. 15. PERIPTYCHIDAE, *s.l.*: shallow lingual slope on protocone. 16. Mesostyle on M^2. 17. Palate deep; hypocone shifted medial to protocone. 18. Hypocone region expanded. 19. P^4 becomes very large; protocone moved laterally between conules. 20. Enamel crenulate; hypocone moved laterally to level of protostyle. 21. Weak occipital exposure of mastoid; lose proximal stapedial artery; large posterior lacerate foramen coalesces with cochlear fissure; postglenoid foramen small or absent; reduced acromion and coracoid processes on scapula; lose clavicle; ventral border of ulna straight or concave; reduced deltoid crest of humerus. 22. CETE: blunt, conical protocones. 23. M^3 small with small hypocone; posterolingual accessory cusp added to P^4; M^3 parastyle small and paracone large; P_4 paraconid deflected posteriorly. 24. M^3 metacone small. 25. $M^{1,2}$ parastyle lost. 26. Paracone larger; all cusps tall and conical; overall size increase. 27. TRIISODONTINAE: accessory cusp anterior to entoconid on M_1. 28. ACREODI: trigonid high; entoconid very reduced. 29. HAPALODECTINI: lower cheek teeth narrow; $M^{1,2}$ parastyles long; $P_{2,3}$ diastema. 30. Mandibular condyle low, with pedicellate glenoid; mandibular angle small. 31. MESONYCHIDAE (amended): mandibular condyle below level of cheek teeth; P^3 protocone moved posteriorly. 32. M^3 very small. 33. M^3 lost. 34. P^2 short; labial cingula on upper molars absent. 35. Horizontal ramus arcuate. 36. Medial portion of lambdoid crest high; $I^{2,3}$ aligned with cheek teeth; elongated premaxilla. CETACEA: all incisors parallel with cheek teeth; medial lambdoid crest semicircular; nasals retracted; protocones small; accessory cusps large. 37. P^4 protocone area broadened. 38. P^4 protocone and metacone enlarged; ventral, slit-like aquaeductus cochleae; reduced medial epicondyle and entepicondylar foramen on humerus; reduced astragalar/cuboid contact. 39. PHENACODONTIDAE (expanded): posterior petrosal epitympanic sinus; mesostyle on upper molars. 40. M^3 posterior cingular cusp expanded. 41. Metaloph added between hypocone and metaconule. 42. PAENUNGULATA (amended): navicular facet of astragalus flattened; 19 or more thoracic vertebrae; M^3_3 larger than M^2_2. 43. DINOCERATA (amended): M_3 metalophid high; V-shaped lophs on M^3; lower incisors multicusped. 44. PYROTHERIA (amended): $M^{1,2}$ fully bilophodont. 45. Cristid between M_3 hypoconid and entoconid. 46. P_4 fully bilophodont with metalophid. 47. Hypolophid forms between entoconid and hypoconid of lower molars; metaconule moves posteriorly and loses connection with protocone. 48. EMBRITHOPODA: M_3 hypoconulid small and merged with posterior cingulum. 49. PANTOMESAXONIA (*sensu* Franz 1924): entoconid and hypoconid opposite one another, at right angles to axis of ramus; lacrimal process in anterior border of orbit; fossa glandis with projecting processus urethralis in penis; ramus anastomosticus; ramus infraorbitalis of stapedial artery in alisphenoid canal; tuber maxillaris. 50. PERISSODACTYLA (*sensu* Owen 1848): lose promontory sulcus; lose stapedial sulcus; extrabullar internal carotid artery; large, bridged tympanohyal; long tympanic process; eustachian sac; tuber maxillaris persists after tooth eruption; reduced first metapodial and fifth metatarsals; enlarged third metapodials. EQUIDAE: foramen ovale and median lacerate foramen confluent; optic foramen and anterior lacerate foramen confluent; lengthened limbs; reduced lower canines. 51. Large parastyle; metaconule merges with metaloph. 52. CERATOMORPHA: paraconule merges with protoloph. 53. Mesostyle and metastylid present; anterior crista obliqua near metaconid. 54. M^3 expanded posteriorly: metastyle elongated. 55. Lower molars W-shaped (crista obliqua connects hypoconid to metaconid); centrale fused to scaphoid.

Discussion

The relationships of ungulates

The Ungulata are a monophyletic group recognizable in several features of their anatomy. They all share:

(1) astragalus with a shorter, more robust head;

(2) foramen for superior ramus of stapedial artery shifted to petrosal or lost;

(3) auditory bulla (if present) composed of ectotympanic;

(4) mastoid foramen lost;

(5) more bunodont teeth with relatively low cusp relief (compared to *Purgatorius*, *Gypsonictops*, or other Late Cretaceous eutherians);

(6) lower molar trigonids shortened anteroposteriorly;

(7) a large, posteriorly-projecting hypoconulid on M_3.

The oldest known ungulate, *Protungulatum*, is as yet described only from teeth and tarsals, so some of these characters cannot be applied to it. More complete material (partially described by Van Valen 1978*b*) might be expected to show these features. Note that Novacek *et al.* (Chapter 3 of this volume) do not find strong evidence for a monophyletic Ungulata.

The artiodactyls appear to be the first of the ungulate groups to diverge, since they lack many of the specializations seen in all other ungulates. *Diacodexis* and *Dichobune*, for example, have very weak third trochanters, and lack the hypocone seen in all other ungulates. In higher artiodactyls, the metaconule takes the position of the hypocone. In some taxa, a lingual cingular cusp that may or may not be homologous with the hypocone also appears. Primitive artiodactyls can also be distinguished from all other ungulates by the following derived dental features: lower molar trigonids are very narrow, with the paraconid and metaconid closely appressed; M^3 metacone enlarged; P^4 with a small protocone crista; P^3 broad. Artiodactyls are also united by a wide variety of unique non-dental features, including:

(1) wide epitympanic recess;

(2) anterior lacerate foramen confluent with foramen rotundum;

(3) enlarged pars facialis of lacrimal, often developing a fossa or vacuity;

(4) enlarged orbitosphenoid, separating frontal from alisphenoid;

(5) paraxonic foot, with double-pulley astragalus, enlarged third and fourth metapodials, symmetrically reduced second and fifth metapodials, and vestigial first metapodials (except in the Tayassuidae) (see also Chapter 9).

At node 3 in Fig. 8.1 are several ungulate groups that developed a true hypocone. These include arctocyonids *sensu stricto*, hyopsodonts, periptychids, and mesonychids. Non-dental evidence suggests that the arctocyonids are the most primitive of these groups. The arctocyonids have long been a paraphyletic 'waste-basket' taxon for many types of primitive ungulates that are not something else. We have relegated the 'triisodontines' (Fig. 8.1, node 22) to the Cete. At the core of this unnatural assemblage, however, is a monophyletic group united by the presence of a complete lingual cingulum on the upper molars. The relationships of selected genera of Arctocyonidae *sensu stricto* are shown in Fig. 8.1. Not all of the known genera are included, since the group is currently undergoing extensive revision and new taxa are being added (Van Valen 1978*b*; Cifelli 1983*b*).

The Arctocyonidae had long been included with the creodonts and carnivores, since they have a somewhat bear-like dentition, and a skull like some primitive carnivores. The lack of a true shearing carnassial, and the presence of ungulate limbs and feet, eventually led to their recognition as ungulates (Ameghino 1901; Patterson and McGrew 1962; Sloan and Van Valen 1965; Van Valen 1966). In addition to a typically ungulate bunodont dentition, which lacks the type of shearing seen in carnivores, arctocyonids have a number of other advanced ungulate features. These include an enlarged third trochanter, a typically ungulate astragalus, and the tympanic aperture of the facial canal that opens anterior to the fenestra ovalis (Cifelli 1982).

Hyopsodonts and periptychids are two groups that were very abundant in the Early Tertiary. They are usually placed in the 'Condylarthra' on the basis of shared primitive similarities. In the past, they have been assigned to many different groups. Periptychids were long associated with pantodonts and excluded from the ungulates, although they were part of Cope's (1881*b*) original definition of the Condylarthra. *Hyopsodus* has been considered a primate, an insectivore, an artiodactyl, or the ancestor of proboscideans or perissodactyls (Gazin 1968). Hyopsodonts and periptychids share a number of derived dental similarities, especially the development of bulbous protocones and paracones on $P^{3,4}$. Some periptychids are even more distinctive in having shallow lingual slopes on the protocones and, in more derived taxa, highly crenulated enamel. The relationships of hyopsodonts have been discussed by Van Valen (1978*b*), Rigby (1980), and Cifelli (1983*b*, pp. 36–8), yet the cladogram of these taxa is still in flux. The periptychids are currently being revised by Archibald (1982; Archibald *et al.* 1983*a*, 1983*b*). In the possession of an inflated tegmen tympani (Cifelli 1982), hyopsodonts and periptychids are more derived than arctocyonids and are thus placed as sister-taxon to higher ungulates (Fig. 8.1, node 13).

Whales have traditionally been placed in their own Order Cetacea, with no apparent relationship to the rest of the Mammalia. In recent years, however, it has become clear that whales are related to a primitive omnivorous group, the Mesonychidae (Van Valen 1966, 1968; Szalay 1969*a*,*b*; Gingerich *et al.* 1983). Because of their bear-like skeletons and pointed teeth, mesonychids were long allied with creodonts and carnivores. Their fully ungulate skeleton (including hooves) and the lack of any carnassial modification strongly indicate that mesonychids are not related to creodonts or carnivores, but are in fact omnivorous ungulates. Several derived features, such as the loss of the proximal stapedial artery, loss of the clavicle, reduced coracoid and acromion processes on the scapula, reduced deltopectoral crest on the humerus, weak occipital exposure of the mastoid, small postglenoid foramen, and a large posterior lacerate foramen that coalesces with the cochlear fissure, suggest that mesonychids are actually quite derived ungulates.

McKenna (1975) resurrected Linnaeus' taxon, Cete, for whales and mesonychids. We would include not only these two groups, but also the triisodontines and some 'oxyclaenine arctocyonids' (*sensu lato*) that share the derived feature of a blunt, conical protocone, which unites the Cete. The more derived cetans (Fig. 8.1, node 28) include *Dissacus*, *Hapalodectes*, mesonychids, and whales, which all had high trigonids and small entoconids. McKenna (1975) suggested Matthew's (1909) taxon 'Acreodi' for the Mesonychidae as a taxon equal in rank to the Cetacea. If the relationships shown in Fig. 8.1 are correct, however, this usage of Acreodi would be the same as the Family Mesonychidae used here, making the term redundant. We prefer to use the taxon as a hierarchical level above the mesonychids and cetaceans, and also include *Dissacus* and *Hapalodectes*.

The ecological gap between whales and other ungulates often seems unbridgable to mammalogists. Still, as living land carnivores and seals have a functional (rather than phylogenetic) intermediate in the otters, land and sea ungulates also have a functional intermediate. The transition can be made by a coastal scavenger that eats fish. If it is accustomed to eating fish washed up on shore, it is a relatively small step to adapt to catching live (probably weak, large, or slow-moving) coastal fish. After an otter-like semiaquatic stage, a whale could evolve. A fair approximation of this functional intermediate for whales is the living brown hyaena (*Hyaena brunnea*). When living along the coast of

South Africa (where it is called the 'strand wolf'), the brown hyaena is a large coastal scavenger (Walker *et al.* 1975).

The phenacodonts were the original nucleus of Cope's (1881*b*) concept of 'Condylarthra'. Cope (1881*a*, 1882) originally placed *Meniscotherium* with the phenacodonts in the 'Condylarthra'. Other authors (reviewed by Gazin 1965) have allied meniscotheres with chalicotheres, proterotheres, hyraxes, and artiodactyls (because of their precociously selenodont teeth). More recent authors (e.g. Gazin 1965) have shown that meniscotheres are most closely related to phenacodonts. They share a number of derived characters, including a well-developed mesostyle and a unique posterior petrosal epitympanic sinus (Cifelli 1982).

Phenacodonts and meniscotheres are the most derived groups formerly referred to the 'Condylarthra'. This is shown by a number of features. In the teeth, they are beginning to show the trend toward squaring and molarization of the premolars that is seen in higher ungulates. *Tetraclaenodon* had broadened the lingual side of the P^4 protocone, and *Phenacodus* and higher taxa have also enlarged the P^4 metacone. A number of non-dental characters clearly indicate that phenacodonts and meniscotheres are actually quite derived ungulates. These features include a reduced astragalus–cuboid contact (beginning the trend toward a serial tarsus) and a reduced medial epicondyle and entepicondylar foramen in the humerus. Phenacodonts and meniscotheres share with higher ungulates a ventral, slit-like aquaeductus cochleae in the petrosal.

McKenna and Manning (1977) first suggested that the uintatheres (Dinocerata), arsinoitheres (Embrithopoda), elephants (Proboscidea), sea cows (Sirenia), desmostylians, hyraxes, and perissodactyls (Fig. 8.1, node 42) formed a natural group of advanced ungulates. They expanded the 'Paenungulata' to include these taxa by adding the Perissodactyla and removing the Pantodonta. These groups all share the enlargement of M^3_3, which is the beginning of the trend toward molars that are approximately equal in size—an advance which made lophodonty possible. In the primitive condition, the molar row tapers posteriorly. They also share a number of derived non-dental features, including a flattened navicular facet on the astragalus, and an increase to 19 or more thoracic vertebrae.

Of the first three groups listed, the uintatheres are clearly the most primitive. A number of features support the monophyly of the Dinocerata, including: a fully developed V-shaped loph on M^3, a high M_3 metalophid, P_3–M_3 metalophids obliquely oriented and posteriorly inclined, metastylids accentuated, paralophids lost, multicusped incisors, and an inframandibular flange (McKenna and Manning 1977; Tong and Lucas 1982). Tong and Lucas (1982) pointed out the derived dental similarities between uintatheres and *Pseudictops*, and suggested that uintatheres might not be ungulates, but it is not only the astragalus that makes uintatheres ungulates. Nearly all the non-dental features found in more primitive ungulates (Fig. 8.1) are also found in uintatheres, particularly in the more primitive forms, such as *Mongolotherium* (Flerov 1957). Although the teeth of *Pseudictops* and uintatheres are somewhat similar, the strong evidence from the skull and postcranial anatomy clearly places the uintatheres within the ungulates.

Arsinoitheres are united with higher ungulates by the common possession of a hypolophid between the entoconid and hypoconid of the lower molars, and a metaconule that shifts posteriorly and loses its connection with the protocone (McKenna and Manning 1977). The Embrithopoda was originally thought to include only the enigmatic Egyptian Oligocene *Arsinoitherium*, a rhino-like beast with paired pointed horns on its nasals and frontals. The group has been considerably expanded with the addition of *Crivadiatherium* from the Late Eocene of Rumania (Radulesco *et al.* 1976), *Phenacolophus* from the Late Palaeocene of Mongolia (McKenna and Manning 1977), other forms from the Late Palaeocene of China (Tong *et al.* 1976), and *Palaeoamasia* from the Late Eocene of Turkey (Sen and Heintz 1979). Apparently, arsinoitheres were an important element of the Early Tertiary large mammal fauna of the Old World, along with pantodonts and uintatheres.

McKenna (1975) united the proboscideans, sirenians, and desmostylians in his new taxon Tethytheria. Although he provided no evidence for this relationship, they are clearly a good monophyletic group, with a number of unique and striking synapomorphies. These were discussed by Domning *et al.* (1986) and Tassy (1981, 1982; Chapter 11 of this volume). The tethytheres are united with hyraxes and perissodactyls by the common possession of lower molars with entoconids and hypoconids opposite one another at right angles to the ramus, forming a more efficient bilophodont molar (McKenna and Manning 1977). This is further supported by the shared presence of a lacrimal process and the unique morphology of the tip of the penis, as discussed by Fischer (1986).

Several authors (Shoshani 1986; de Jong and Goodman 1982; Novacek 1982, 1986; Novacek and Wyss 1986; Chapter 3 of this volume) have suggested that hyraxes are more closely related to tethytheres than they are to perissodactyls. We feel that the anatomical evidence strongly supports the hypothesis that hyraxes and perissodactyls are most closely related. The dental similarities between primitive Fayum hyraxes and certain European 'palaeotheres' are very striking, and suggest that some perissodactyls share many derived dental features with hyraxes. Indeed, the oldest known equoid was named *Hyracotherium* by Owen in 1841 because of this similarity. The list of non-dental features allying hyraxes and perissodactyls is even stronger. In the petrosal region (Cifelli 1982) they share a ventral hiatus fallopii, a large bridged tympanohyal, a long tympanic process, and a tympanic aperture of the canalis facialis at the fenestra ovalis. They both lose the stapedial sulcus, and instead share an extrabullar path of the internal carotid artery (Wible 1986; Fischer 1986). By far the most striking feature of hyraxes and perissodactyls is the inflated eustachian sac (Fischer 1986, 1988). Fischer (1986, 1988) lists additional characters of the course of the sternoscapular muscle, the details of the ungual phalanges, and the development of the tuber maxillare, which strongly suggest that hyraxes are perissodactyls. The feet of the hyrax are nearly identical with those of *Hyracotherium* and other primitive perissodactyls. Both groups have greatly reduced or have lost the first metacarpals, and first and fifth metatarsals. They emphasize the third metapodial, producing a strongly mesaxonic foot. Since Owen (1848) originally included *Hyrax* in his definition of the Perissodactyla, we suggest that the 'Order Hyracoidea' be demoted to a subgroup of the Perissodactyla.

The Perissodactyla has long been recognized as a monophyletic group. Several features support this relationship. In the skull, only perissodactyls (except certain hyraxes, advanced tapirs, and rhinos) have lacrimals that broadly contact the nasals. In the $P^{3,4}$ of perissodactyls, the metacone is equal in size to the paracone. There are a number of derived postcranial similarities. The most striking is the concave navicular facet on the astragalus (and corresponding convex astragalar facet on the navicular). Perissodactyl and hyrax fibulae have lost their articulation with the calcaneum, and the third metapodials are even more enlarged than in hyraxes.

Other possible ungulates

Space limitations prevent us from more fully discussing several problematic groups. Aardvarks have long been allied with ungulates (Patterson 1975), but the evidence is ambiguous. They share a confusing mix of characters seen in ungulates, but also lack many key features. In their skull and ear features, they are clearly more primitive than any ungulate (Thewissen 1985; Novacek 1986; Novacek and Wyss 1986; Wible 1987). Molecular evidence (discussed below) tends to place aardvarks with other 'paenungulates' (and with tupaiids), where there are data available. The posteriorly inclined coronoid process and shallow mandibular symphysis of the peculiar pantodont *Pantolambdodon* may suggest a relationship of aardvarks to one group of pantodonts. Further resolution of this problem requires a broader study than our space permits.

Because they are large and heavy-bodied, the pantodonts have frequently been placed with the ungulates. At one time, for example, they were grouped with the periptychids and the uintatheres in the Amblypoda. Simpson (1937) and Patterson (1939) effectively demolished the Amblypoda, but proposed no alternative for pantodont relationships. Pantodonts do share a few features that are found in the ungulates, such as an enlarged femoral third trochanter and hooves in some of the more derived genera. Szalay (1977, pp. 354–5) argued that the pes of *Pantolambda* is like that of ungulates, but gives no specific features to support this contention. Our examination of primitive pantodont feet produced no derived ungulate features, only shared primitive similarity. More importantly, the teeth of pantodonts are specialized in a completely different manner from primitive ungulates. Instead of the transverse shear and low cusp relief that is diagnostic of primitive ungulates, pantodonts have specialized tribosphenic teeth with a high, W-shaped ectoloph formed on a broad stylar shelf for vertical shearing. The hypocone, when present, is never large or high enough to produce the typically low, rectangular ungulate upper molar. For the present, we feel that *Deltatherium* is probably the closest sister-taxon of pantodonts, and suspect that pantodonts are closer to Ferae than Ungulata. The few derived features that pantodonts share with ungulates are probably convergent.

Schoch (1986) has provided convincing evidence that another Palaeogene large mammal group, the taeniodonts, is only distantly related to ungulates. Schoch indicates a relationship of taeniodonts to the leptictimorph group, including palaeoryctines, leptictids, and pantolestids (Schoch 1986, pp. 171–3).

South American ungulates

The enormous radiation of endemic ungulates in South America has long fascinated palaeontologists. Most authors thought that South American ungulates originated from a single primitive stock, and that they formed a monophyletic group. McKenna (1975) formalized this concept in proposing the 'Meridiungulata', although he provided no anatomical evidence for this taxon. A complete review of all the evidence is inappropriate here, but such work is currently in progress (Cifelli 1983*a*,*b*; *pers. comm.*). From Cifelli's work, it appears that didolodonts, litopterns, and notoungulates share derived features of the dentition and feet with hyopsodonts. Alternatively, *Didolodus* may be related to periptychids, as suggested by the resemblance of its teeth to those of *Ectoconus*. Likewise, primitive notoungulates show similarities to the enigmatic Chinese ungulate *Lantianius*. Astrapotheres and trigonostylopoids share more features with uintatheres than with any other group. The bizarre Palaeocene mammal *Carodnia* has frequently been united with pyrotheres, and both have derived dental similarities with uintatheres. For this reason, *Carodnia* and the pyrotheres have been united with the uintatheres here in an expanded Dinocerata (Fig. 8.1, node 43). *Carodnia*, however, has a peculiar type of serial tarsus, unlike that of uintatheres and pyrotheres. However, the serial tarsus has often been independently derived, as discussed above. At present, it appears that South American ungulates may be related to at least two different ungulate groups, the hyopsodonts (didolodonts, litopterns, notoungulates) and the uintatheres (astrapotheres, trigonostylopoids, pyrotheres, and maybe *Carodnia*). This suggests that the South American ungulates are not a monophyletic group. The 'Meridiungulata' appears to be merely a geographic construct (like the 'subungulates'), with no phylogenetic basis.

Comparison with molecular phylogenies

In the last decade, a new body of phylogenetic information has emerged from molecular biology. From crude immunological distances, to more sophisticated amino acid sequencing, to currently developing DNA analysis, molecular studies have provided a new source of data to test phylogenetic hypotheses based on traditional morphological features. As clearly shown by Goodman *et al.* (1979, 1982), however, the molecular evidence is no magical 'black box' that delivers unambiguous truth (see also Chapter 3 of this volume). Molecular sequences are also subject to parallelism and convergence, as well as insidious gene duplications (GD), and events that prevent gene expression (GE). For example, Joysey (*pers. comm.* 1985) has shown that the blood proteins of South American didelphids and primates show amazing convergence, which he attributes to some ecological crisis affecting all the mammals of this island continent in a similar fashion. This may explain some of the molecular similarities between endemic African forms, such as hyraxes and tethytheres.

In many cases, there are so few molecular differences that several possible arrangements are equally parsimonious. Phylogenetically-meaningful molecular differences can be overprinted by parallelism, back mutation, and other phylogenetic 'noise' (Goodman *et al.* 1982, p. 119). As a result, most molecular phylogenies are not simply the most-parsimonious arrangement of nucleotide replacements (NR) for a given array of taxa. Once a most-parsimonious tree has been produced and tested, a host of assumptions is added to the concoction. These *a priori* assumptions (Goodman *et al.* 1979, p. 139) modify the clustering so that certain assumed monophyletic groups automatically become closely related. For example, in nearly all trees, the major orders of mammals are assumed to be monophyletic, and the Perissodactyla and Artiodactyla are automatically clustered. Within the artiodactyls, ruminants and tylopods are grouped together, excluding the suines. These *a priori* assumptions are often hard to decipher from the final, most-parsimonious tree given in the last figure of a paper, and sometimes their influence on the data can be misleading.

Consider, for example, the development of the myoglobin sequence. First presented by Romero-Herrera *et al.* (1973), the data set emphasized primates and included the horse, ox, sheep, and three cetaceans; all of these ungulate groups clustered together. A larger sample of these three orders (Goodman *et al.* 1979, figs. 16 and 17) provided the same arrangement, although three 'gene expression' events were required to hold ruminants together, and pigs clustered with insectivores. Goodman *et al.* (1982, figs. 1 and 2) added elephants to the pool; they clustered with the rest of the ungulates, as did the primate *Lepilemur*. Dene *et al.* (1983) added the aardvark to the arrangement, and the most-parsimonious NR tree with no *a priori* adjustments (Dene *et al.* 1983, fig. 3) clusters elephants with perissodactyls, followed by artiodactyls and then by whales. Except for *Lepilemur*, which once again clusters with elephants, this arrangement of monophyletic ungulates resembles ours. When *a priori* expectations from other protein sequences (primarily lens α-crystallin, as discussed below) were added to the pool, however, Dene *et al.* (1983, fig. 4) produced a completely different arangement, with elephants widely separated from ungulates and placed near the base of the Eutheria.

Other protein sequences produced comparable results. α-haemoglobin (Goodman *et al.* 1979, figs. 6–9) clustered artiodactyls and perissodactyls with few GD or GE events required, but whales, elephants, and hyraxes were not sampled. β-haemoglobin (Goodman *et al.* 1979, figs. 10–13) cleanly clustered most artiodactyls and the horse, but the llama and elephant fell outside the ungulates. Judicious branch-swapping using GE and GD events where necessary finally placed the pig and llama within the artiodactyls, with elephants just outside the ungulates. We wonder if the further assumption that the elephant is also an ungulate would have produced an almost equally parsimonious arrangement. The β-haemoglobin tree of Goodman *et al.* (1982, fig. 6) showed similar results, although hedgehog and armadillo also cluster near the elephant at a node just more primitive than other ungulates. Fibrinopeptides A and B (Goodman *et al.* 1982) place the elephants nearest the artiodactyls.

The most-parsimonious arrangement of cytochrome *c* (Baba *et al.* 1981; Goodman *et al.* 1982, fig. 3) clustered whales and artiodactyls, and placed perissodactyls on the next most remote node. This arrangement required no GE or GD events. No data were available for hyraxes or elephants. Pancreatic ribonucleases (Beintema and Lenstra 1982) clustered artiodactyls and whales, with perissodactyls at the next most remote node; again, no data were available for elephants or hyraxes. From all the systems so far discussed, it is clear that artiodactyls, perissodactyls, and whales form a monophyletic group, and in most cases elephants cluster within or just outside this group. Goodman *et al.* (1982, p. 148) readily concede the ambiguous position of the Proboscidea.

Yet the most recent papers of the molecular phylogeneticists (Goodman *et al.* 1982, 1985; Miyamoto and Goodman 1986; Shoshani 1986) have strongly argued that elephants and hyraxes form a 'paenungulate' group not closely related to other ungulates, and have gone so far as to reclassify the mammals this way (Miyamoto and Goodman 1986). The basis for this abrupt switch appears to be the interpretation of yet another protein, eye lens protein α-crystallin A (de Jong *et al.* 1977; de Jong and Goodman 1982; de Jong 1982). This protein phylogeny (de Jong 1982, fig. 11) clusters aardvark, hyrax, and manatee with elephant as another outgroup near the base of the eutherian mammals, and places whales plus perissodactyls as closest sister-taxa, joined by the artiodactyls to form an ungulate group at the derived end of the phylogram.

On closer inspection, however, there appear to be some problems with the analysis. For example, de Jong (1982, table 2) shows that cetaceans, perissodactyls (except tapirs), proboscideans, hyraxes, manatees, aardvarks, and opossums all share methionine at locus 150. Since its distribution indicates that it is primitive for mammals, it would seem that clustering these groups would place them together at the base of the Eutheria. Instead, the algorithm uses methionine to valine substitution at locus 150 to set the 'paenungulates' in a node more primitive than all other eutherians (including edentates) and then requires a back substitution to account for the presence of methionine in whales and perissodactyls further up the tree. The separation of 'paenungulates' from other ungulates is supported by only two unique substitutions: threonine to alanine at locus 4, and aspartic acid to glutamic acid at locus 91. The argument of de Jong and Goodman (1982, p. 268) against a 'paenungulate'–ungulate grouping on the grounds that it requires 'five additional NRs' seems less impressive when the NRs that separate these groups in their arrangement include two changes in locus 150 mentioned above (which unite, rather than separate the groups), one back mutation, and two that occur in parallel elsewhere. Only two unique substitutions support this separation.

So far, we have four proteins that support some sort of proboscidean–ungulate affinity (β-haemoglobin, fibrinopeptides A and B, myoglobin), one that separates elephants and hyraxes from other ungulates (lens α-crystallin), and the rest have not been sampled for all the relevant groups. Yet when all of these sequences are combined, the resulting phylograms (Goodman *et al.* 1985; Miyamoto and Goodman 1986) widely separate elephants from other ungulates. The lens α-crystallin has a disproportionate effect on the analysis, even though it is only one of seven proteins sampled. Apparently this is because lens α-crystallin has been sampled for a wider array of taxa than any other protein. When other proteins have not been sampled for the relevant taxa, they cannot outweigh the disproportionate effect of this one protein.

Finally, Weitz (1953) and Shoshani (1986) used the cruder immunological distance method to determine relative phylogenetic distances between taxa. Their results suggest that hyraxes and elephants are immunologically close, but horses are much less close. Since immunological distance cannot be broken down into discrete steps and analysed as above, it is difficult to assess the meaning of these data. It is not clear that immunological distance is always related to time of phylogenetic separation. Much of the similarity could be symplesiomorphic, and the horse could be different because it has diverged from a common ancestor with elephants much more rapidly than hyraxes have. This is analogous to uniting crocodilians with turtles because they share an overwhelming number of primitive similarities and only a few derived characters with birds. As has been shown many times, however, the fact that birds have diverged much more from their most recent common ancestor than crocodilians, does not justify placing crocodilians with turtles on shared primitive similarity. In this context, it is interesting to note that Buettner-Janusch *et al.* (1964) observed little similarity between hyraxes and elephants in an electrophoretic analysis of plasma proteins and haemoglobin.

In conclusion, we find no convincing evidence from the molecular data against our hypothesis that hyraxes and perissodactyls are sister-taxa, and that elephants belong with other ungulates in a monophyletic group. Indeed, it appears that four proteins (myoglobin, β-haemoglobin, and fibrinopeptides A and B) support this relationship, and only one (lens α-crystallin) appears to contradict it. The phylogenetic diagrams of Goodman *et al.* (1985) and Miyamoto and Goodman (1986) obscure this by throwing all proteins together, when many of the proteins have not been sampled for the relevant taxa (except lens α-crystallin). In their discussions, Goodman *et al.* (1982, 1985) and de Jong and Goodman (1982) have apparently been strongly influenced by the long tradition of placing hyraxes and elephants together, separate from ungulates (Simpson 1945; Thenius 1969) and have only recently become aware of the growing body of evidence supporting hyrax–perissodactyl affinities (McKenna 1975; Fischer 1986, 1988) and ungulate monophyly (Novacek 1986; Novacek and Wyss 1986). Since their methods are explicitly dependent on *a priori* assumptions, we would be interested to see what new, perhaps even more parsimonious, solutions could be produced with different assumptions.

Other recent phylogenetic hypotheses

Novacek (1982, 1986) and Novacek and Wyss (1986) have provided the first testable phylogenetic analysis of the living ungulate groups, which is explicitly based on shared derived characters (see also Chapter 3 of this volume). Their arrangement independently corroborated our hypothesis, which was developed in 1985 and earlier, without knowledge of their work. Subsequently, we have incorporated many of their characters in our hypothesis. We differ from them only in that we place hyraxes closer to perissodactyls than to tethytheres. Our disagreements with their interpretation of the characters that support the 'paenungulates' are discussed above and by Fischer (1986, 1988). Although a few of their characters appear strong, we are more impressed with the uniqueness of the characters shared by hyraxes and perissodactyls (especially the eustachian sac), and by the overwhelming preponderance of characters that support this relationship. On the grounds of both uniqueness of characters and of parsimony, then, hyraxes are clearly closer to perissodactyls than to elephants.

Shoshani (1986) proposed a phylogeny based on morphological characters that was a condensed version of his doctoral dissertation. Unfortunately, in condensing the work for the PAUP phylogenetic program and for publication, there are many errors that invalidate the hypothesis. For example, the clavicle (node F2) is *not* lost in primitive artiodactyls, aardvarks, or most 'condylarths'. Node Q5 implies that artiodactyls have an odd number of digits in their feet. Some of the errors occur because only living members of the group were examined. For example, the astragalar foramen (node N5) is present in some primitive members of clade C, but is lost in the derived members. Besides, this feature is primitive for mammals. The tibia and fibula (node P2) and radius and ulna (node R8) are fused only in the more derived members of the perissodactyls and artiodactyls. Fewer than 14 thoracic vertebrae (node M3) is primitive for mammals (Gregory 1910). There are many more examples of the above, but space does not permit us to list them all here. Shoshani's clades B and C (his fig. 3) include most of the ungulate groups, but nearly all the characters at nodes H and I ('Paenungulata' and Tethytheria) also occur in nodes Q and R (notoungulates, perissodactyls, and artiodactyls). Elsewhere, Shoshani indicates that it is only slightly less parsimonious to join clades B and C, eliminating these unnecessary parallelisms by postulating a monophyletic Ungulata. That hypothesis clearly approaches our own.

'Condylarthra' and other paraphyletic 'waste-baskets'

Unlike polyphyly (the grouping of taxa on the basis of shared convergent characters), paraphyly (the grouping of taxa based on shared primitive characters) has been a persistent problem in ungulate systematics. A few polyphyletic taxa were proposed in early classifications. Rhinos were united with rodents in an early version of the Glires on the basis of enlarged incisors. Elephants were grouped with hippos in an early Pachydermata on the basis of thick skin. Mercifully, these taxa were quickly dropped. Paraphyletic taxa, on the other hand, have increased within the Ungulata at an alarming rate.

In recent years, systematists have increasingly rejected the idea that paraphyletic taxa have any real meaning in nature, or are useful in phylogenetic analysis (e.g. Patterson 1977, 1981, 1982). The supposed 'biological reality' of 'adaptively unified groups' (Mayr 1974, 1981; Van Valen 1971, 1978*a*) does not stand up to the close scrutiny of either genetic or anatomical analysis (Patterson 1982). Paraphyletic groups are 'non-groups', defined by the common absence of derived characters, and few useful statements can be made about them. Even as non-phylogenetic, ecological groupings, paraphyletic taxa rarely provide useful information because of their emphasis on the *absence* of important (often ecologically significant) features.

The Mammalia, like most major taxa, has been subdivided into a mixture of monophyletic and paraphyletic groups. Most of the orders of mammals are clearly monophyletic, and were recognized as such from the time of Linnaeus. Bats, whales, carnivores, elephants, rabbits, and rodents, for example, have so many unique derived characters that even schoolchildren recognize them as natural groups. Unfortunately, many other groups have played the role of paraphyletic waste-baskets. Although there is a monophyletic core to the Insectivora, the Lipotyphla (Gregory 1910; McDowell 1958; McKenna 1975), many insectivorous mammals with no special relationships to the Lipotyphla have been placed in an expanded 'Insectivora' because of shared primitive eutherian features (Van Valen 1978*a*).

The Ungulata has been burdened with a large number of paraphyletic groups. Apparently, the psychological need for 'groundmass' taxa (to use Mayr's term) is so great that for every natural monophyletic group, there is an equal and opposite paraphyletic non-group. The most serious problem for ungulate systematics is the paraphyletic grouping of primitive ungulates—the 'Condylarthra'—which has resisted several attempts to break it (McKenna 1975; Cifelli 1983*a*,*b*). Cope (1881*b*) originally proposed the Condylarthra as a subunit of the Perissodactyla to include phenacodonts (*Phenacodus* and *Tetraclaenodon*), periptychids (*Anisonchus* and *Periptychus*), and a hyopsodont (*Mioclaenus*). He considered condylarths to be 'the ancestral type of the known Perissodactyla' (Cope 1881*b*, p. 1018). When it became clear that *Phenacodus* had feet quite unlike those of perissodactyls, Cope (1882) removed the 'condylarths' from the Perissodactyla. In 1884, he expanded the 'Condylarthra' to include additional periptychids, an arctocyonid (*Anacodon*), a meniscothere (*Meniscotherium*), a taeniodont (*Conoryctes*), and an artiodactyl (*Diacodexis*). With the revised contents, Cope gave the group a new meaning, as the 'most primitive type of the Ungulata' (Cope 1884, p. 793). Over the years, the concept of Condylarthra has expanded to include a great diversity of phenacodonts, meniscotheres, didolodonts, periptychids, hyopsodonts, mesonychids, and arctocyonids, with the last treated as the central 'stem group' (Simpson 1945; Van Valen 1978*b*; Cifelli 1983*b*). To retain Cope's name for a monophyletic group, the Condylarthra would have to be a synonym of nearly the entire Ungulata to fit his original concept, or a synonym of nearly the entire Eutheria to fit his 1884 concept. 'Condylarthra' has taken on such significance as a grouping of primitive ungulates that it has hampered the investigation of the interrelationships of the major ungulate groups for over a century.

It is much better to abandon the term 'Condylarthra' altogether, and with it all the erroneous connotations it has acquired. There are many good reasons for this. Although many 'condylarths' share primitive mammalian, eutherian, and ungulate characters, the far more important derived ungulate characters have been obscured by the use of a waste-basket term. Even some of these primitive similarities hide significant distinctions. General statements about 'condylarths' emphasize primitive ungulate morphology and hide important differences between early ungulate lineages. When we refer to 'condylarths', are we talking about arctocyonids, phenacodonts, meniscotheres, periptychids, or hyopsodonts? Each of these groups is different in detail from the others, likely to have major ecological differences as well. Despite claims to the contrary (Van Valen 1978*a*), 'condylarths' are not a clear example of an 'adaptively unified group'. For example, the teeth of *Arctocyon* suggest an omnivore, while *Meniscotherium* was apparently a specialized herbivore that lived in herds.

A more serious problem with the persistence of a waste-basket taxon like 'Condylarthra' is its effect on the people who do not know the fossils first-hand. The waste-basket term not only obscures derived features, but it also hinders understanding by suggesting false relationships, and prevents serious discussion of relationships between taxa. Like the emperor's new clothes, horizontal groups obtain a reality in people's eyes they never really had. Shoshani's (1986) phylogenetic analysis vividly demonstrates the pitfalls of the 'Condylarthra'. By treating the mesonychids as 'condylarths' (along with *Arctocyon* and *Phenacodus*, to which they are only distantly related) rather than as a separate group, his analysis placed the mesonychids and the whales in completely different clades. Most of the characters Shoshani used for 'condylarths' are not true of the majority of the taxa concerned, since there are no derived characters that unite the group.

As this study has shown, there was an important substructure of nested, derived similarity hiding beneath the obscuring blanket of the 'Condylarthra' all along. It is hoped that the framework presented will begin to clear away the vagueness and confusion, and provide meaningful biological statements for further testing of the natural groups of ungulates. Sometimes it is useful to make statements about generalized or primitive ungulates, but no formal taxonomic term is required for such a concept. 'Primitive' or 'archaic ungulate' or some similar term is much more evocative than 'condylarth', and carries no misleading phylogenetic connotations. The time has come to relegate 'Condylarthra' to the waste-basket of forgotten and obsolete taxa, and move on to problems regarding *real* monophyletic taxa within the ungulates.

Acknowledgements

We thank Drs J. D. Archibald, R. L. Cifelli, J. Czelusniak, W. W. de Jong, D. P. Domning, A. W. Gentry, M. Goodman, J. J. Hooker, C. Janis, M. C. McKenna, M. J. Novacek, J. Shoshani, P. Tassy, L. Van Valen, J. R. Wible, and A. Wyss for many helpful comments and criticisms. Prothero acknowledges the support of a grant from the donors of the Petroleum Research Fund of the American Chemical Society, and from NSF grant EAR87-08221 during the preparation of this manuscript. Special thanks are due to Dr E. S. Gaffney who, as a

Columbia University professor in 1973, first introduced Manning to cladistic methodology.

References

Ameghino. F. (1901). Notices preliminaires sur des mammifères nouveaux des terrains crétacés de Patagonie. *Bol. Acad. Nac. Cienc. Cordoba, Argent.* **16**, 350–426.

Andrews, C. W. (1906). *A Descriptive Catalogue of the Tertiary Vertebrata of the Fayum, Egypt*, 324 pp, 26 pl. British Museum, London.

Archibald, J. D. (1982). A study of the Mammalia and geology across the Cretaceous–Tertiary boundary in Garfield County, Montana. *Univ. Calif. Publ. Geol. Sci.* **122**, 1–286.

Archibald, J. D., Rigby, J. K., Jr., and Robison, S. F. (1983*a*). Systematic revision of *Oxyacodon* (Condylarthra, Periptychidae) and a description of *O. ferronensis* n. sp. *J. Paleontol.* **57**, 53–72.

Archibald, J. D., Schoch, R. M., and Rigby, J. K., Jr. (1983*b*). A new subfamily, Conacodontinae, and a new species, *Conacodon kohlbergeri*, of the Periptychidae (Condylarthra, Mammalia). *Postilla* **191**, 1–24.

Baba, M. L., Darga, L. L., Goodman, M., and Czelusniak, J. (1981). Evolution of cytochrome *c* investigated by the maximum parsimony method. *J. Mol. Evol.* **17**, 197–213.

Beintema, J. J. and Lenstra, J. A. (1982). Evolution of mammalian pancreatic ribonucleases. In *Macromolecular Sequences in Systematic and Evolutionary Biology* (ed. M. Goodman), pp. 43–74. Plenum Press, New York.

Blainville, H. M. de (1816). Prodrome d'une nouvelle distribution systématique du regne animal. *Bull. Sci. Soc. Philom. Paris, Ser. 3* **3**, 105–24.

Buettner-Janusch, J., Buettner-Janusch, V., and Sale, J. B. (1964). Plasma proteins and haemoglobins of the African elephant and the hyrax. *Nature (London)* **201**, 510–11.

Cifelli, R. L. (1982). The petrosal structure of *Hyopsodus* with respect to that of some other ungulates, and its phylogenetic implications. *J. Paleontol.* **56**, 795–805.

Cifelli, R. L. (1983*a*). Eutherian tarsals from the Late Paleocene of Brazil. *Am. Mus. Novit.* **2761**, 1–31.

Cifelli, R. L. (1983*b*). The origin and affinities of the South American Condylarthra and early Tertiary Litopterna (Mammalia). *Am. Mus. Novit.* **2772**, 1–49.

Coombs, M. C. (1983). Large mammalian clawed herbivores: a comparative study. *Trans. Am. Philos. Soc.* **73**(7), 1–96.

Cope, E. D. (1881*a*). A new type of Perissodactyla. *Am. Nat.* **15**, 1017–18.

Cope, E. D. (1881*b*). On some Mammalia of the lowest Eocene beds of New Mexico. *Proc. Am. Philos. Soc.* **19**, 484–95.

Cope, E. D. (1882). New characters of the Perissodactyla Condylarthra. *Am. Nat.* **16**, 334.

Cope, E. D. (1884). The Condylarthra. *Am. Nat.* **18**, 790–805.

Cope, E. D. (1897). The position of the Periptychidae. *Am. Nat.* **31**, 335–6.

de Jong, W. W. (1982). Eye lens protein and vertebrate phylogeny. In *Macromolecular Sequences in Systematic and Evolutionary Biology* (ed. M. Goodman), pp. 75–114. Plenum Press, New York.

de Jong, W. W., Gleaves, J. T., and Boulter, D. (1977). Evolutionary changes of α-crystallin and the phylogeny of mammalian orders. *J. Mol. Evol.* **10**, 123–35.

de Jong, W. W. and Goodman, M. (1982). Mammalian phylogeny studied by sequences analysis of the eye lens protein α-crystallin. *Z. Säugertierkd.* **47**, 257–76.

Dene, H., Goodman, M., Walz, D., and Romero-Herrera, A. E. (1983). The phylogenetic position of aardvark (*Orycteropus afer*) as suggested by its myoglobin. *Hoppe Seyler's Z. Physiol. Chem.* **364**, 1585–95.

Domning, D. P., Ray, C. E., and McKenna, M. C. (1986). Two new Oligocene desmostylians and a discussion of tethytherian systematics. *Smithson. Contrib. Paleobiol.* **59**, 1–56.

Fischer, M. S. (1986). Die Stellung der Schliefer (Hyracoidea) im phylogenetischen System der Eutheria. *Courier Forsch.-Inst. Senckenberg* **84**, 1–132.

Fischer, M. S. (1988). Hyracoids as sister-group of perissodactyls. In *The Evolution of the Perissodactyls* (eds. D. R. Prothero, R. M. Schoch, and J. L. Franzen). *Courier Forsch.-Inst. Senckenberg* (in press).

Flerov, K. K. (1957). Dinocerata of Mongolia. *Tr. paleontol. Inst. Akad. Nauk SSSR* **67**, 1–82.

Franz, V. (1924). *Die Geschichte der Organismen*, 948 pp. G. Fischer, Jena.

Gazin, C. L. (1965). A study of the early Tertiary condylarthran mammal *Meniscotherium. Smithson. Misc. Collect.* **149**(2), 1–98.

Gazin, C. L. (1968). A study of the Eocene condylarthran mammal *Hyopsodus. Smithson. Misc. Collect.* **153**(4), 1–90.

Gingerich, P. D., Wells, N. A., Russell, D. A., and Shah, S. M. I. (1983). Origin of whales in epicontinental remnant seas: new evidence from the early Eocene of Pakistan. *Science, N.Y.* **220**, 403–6.

Goodman, M., Czelusniak, J., and Beeber, J. E. (1985). Phylogeny of primates and other eutherian orders: a cladistic analysis using amino acid and nucleotide sequence data. *Cladistics* **1**, 171–85.

Goodman, M., Czelusniak, J., Moore, G. W., Romero-Herrera, A. E., and Matsuda, G. (1979). Fitting the gene lineage into its species lineage: a parsimony strategy illustrated by cladograms constructed from globin sequences. *Syst. Zool.* **28**, 132–63.

Goodman, M., Romero-Herrera, A. E., Dene, H., Czelusniak, J., and Tashian, R. E. (1982). Amino acid sequence evidence on the phylogeny of primates and other eutherians. In *Macromolecular Sequences in Systematic and Evolutionary Biology* (ed. M. Goodman), pp. 115–91. Plenum Press, New York.

Gregory, W. K. (1910). The orders of mammals. *Bull. Am. Mus. Nat. Hist.* **27**, 1–524.

Gregory, W. K. (1920). Studies in comparative myology and osteology: no. IV. A review of the evolution of the lacrimal in vertebrates with special reference to that of the mammals. *Bull. Am. Mus. Nat. Hist.* **42**, 95–263.

Kellogg, R. (1936). A review of the Archaeoceti. *Publ. Carnegie Inst. Wash.* **482**, 1–366.

Kumar, K. and Sahni, A. (1986). *Remingtonocetus harudiensis*, new combination, a Middle Eocene archaeocete (Mammalia, Cetacea) from western Kutch, India. *J. Vertebr. Paleontol.* **6**, 326–49.

Lazarus, D. B. and Prothero, D. R. (1984). The role of stratigraphic and morphologic data in phylogeny reconstruction. *J. Paleontol.* **58**, 163–72.

Linnaeus, C. (1758). *Systema naturae per regna tri naturae, secundum classis, ordines, genera, species cum characteribus, differentiis, synonymis locis*, 10th edn., Vol. 1, 824 pp. Laurentii Salvii, Stockholm.

MacFadden, B. J. (1976). Cladistic analysis of primitive equids, with notes on other perissodactyls. *Syst. Zool.* **25**, 1–14.

Matthew, W. D. (1897). A revision of the Puerco fauna. *Bull. Am. Mus. Nat. Hist.* **9**, 249–323.

Matthew, W. D. (1909). The Carnivora and Insectivora of the Bridger Basin, Middle Eocene. *Mem. Am. Mus. Nat. Hist.* **9**, 289–567.

Matthew, W. D. (1937). Paleocene faunas of the San Juan Basin, New Mexico. *Trans. Am. Philos. Soc.* **30**, 1–510.

Mayr, E. (1974). Cladistic analysis or cladistic classification? *Z. Zool. Syst. Evol.* **12**, 94–128.

Mayr, E. (1981). Biological classification: toward a synthesis of opposing methodologies. *Science, N.Y.* **214**, 510–16.

McDowell, S. B. (1958). The Greater Antillean insectivores. *Bull. Am. Mus. Nat. Hist.* **115**, 113–214.

McKenna, M. C. (1975). Toward a phylogenetic classification of the Mammalia. In *Phylogeny of the Primates: a Multidisciplinary Approach* (eds. W. P. Luckett and F. S. Szalay), pp. 21–46. Plenum Press, New York.

McKenna, M. C., Engelmann, G. F., and Barghoorn, S. F. (1977). Review of 'Cranial anatomy and evolution of the early Tertiary Plesiadapidae (Mammalia, Primates)' by Philip D. Gingerich. *Syst. Zool.* **26**, 233–8.

McKenna, M. C. and Manning, E. (1977). Affinities and palaeobiogeographic significance of the Mongolian Paleogene genus *Phenacolophus*. *Géobios, Mém. Spéc.* **1**, 61–85.

Miyamoto, M. M. and Goodman, M. (1986). Biomolecular systematics of eutherian mammals: phylogenetic patterns and classification. *Syst. Zool.* **35**, 230–40.

Novacek, M. J. (1982). Information for molecular studies from anatomical and fossil evidence on higher eutherian phylogeny. In *Macromolecular Sequences in Systematic and Evolutionary Biology* (ed. M. Goodman), pp. 3–41. Plenum Press, New York.

Novacek, M. J. (1986). The skull of leptictid insectivorans and the higher-level classification of eutherian mammals. *Bull. Am. Mus. Nat. Hist.* **183**, 1–112.

Novacek, M. J. and Wyss, A. R. (1986). Higher-level relationships of the recent eutherian orders: morphological evidence. *Cladistics* **2**, 257–87.

Osborn, H. F. (1898). Evolution of the Amblypoda. *Bull. Am. Nat. Hist.* **10**, 169–218.

Owen, R. (1841). Description of the fossil remains of mammal (*Hyracotherium leporinum*) and of a bird (*Lithornis vulturinus*) from the London Clay. *Trans. Geol. Soc. London* **2**(6) 1, 203–8.

Owen, R. (1848). Description of teeth and portions of jaws of two extinct anthracotheroid quadrupeds (*Hyopotamus vectianus* and *Hyop. bovinus*) discovered by the Marchioness of Hastings in the Eocene deposits on the N.W. coast of the Isle of Wight: with an attempt to develop Cuvier's idea of the classification of Pachyderms by the number of their toes. *Q. J. Geol. Soc. London* **4**, 103–41.

Patterson, B. (1939). New Pantodonta and Dinocerata from the Upper Paleocene of western Colorado. *Field Mus. Nat. Hist., Geol. Ser.* **5**, 351–84.

Patterson, B. (1975). The fossil aardvarks (Mammalia: Tubulidentata). *Bull. Mus. Comp. Zool.* **147**, 185–237.

Patterson, B. and McGrew, P. O. (1962). A new arctocyonid from the Paleocene of Wyoming. *Breviora* **174**, 1–10.

Patterson, C. (1977). The contribution of paleontology to teleostean phylogeny. In *Major Patterns in Vertebrate Evolution* (eds. M. K. Hecht, P. C. Goody, and B. M. Hecht), pp. 579–643, NATO Advanced Study Inst., Series A, 14. Plenum Press, New York.

Patterson, C. (1981). Significance of fossils in determining evolutionary relationships. *Ann. Rev. Ecol. Syst.* **12**, 195–223.

Patterson, C. (1982). Cladistics and classification. *New Sci.* **94**, 303–6.

Radinsky, L. B. (1963). The perissodactyl hallux. *Am. Mus. Novit.* **2145**, 1–8.

Radinsky, L. B. (1966). The adaptive radiation of the phenacodontid condylarths and the origin of the Perissodactyla. *Evolution* **20**, 408–17.

Radulesco, C., Iliesco, G., and Iliesco, M. (1976). Découverte d'un Embrithopode nouveau (Mammalia) dans le Paléogene de la dépression de Hateg (Roumanie) et considérations générales sur la géologie de la région. *Neues Jahrb. Geol. Paläontol., Monatsh.* **11**, 690–98.

Rigby, J. K., Jr. (1980). Swain Quarry of the Fort Union Formation, Middle Paleocene (Torrejonian), Carbon County, Wyoming, geologic setting and mammalian fauna. *Evol. Monogr.* **3**, 1–178.

Romero-Herrera, A. E., Lehmann, H., Joysey, K. A., and Friday, A. E. (1973). Molecular evolution of myoglobin and the fossil record: a phylogenetic synthesis. *Nature (London)* **246**, 389–95.

Rose, K. D. (1982). Skeleton of *Diacodexis*, oldest known artiodactyl. *Science, N.Y.* **216**, 621–3.

Rose, K. D. (1985). Comparative osteology of North American dichobunid artiodactyls. *J. Paleontol.* **59**, 1203–26.

Rose, K. D. (1987). Climbing adaptations in the early Eocene mammal *Chriacus* and the origin of Artiodactyla. *Science, N.Y.* **236**, 314–16.

Russell, D. E. (1964). Les mammifères paléocènes d'Europe. *Mém. Mus. Natl. Hist. Nat., Paris, Ser. C* **13**, 1–324.

Russell, D. E., Thewissen, J. G. M., and Sigogneau-Russell, D. (1983). A new dichobunid artiodactyl (Mammalia) from the Eocene of north-west Pakistan. *Proc. K. Ned. Akad. Wet. B* **86**, 285–99.

Schaeffer, B. (1947). Notes on the origin and function of the artiodactyl tarsus. *Am. Mus. Novit.* **1356**, 1–24.

Schoch, R. M. (1986). Systematics, functional morphology and macroevolution of the extinct mammalian order *Taeniodonta. Bull. Peabody Mus. Nat. Hist.* **42**, 1–307.

Scott, W. B. (1888). On some new and little known creodonts. *J. Acad. Nat. Sci. Phila.* **2**, 155–85.

Sen, S. and Heintz, E. (1979). *Palaeoamasia kansui* Ozansoy 1966, embrithopode (Mammalia) de l'Éocène d'Anatolie. *Ann. Paléontol. Vertébr.* **65**, 73–91.

Shoshani, J. (1986). Mammalian phylogeny: comparison of morphological and molecular results. *Mol. Biol. Evol.* **3**, 222–42.

Simpson, G. G. (1937). The Fort Union of the Crazy Mountain Field, and its mammalian faunas. *Bull. U.S. Natl. Mus.* **169**, 1–287.

Simpson, G. G. (1945). The principles of classification and a classification of mammals. *Bull. Am. Mus. Nat. Hist.* **85**, 1–350.

Simpson, G. G. (1953). *The Major Features of Evolution*, pp. 1–434. Columbia University Press, New York.

Sloan, R. E. (1970). Cretaceous and Paleocene terrestrial communities of western North America. *Proc. N. Am. Paleontol. Conv.* **E**, 427–53.

Sloan, R. E. (1987). Paleocene and latest Cretaceous mammal ages, biozones, magnetozones, rates of sedimentation and evolution. *Geol. Soc. Am. Spec. Pap.* **209**, 165–200.

Sloan, R. E. and Van Valen, L. (1965). Cretaceous mammals from Montana. *Science, N.Y.* **148**, 220–27.

Szalay, F. S. (1969*a*). The Hapalodectinae and a phylogeny of the Mesonychidae (Mammalia, Condylarthra). *Am. Mus. Novit.* **2361**, 1–26.

Szalay, F. S. (1969*b*). Origin and evolution of function of the mesonychid condylarth feeding mechanism. *Evolution* **23**, 703–20.

Szalay, F. S. (1977). Phylogenetic relationships and a classification of the eutherian Mammalia. In *Major Patterns in Vertebrate Evolution* (eds. M. K. Hecht, P. C. Goody, and B. M. Hecht), pp. 315–74, NATO Advanced Study Inst., Series A, 14. Plenum Press, New York.

Tassy, P. (1981). Le crâne de *Moeritherium* (Proboscidea, Mammalia) de l'Éocène de Dor el Talha (Libye) et le problème de la classification phylogénétique du genre dans les Tethytheria McKenna 1975. *Bull. Mus. Natl. Hist. Nat., Paris, Ser. 4* **C3**, 87–147.

Tassy, P. (1982). Les principales dichotomies dans l'histoire des Proboscidea (Mammalia): une approche phylogénétique. *Géobios, Mém. Spéc.* **6**, 225–45.

Thenius, E. (1969). *Phylogenie der Mammalia.* De Gruyter, Berlin.

Thewissen, J. G. M. (1985). Cephalic evidence for the affinities of the Tubulidentata. *Mammalia* **49**, 257–84.

Tong, Y., Zhang, Y., Wang, B. and Ding, S. (1976). The lower Tertiary of the Nanxiang and Chijian basins. *Vertebr. PalAsiat.* **14**, 16–25.

Tong, Y. and Lucas, S. G. (1982). A review of Chinese uintatheres and the origin of the Dinocerata (Mammalia, Eutheria). *Proc. 3rd N. Am. Paleontol. Conv.* **2**, 551–6.

Van Valen, L. (1966). Deltatheridia, a new order of mammals. *Bull. Am. Mus. Nat. Hist.* **132**, 1–126.

Van Valen, L. (1968). Monophyly or diphyly in the origin of whales. *Evolution* **22**, 37–41.

Van Valen, L. (1971). Adaptive zones and the orders of mammals. *Evolution* **25**, 420–28.

Van Valen, L. (1978*a*). Why not to be a cladist. *Evol. Theory* **3**, 285–99.

Van Valen, L. (1978*b*). The beginning of the Age of Mammals. *Evol. Theory* **4**, 45–80.

Walker, E. P. *et al.* (1975). *Mammals of the World.* 3rd edn., 1500 pp. The Johns Hopkins University Press, Baltimore.

Weitz, B. (1953). Serological relationships of hyrax and elephant. *Nature (London)* **171**, 261.

Wible, J. R. (1986). Transformations in the extracranial course of the internal carotid artery in mammalian phylogeny. *J. Vertebr. Paleontol.* **6**, 313–25.

Wible, J. R. (1987). The eutherian stapedial artery: character analysis and implications for superordinal relationships. *Zool. J. Linn. Soc.* **91**, 107–35.

9. The phylogeny of the Artiodactyla

A. W. GENTRY and J. J. HOOKER
Department of Palaeontology, British Museum (Natural History), London, UK

Abstract

Cladistic analysis of artiodactyls was undertaken using dental and skeletal characters. Much homoplasy was found, resulting in some unresolved polychotomies. Despite this, certain more robust monophyletic groupings were disclosed that modify the current schemes of artiodactyl classification. Among them, most 'Dichobunidae' (paraphyletic), including a new subfamily, are attached to the infraorder Ruminantia to form a new suborder; the remaining 'Dichobunidae' (the 'Uintan homacodontines', here named as a new family) and the infraorder Oreodonta (including Mixtotheriidae) are referred to the suborder Tylopoda, supporting the concepts of Scott and of Romer versus the Neoselenodontia of Webb and Taylor; the 'Palaeodonta' and 'Ancodonta' are shown to be paraphyletic, and their components are spread amongst the other groups; the 'Anthracotheriidae' also become paraphyletic by the nesting of the Hippopotamidae within them; the old classic primary division of artiodactyls into Bunodontia and Selenodontia is supported. The genus *Diacodexis* appears to be paraphyletic, different species being related to different subordinal groups of the Selenodontia, and thus is not the stem artiodactyl. Within the ruminants, Antilocapridae emerged closer to Bovidae than to Cervidae, contrasting with several recent studies. Giraffoidea were the next most primitive sister-group. Below them, the cervoid groups were paraphyletic but some problems with their analysis are discussed. A convenient higher grouping, perhaps best taken as Pecora at a sub-infraordinal rank, would exclude *Gelocus*. Three independent evolutions of horns are postulated among pecorans.

The Phylogeny and Classification of the Tetrapods, Volume 2: Mammals (ed. M. J. Benton), Systematics Association Special Volume No. 35B, pp. 235–72. Clarendon Press, Oxford, 1988.

Introduction

Cuvier (1822) and Blainville (1839–64) were the first to recognize the cogency of the mammalian group now known as the Order Artiodactyla. Owen (1848) demonstrated that the typical ruminant cannon bone was not an intermediate stage in toe bone fusion leading to the horse, but reflected a different foot symmetry, the longitudinal axis passing between two toes (paraxony) instead of along the central one (mesaxony). He also gave the group its formal name and ordinal rank, thus finally dismembering that old unnatural group, the Pachydermata (non-ruminating ungulates).

The artiodactyls are a very distinct group. Basically, they are ungulates with a paraxonic foot through digits III and IV and an astragalus with a trochleated distal articulation (Schaeffer 1948). These and other diagnostic characters were dealt with in Chapter 8 of this volume. An interesting addition is the formation of a partial double mesocylix in the distal deciduous premolar(s) (except P^4) by: mesial prolongation of the paraconid and paracristid parts of the lower premolar trigonids to occlude between the paracone and metacone of the upper premolar next in front; and mesial prolongation of the upper premolar parastyle to occlude between the protoconid and hypoconid of its lower homologue. In more advanced artiodactyls, this trend can affect more mesial teeth and also the permanent premolars (Fig. 9.1). A similar trigonid prolongation and its occlusion with the mesocylix in front occurs also in phenacodonts (West 1971), proboscideans, desmostylians and sirenians (Domning *et al.* 1986), advanced ceratomorph perissodactyls (Butler 1951), and Leptictidae (West 1972). However, in these, the parastyle occludes in front of the protoconid instead of between the protoconid and hypoconid. A similar pattern affects the molars of docodonts (Kermack *et al.* 1987).

Other characteristic artiodactyl features evolved within the order are mentioned by many authors (e.g. Flower and Lydekker 1891; Viret 1961), and most of the hard structures are dealt with herein.

Fossil artiodactyls have been recognized as such with ease, although primitive forms caused problems until their foot structure became known. Relationships of the order have been problematical. The primitive 'arctocyonids' *Chriacus*, *Tricentes*, and *Carcinodon* bear strong dental resemblances to primitive artiodactyls but not in foot structure (Van Valen 1971; Rose 1987). We shall consider only classification and phylogeny within the group. It was concluded in Chapter 8 that the Artiodactyla are the sister-group of all other ungulates.

AWG is mainly responsible for the ruminants herein, and JJH for all other forms.

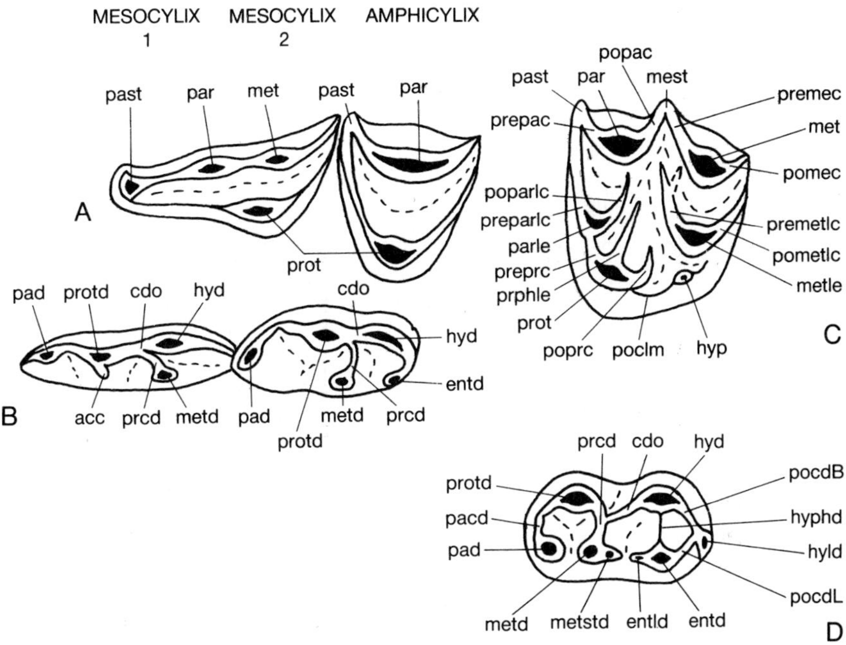

Fig. 9.1. Dental nomenclature of artiodactyl cheek teeth. Mesial side towards left of page, buccal side towards top. A, left P^3+P^4; B, right P_3+P_4; C, left upper molar; D, right lower molar. Abbreviations, A, C: hyp—hypocone, mest—mesostyle, met—metacone, metle—metaconule, par—paracone, parle—paraconule, past—parastyle, poclm—postcingulum, pomec—postmetacrista, pometlc—postmetaconule crista, popac—postparacrista, poparlc—postparaconule crista, poprc—postprotocrista, premec—premetacrista, premetlc—premetaconule crista, prepac—preparacrista, preparlc—preparaconule crista, preprc—preprotocrista, prphle—protolophule, prot—protocone. B, D: acc—accessory crest, cdo—cristid obliqua, entd—entoconid, entld—entoconulid, hyd—hypoconid, hyld—hypoconulid, hyphd—hypolophid, metd—metaconid, metstd—metastylid, pacd—paracristid, pad—paraconid, pocd B & L—postcristid buccal and lingual, prcd—protocristid, protd—protoconid. Centrocrista occupies postparacrista + premetacrista.

Previous classifications

Historical coverage

For about 80 years following the works of Blainville and Owen, a primary division of the order into the Non-ruminantia (or Bunodontia) and Ruminantia (or Selenodontia) was usually maintained, founded on living taxa (e.g. Flower and Lydekker 1891; Matthew 1929). Details are given in Fig. 9.2, in which the Pecora comprise Cervidae, Giraffidae, Antilocapridae, and Bovidae. Fossils often appeared to combine characters defining two or more of these groups in a random way, and were considered to confuse rather than clarify the situation. At this

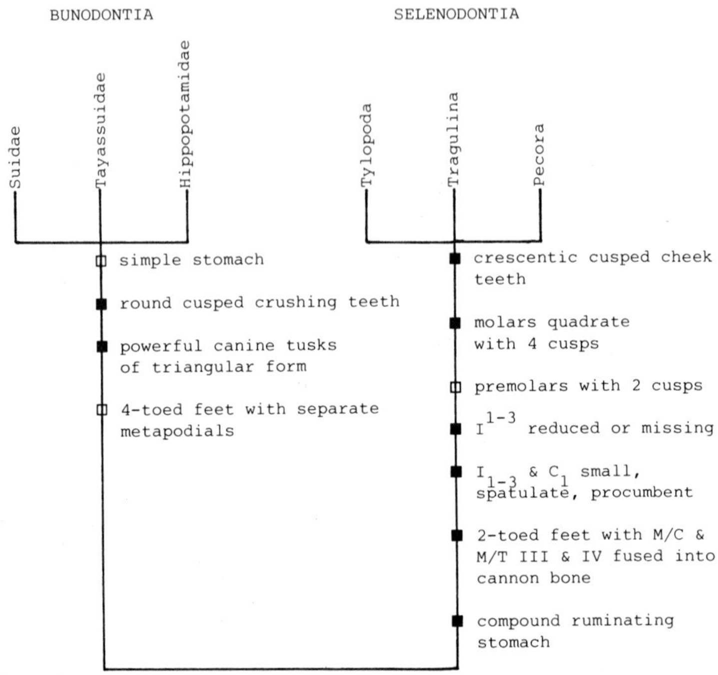

Fig. 9.2. An interpretation of Matthew's (1929) classification of modern artiodactyls. The polarities of his characters are indicated by open (primitive) and solid (advanced) squares.

stage, however, many of the fossils were very poorly known, and advanced and primitive character states were mixed in the classification. Sometimes certain of the fossils were placed in a third primary division, Bunoselenodontia (Zittel 1925, p. 171).

The problems were recognized by Matthew (1929), who reduced the former hierarchical structure and distributed 17 families among five primary divisions: Palaeodonta, Hyodonta, Ancodonta, Tylopoda, and Pecora. He made no attempt to relate these divisions, despite some evident shared characters (Fig. 9.3). Sudre (1978) has pointed out that Matthew's classification has predominated subsequently (e.g. Viret 1961; Romer 1966), although Simpson (1945) reduced the five divisions to three suborders; Suiformes, Tylopoda, and Ruminantia. Even Matthew himself was unhappy about how natural some of his groups were, e.g. Ancodonta. Moreover, Scott (1940, p. 365) wrote of the Palaeodonta, 'This is a provisional group of ancient and primitive genera, united chiefly by their primitive character and their high antiquity . . . The association [of Leptochoeridae and Entelodontidae] seems rather forced and the group is somewhat like a waste-basket, into which are thrown the families for which there is no other place.'

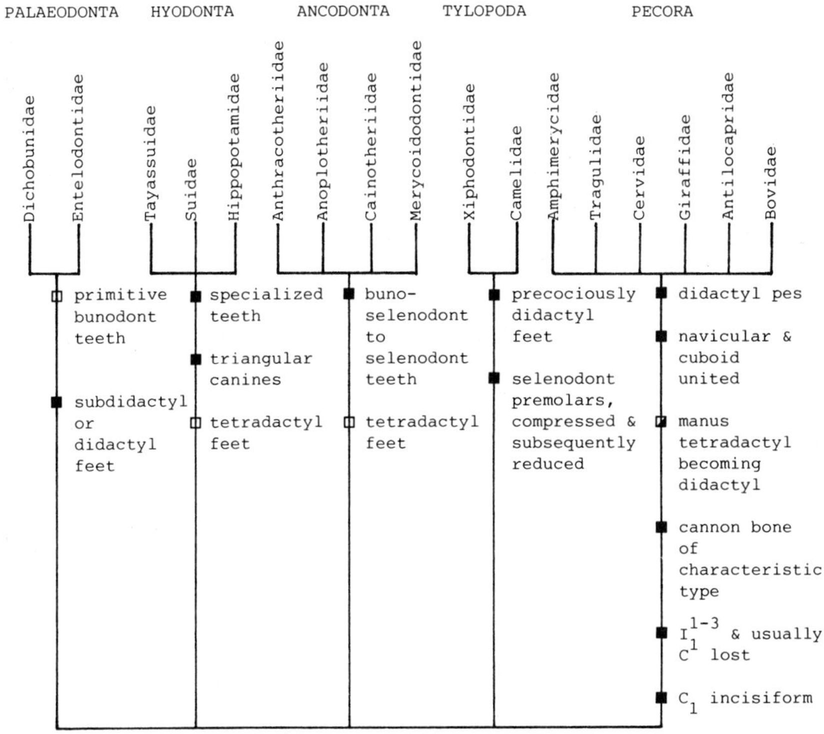

Fig. 9.3. A representation of Matthew's (1929) classification of modern and fossil artiodactyls. Symbols as for Fig. 9.2.

Other previous classifications of the Artiodactyla have been reviewed by Simpson (1945). Within the last decade, specific use has been made of advanced versus primitive character states in a cladistic analysis of hornless ruminants by Webb and Taylor (1980). This resulted in an increase of hierarchical grouping and a clarification of relationships, complementing the work of Flower (1875, p. 189).

Specific problems

Paraphyly, already mentioned for the Ancodonta and Palaeodonta, also affects the Dichobunoidea, comprising Dichobunidae *s.l.*, Choeropotamidae, and Cebochoeridae (see Viret 1961); the Helohyidae (see Gazin 1955; Coombs and Coombs 1977*a*,*b*), comprising especially *Helohyus*, *Gobiohyus*, *Achaenodon*, and the genera sometimes separated as Raoellidae (Sahni *et al.* 1981); the Anoplotheriinae and Dacrytheriinae (see Sudre 1978, where they are raised to family status); the Dichobunidae *s.l.* and the genus *Diacodexis*.

Certain groups can be recognized with ease back to the Oligocene, where they appear abruptly, e.g. Suoidea. Others have a very poor fossil record, e.g. Hippopotamidae. Such groups, because of their many autapomorphies, are difficult to relate to other artiodactyl groups. In particular, there is dispute over whether Hippopotamidae are closer to Anthracotheriidae (Colbert 1935; Viret 1961) or to Tayassuidae (Pickford 1983). In a similar vein, it is unclear whether Giraffidae, and independently Antilocapridae, are closer to Bovidae (Leinders 1984) or to Cervidae (Pilgrim 1941, p. 158).

Webb and Taylor (1980) have proposed the clade Neoselenodontia for the Tylopoda plus Ruminantia, but excluding Oreodonta, Cainotheriidae, and Anoplotheriidae, which Romer (1966, following others) included in an enlarged concept of the Tylopoda. Of relevance here also is the disputed relationships of European Amphimerycidae with Ruminantia (Matthew 1929; Sudre 1978) or with European Xiphodontidae (as tylopods; Webb and Taylor 1980), and secondly of Xiphodontidae with North American Camelidae (Matthew 1929; Webb and Taylor 1980; Dechaseaux 1967; Sudre 1978) or with European Anoplotheriidae (as non-tylopodan Bunoselenodontia; Viret 1961; Stehlin 1906–10). This set of examples raises the question of whether groups with certain morphological similarities on different continents (especially in the Eocene) were simply evolving convergently, were the result of multiple dispersal events, or had evolved in parallel from an already differentiated primitive artiodactyl group of low diversity, spread by an initial major dispersal event around the Palaeocene–Eocene boundary.

Some character polarities in early artiodactyls are also a problem because of ignorance of a close sister-group. They have been highlighted by the discovery of a partial skeleton of *Diacodexis metsiacus* from the North American Early Eocene. The long slender limbs of this animal and its tibia–fibula fusion led Rose (1982) to consider it relatively advanced. Rose (1985), however, considered it more likely that long limbs characterized the primordial artiodactyl and that the later short-limbed forms reverted from this state. Outgroup comparison supports his 1982 conclusion, the later one being more the result of his assumption that *Diacodexis* is the primordial artiodactyl.

Methodology

A cladistic analysis was undertaken in order to attempt a solution to these problems.

Choice of taxa

A consistent method was needed to reduce the large number of artiodactyl taxa to a manageable quantity. We have concentrated on major modern and fossil groups, usually families, defined by their most primitive definite members. Where there is no single clear primitive member, the most primitive state of a given character was chosen from the most relevant of the several relatively primitive genera available. Moreover, the most primitive members were often known only from teeth, so postcranial information had to be assessed from more advanced members. Some subjectivity in group constitution is introduced and some potential information on relationships is removed when advanced members of certain groups have characters that define other groups. To reduce this subjectivity, only well-defined groups were left at family level and other 'looser' or probably paraphyletic groups were split into smaller component units. Certain groups considered difficult to define (e.g. Cervidae, Helohyidae, Anthracotheriidae, Dichobunidae, and *Diacodexis*) were split into several relatively well-presented genera or species to see if they had any bearing on stem relationships of major groups. Other apparently paraphyletic groups (e.g. Dacrytheriinae and Anoplotheriinae) were resolved mainly by removing genera either to a level of *incertae sedis* (e.g. *Tapirulus*, see below) or to other groups to which they were considered closely related, e.g. *Leptotheridium* to the Xiphodontidae (Stehlin 1906–10) or *Robiacina* to the Cainotheriidae (Sudre 1977; so that the distal shift of molar protocone and metastylid demonstrated by Hürzeler (1936) took place within the family).

Haplobunodontidae were lumped with Choeropotamidae following a current study (Hooker, *in prep.*), and Leptotragulidae were lumped with Protoceratidae following Golz (1976) and Patton and Taylor (1973). The Sanitheriidae, now raised to family rank, are considered still as most closely related to the Suidae by Pickford (1984). They have not been treated separately herein as they add nothing to the relationships of the Suoidea. Some poorly known taxa were omitted, such as the primitive genera *Wasatchia*, *Bunophorus*, *Chorlakkia*, *Lantianius*, *Hexacodus*, and *Texodon*.

Choice of characters

A range of characters as objective as possible was chosen, covering dentition, skull, and postcranial elements. However, characters known in relatively few taxa, and those of uncertain polarity (e.g. various characters of the otic region, see Coombs and Coombs 1982), were omitted.

Character polarity

Character polarities were worked out by outgroup comparison mainly using the 'Arctocyonidae' *s.l.*, and where relevant were extended to other eutherians backed up in cases of doubt by the concept of simple to complex and by known evolutionary sequences (e.g. Krishtalka and Stucky 1985; Stehlin 1906–10; Sudre 1978). Multiple states were formulated for many of the characters.

Characters

Two data matrices were constructed, one for artiodactyls as a whole and a second for 'undoubted' Ruminantia, i.e. omitting Amphimerycidae and Hypertragulidae. Characters were assessed wherever possible by direct observation of specimens. In other cases, use was made of the published literature. References for data matrix 1 are: Bergounioux and Crouzel (1967), Colbert (1938, 1941), Coombs and Coombs (1977*a*,*b*, 1982), Cuvier (1822), Dechaseaux (1965, 1967, 1974), Depéret (1917), Edwards (1976), Flower and Lydekker (1891), Franzen (1981, 1983), Franzen and Krumbiegel (1980), Gazin (1955), Golz (1976), Heller (1934), Hürzeler (1936), Kowalevsky (1873, 1874), Krishtalka and Stucky (1985), Macdonald (1955), Patton and Taylor (1971, 1973), Pearson (1927), Peterson (1919), Pickford (1983), Pilgrim (1928), Pilgrim and Cotter (1916), Rose (1985), Russell *et al.* (1983), Schaeffer (1948), Schlosser (1886), Scott (1940), Sinclair (1914), Stehlin (1899–1900, 1906–10), Stock (1934), Sudre (1978), Sudre *et al.* (1983), Thewissen *et al.* (1983), Tobien (1985), Viret (1961), Webb and Taylor (1980), West (1980, 1984), Wortman (1898). References for data matrix 2 are: Black (1978), Bouvrain and Geraads (1985), Colbert (1941), Filhol (1877, 1881, 1882), Flower (1875), Frick (1937), Furlong (1927), Hendey (1978), Matthew (1908), Mayet (1908), Schlosser (1902), Scott (1940), Sigogneau (1968), Stehlin (1937, 1939), Sudre (1984), Webb and Taylor (1980), Whitworth (1958). C, canine; I, incisor; M, molar; P, premolar; $\overline{M}$, lower molar; $\underline{M}$, upper molar; M^1, upper molar 1; M_1, lower molar 1.

The list of characters for data matrix 1 is:

(1) cheek tooth cresting significantly reduced;

(2) molar crown height semi-hypsodont;

(3) $\underline{M}$ preparacrista mesiobuccally orientated (primitive state: mesiodistal);

(4) $\underline{M}$ preparacrista mesiolingually orientated (primitive state: mesiodistal);

(5) $\underline{M}$ postparacrista distobuccally orientated (primitive state: mesiodistal);

(6) $\underline{M}$ postparacrista distolingually orientated (primitive state: mesiodistal);

(7) $\underline{M}$ premetacrista mesiobuccally orientated (primitive state: mesiodistal);

(8) $\underline{M}$ premetacrista mesiolingually orientated (primitive state: mesiodistal);

(9) $\underline{M}$ postmetacrista mesiobuccally orientated (primitive state: mesiodistal);

(10) $\underline{M}$ postmetacrista distolingually orientated (primitive state: mesiodistal);

(11) $\underline{M}$ parastyle weak;

(12) $M^{1,2}$ distobuccal corner obtuse (primitive state is a right angle);

(13A) $\underline{M}$ mesostyle loop an open V; (B) a narrow U; (C) occluded/lost leaving an isolated mesostyle (primitive state is a straight centrocrista and no mesostyle);

(14A) $\underline{M}$ paraconule reduced; (B) lost;

(15) $\underline{M}$ paraconule enlarged;

(16) $\underline{M}$ paraconule displaced lingually towards protocone;

(17) $\underline{M}$ paraconule extended distobuccally;

(18) $\underline{M}$ postparaconule crista reduced;

(19A) $\underline{M}$ postprotocrista broken; (B) lost;

(20A) $\underline{M}$ postprotocrista strengthened but still orientated distally; (B) orientated buccally;

(21) $\underline{M}$ protolophule prominent/elongated;

(22) $\underline{M}$ protostyle present;

(23) $\underline{M}$ metaconule large;

(24A) $\underline{M}$ premetaconule crista buccal end lost; (B) new buccal extension directed towards mesostyle (primitive state: buccal end passing up side of metacone);

(25A) $\underline{M}$ hypocone medium-sized; (B) same size as protocone; (C) fused to metaconule (primitive state: hypocone absent or very small);

(26) $\underline{M}$ lingual cingulum tends to be absent;

(27) M^3 longer than M^2;

(28) $\overline{M}$ paraconid indistinct, fused to metaconid;

(29A) $\overline{M}$ paracristid deeply but narrowly notched; (B) broadly excavated lingual to paraconid;

(30A) $\overline{M}$ metaconid slightly in front of protoconid; (B) well in front of protoconid;

(31) $\overline{M}$ protocristid deeply notched;

(32) $\overline{M}$ metastylid present;

(33) $\overline{M}$ entoconulid present;

(34) $M_{1,2}$ hypoconulid displaced lingually of midline;

(35) $\overline{M}$ hypolophid present;

(36A) $M_{1,2}$ entoconid opposite hypoconid; (B) slightly in front of hypoconid; (C) well in front of hypoconid (primitive state: behind hypoconid);

(37) $M_{1,2}$ hypoconulid displaced distally;

(38) $M_{1,2}$ postcristid lingual arm absent;

(39) M_3 reduced;

(40) P^4_4 metacone and cuspate hypoconid present;

(41) $P^{34}_{3,4}$ inflated;

(42A) P_3 length greater than 120% M_1 length; (B) greater than 150% M_1 length;

(43) P_4 hypoconid occluding between P^4 parastyle and paracone;

(44) P^4 postprotocrista high, joining back of paracone (primitive state is low, directed towards the distobuccal corner);

(45) P^3 parastyle absent;

(46) P^3 parastyle as large and distinct as paracone;

(47) P_4 paraconid absent;

(48) P_4 metaconid present;

(49A) P_4 protocristid independent of cristid obliqua; (B) partly confluent with cristid obliqua (primitive state: no protocristid);

(50) P_3 paraconid absent;

(51A) P_3 metaconid present in a mesial position; (B) in a distal position;

(52A) P_3 protocristid independent of cristid obliqua; (B) partly confluent with cristid obliqua (primitive state: no protocristid);

(53) P^1 absent;

(54) P^1 complex, premolariform;

(55) P^1 caniniform;

(56) P_1 absent;

(57) P_1 complex, premolariform;

(58) P_1 caniniform;

(59) $\underline{C}$ reduced;

(60) $\underline{C}$ apically elongated as a long thin sabre in males;

(61) $\underline{C}$ growing laterally (primitive state: growing occlusally);

(62) $\underline{C}$ premolariform;

(63) $\overline{C}$ with triangular cross-section;

(64A) $\overline{C}$ same size as incisors; (B) also spatulate (primitive state: larger than incisors);

(65A) I^{1-3} reduced; (B) absent;

(66) I^3 larger than $I^{1,2}$;

(67) face short;

(68) postorbital bar closed;

(69) dentary and zygomatic protuberances present;

(70) squamosal with pointed dorsal process;

(71) occiput high and narrow, flared dorsally;

(72) mastoid process of periotic not exposed on braincase surface;

(73) mastoid process orientated occipitally (primitive orientation: lateral);

(74) glenoid convex for articulation with concave dentary condyle;

(75) medial glenoid pit present (see Pearson 1927, p. 448);

(76) tympanic bulla enlarged;

(77A) tympanohyal pit emarginating bulla posterolaterally; (B) posteriorly (primitive position is posterolateral but not emarginating bulla);

(78) horizontal ramus of dentary deepening posteriorly;

(79) dentary condyle at about level of tooth row (primitive position: significantly above);

(80) ventral edge of dentary essentially without concavity near M_3;

(81) coronoid process of dentary reduced;

(82) angular process of dentary retracted posteroventrally;

(83) angular process of dentary downturned;

(84) angular process of dentary with dorsal hook;

(85) angular process plus condyle of dentary projected posteriorly;

(86A) angular process of dentary expanded dorsally slightly; (B) markedly;

(87) deep concavity between condyle and coronoid of dentary;

(88) distal humerus: medial border of trochlea approximately parallel with long axis of bone;

(89A) distal humerus: entepicondylar process reduced; (B) lost;

(90A) distal humerus: intercondylar ridge narrow, low; (B) narrow, high (primitive state: broadly rounded);

(91) distal humerus: lateral border of trochlea raised into a sharp ridge bordering a groove;

(92) distal humerus: medial facet convex (primitive state: concave);

(93) distal humerus: trochlea a near-parallel-sided cylinder (primitive state: a cone tapering laterally);

(94) distal humerus: lateral epicondyle projected posteriorly;

(95A) radius and ulna fused distally; (B) fused completely;

(96) magnum and trapezoid fused in carpus;

(97) metacarpal II reduced to splints/nodules;

(98) metacarpal V reduced to splints/nodules;

(99) tibia and fibula fused distally;

(100) tibia and fibula fused proximally;

(101) astragalus: distal trochlear ridge parallel with long axis;

(102) astragalus: astragalocalcaneal facet displaced from proximal end;

(103) astragalus: proximal part of astragalocalcaneal facet orientated to form transverse ledge;

(104) astragalus: lateral side flat (primitive state: excavated);

(105) astragalus: lateral edges of proximal and distal trochleae parallel;

(106) cuboid and navicular fused—associated with weakening of distal trochlear ridge of astragalus;

(107) ectocuneiform and mesocuneiform fused;

(108) cubonavicular and ectomesocuneiform fused;

(109) metatarsal II reduced to splint/nodule;

(110) metatarsal V reduced to splint/nodule;

(111) metatarsals III and IV fused proximally;

(112) keels on distal metapodials extending anteriorly;

(113) axis vertebra: atlantoid facet extending half-way up neural arch (primitive state: restricted to below neural arch);

(114) axis vertebra: atlantoid facet transverse in posterodorsal view (primitive state: swept back);

(115) axis vertebra: odontoid process forming an incipient spout (primitive state is a peg);

(116) vertebrarterial canal of cervical vertebrae 3–6 exits posteriorly inside neural canal.

The dental terminology diagram explains many of the above tooth characters (Fig. 9.1). Some characters need further explanation:

Bunodonty and selenodonty: These were considered too generalized to be treated as single characters and are divided among characters 1, 3–11, 13, 17, 20, and 24.

$\overline{M}$ *paracristid (character 29):* Primitively this is high and only slightly notched, a state correlated with the presence of a distinct paraconid. As the paraconid reduces in size and moves nearer the metaconid, the paracristid, acting as front wall of the trigonid, is breached by deepening of the notch (29A). As the paraconid reduces still further (28), the paracristid may stay in this condition or its notch may broaden, so that

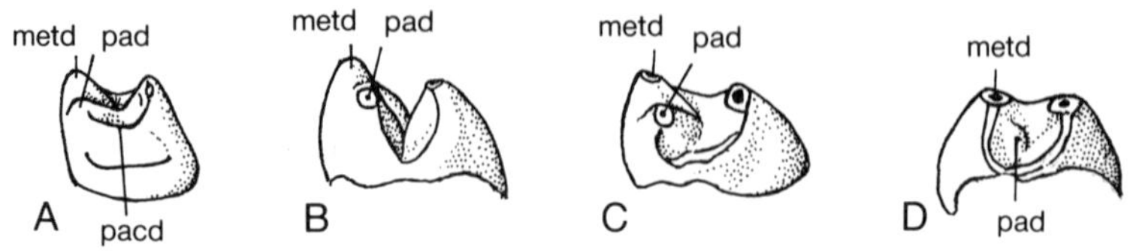

Fig. 9.4. Mesial views of left lower molars showing characters 28 and 29. A, *Diacodexis*, M 42177; B, *Homacodon*, M 37888; C, *Cuisitherium*, M 49466 (described as a dacrytheriid (sic) by Sudre *et al.* (1983), but which is probably a primitive choeropotamid); D, *Anthracothema*, M 13170. A–C show primitive state for character 28. D shows advanced state for character 28. A shows primitive state for character 29. B shows character 29A. C–D show character 29B. A–C (casts) are ×4.5; D is ×0.8. pad=paraconid; metd=metaconid; pacd=paracristid. Register nos. are of the British Museum (Natural History).

it forms a widely scalloped opening, the lingual side having parted company with the paraconid (29B). Understanding this trend is important in homologizing the vestigial state of the paraconid, especially in anthracotheres and suoids (Fig. 9.4).

$\overline{M}$ metaconid position (character 30): Primitively this is either opposite the protoconid or, usually when there is still a distinct paraconid, slightly behind. It is sometimes, but not always, correlated with entoconid position (36). Directly correlated with either or both, however, are the positions in the $\underline{M}$s of protocone and metaconule relative to paracone and metacone, respectively, and so these are not covered separately here. The most advanced states of both characters 30 and 36 produce molars with a very distinct mesiolingual–distobuccal obliquity.

Interaction of $P_{3,4}$ protocristid and cristid obliqua (characters 49, 52): When the metaconid on these teeth is mesially situated, near the protoconid, the protocristid links it directly with the protoconid. Displacement of the metaconid distally (usually P_3 character 51B) is often, but not always (e.g. *Mixtotherium*), associated with a shift in the mesiobuccal end link of the protocristid. This moves distally so that it forms a T-junction with the cristid obliqua, sometimes as far back as the hypoconid (49B, 52B). Whether the crest adjoining the protoconid now represents part of the protocristid or cristid obliqua is a semantic point. Nevertheless, in terms of position, the part that is definitely protocristid resembles a hypolophid and the metaconid resembles an entoconid (referred to as such by, for example, Webb and Taylor 1980). Their origin is not in doubt though, and the cusp that buds off the P_3 protoconid in leptomerycids and pecorans is a neoformed structure not homologous with the metaconid, although referred to as such by, for example, Webb and Taylor (1980) (see Fig. 9.1).

Downturned angular process of dentary (character 83): This is typically present in hippopotamuses and anthracotheres. It has been recognized also in some helohyids by Coombs and Coombs (1977*a*). In the latter, along with other primitive mammals, there is a concavity in the ventral margin just in front of the angular process. This often flattens out (80) without affecting the shape of the angle, and the specific downturning is superimposed on this state. Figure 9.5 shows these and other states of the back of the dentary.

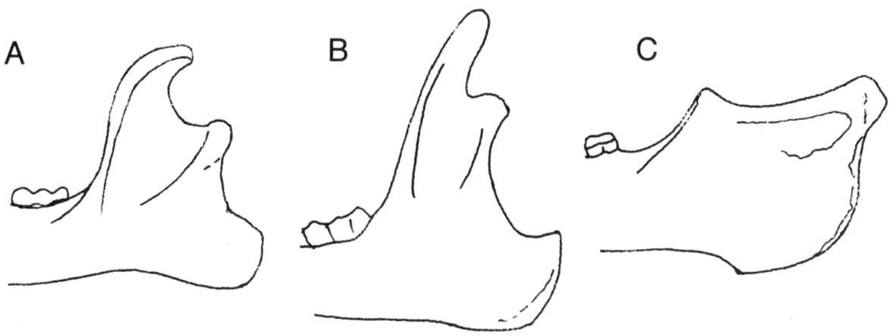

Fig. 9.5. Lateral views of backs of left dentaries. A, *Haplobunodon*, 29713; B, *Poebrotherium* (a primitive camelid); and C, *Archaeotherium* (entelodont); showing characters 79–84, 86. A shows the primitive state for all characters. B shows characters 80, 84, 86A. C shows characters 79–83. B and C are from Scott (1940). A is ×0.55; B is ×0.40; C is ×0.13. Register no. is of the British Museum (Natural History).

The list of characters for data matrix 2 is:

(1) two lacrimal orifices on orbit rim (primitive state is single or no orifice);

(2) horns present;

(3) compact bone of horn core cortex forms a wide layer (primitive state: narrow);

(4) independent ossicone present as whole or part of horn core;

(5A) horn cores with sheath; (B) sheath deciduous;

(6) horn core itself (=antler) deciduous;

(7) occipital horn present;

(8) horns forked;

(9A) foramen ovale large; (B) very large;

(10) tympanohyal pit emarginating lateral side of bulla (primitive state is equivalent to character 77A of data matrix 1);

(11) auditory bulla hollowed internally (primitive state is cancellous);

(12) postglenoid process absent;

(13) cheek teeth hypsodont;

(14A) I^{1-3} much reduced; (B) lost;

(15) $\underline{C}$ reduced or absent;

(16) $\underline{C}$ elongated in males;

(17) $\underline{C}$ bi-lobed;

(18) premolar row elongated;

(19) P_1 absent;

(20) P_1 caniniform;

(21A) hypoconid on $P_{2,3}$ slightly salient; (B) not salient;

(22) distolingual crest on P_4 present;

(23) P_4 with transverse ridge from entoconid to cristid obliqua;

(24) $\underline{M}$ lingual cingulum absent;

(25) $\underline{M}$ entostyle present;

(26) $\underline{M}$ crest labial to postprotocrista present;

(27) $\underline{M}$ postprotocrista orientated buccally (primitive orientation: distally);

(28) $\overline{M}$ metastylid present;

(29) $\overline{M}$ '*Palaeomeryx*-fold' (diagonally on distolabial face of protoconid) present;

(30) $\overline{M}$ metaconid forward of level of protoconid;

(31) $\overline{M}$ postcristid becomes connected to lingual side of tooth in early wear (primitive state: connection not made until later wear);

(32) $\overline{M}$ with lingual fold on metaconid;

(33A) axis vertebra: atlantoid facet extending half-way up neural arch; (B) extending higher;

(34A) axis vertebra: odontoid process transitional to spout-shaped; (B) spout-shaped;

(35) limbs much elongated;

(36) astragalus proximal and distal parts in line;

(37) metatarsals III and IV fused from proximal end for most or entire length;

(38A) metatarsals III and IV fused with anterior longitudinal gully open distally; (B) bridged;

(39) metapodial keels on distal condyles complete anteriorly;

(40) metacarpals III and IV fused from proximal end for most or entire length;

(41) metacarpals II and V incomplete.

Two of the above characters need further explanation:

Tympanohyal pit (character 10): Primitively, the tympanohyal pit lies posterolateral to the auditory bulla and scarcely impinges on it. In most cervids there has been expansion in the tympanic neck region, but this still leaves the tympanohyal pit in its primitive position relative to the bulla. Emargination of the bulla by the pit may appear stronger, possibly because of the customarily small inflation of the bulla in cervids. In bovids, giraffids, and antilocaprids there has been an expansion of the bulla posteromedially, which leaves the tympanohyal pit lying lateral to the bulla; this is taken as the advanced condition. Garrod (1877*a*, p. 14) made observations on this character in deer.

Postprotocrista on upper molars (characters 26, 27): Buccal orientation of the postprotocrista is one of the characters whereby selenodonty is manifested in ruminants. Among early cervoids, however (see below), the postprotocrista retains its primitive orientation, while a new crest appears between the postprotocrista and the protolophule. This new crest fuses to the postprotocrista and takes over the function of the latter's buccal arm in other groups. See illustration of *Palaeomeryx* in Qiu *et al.* (1985, fig. 3).

Analysis

Two data matrices were constructed: one for artiodactyls as a whole and a second for 'undoubted' Ruminantia (i.e. omitting Amphimerycidae and Hypertragulidae). Both were computer-analysed using Swofford's (1984) PAUP program, with global branch-swapping (MULPARS) and reduced reversal (DELTRAN) options. The data were rooted in a hypothetical ancestor represented by the primitive state for all the characters.

Entelodontidae	1000000000100000012000012100201000011110010010100100000000000000000001100100000011111000001201012011001109001011000000
Tayassuidae	1000000000100000012000123101200000111100000010011100000000000000100000011100000000100000011201002000000101000001111010
Suidae	1000000000100000012000123101200000111100000010000000000000000101000000111000100011000000122110010001101010000000011000
Cebochoeridae	0000000000100000010010110100200000111100000000110100001001000001000000010000110110000109999999999999999999999999999
Achaenodon	1000000000100101012000110011200000111100000010100100100100000000000000010009901101000019999999990099999999900099999
Helohyus	0000000000100100000000100010200000011000000009910010000000000000000000900000009000090090999999999999900900009999999999
Raoellidae	1001000001100100000000100001201000111100000010100100000000099909990900099999900000001009999999999999909909099999999999
Gobiohyus	0000000000000000000000100010100000111100000010110100000000000000099090000009990010000000099999999999999999999999999999
Choeropotamidae	0000101000011000012010100000200000011000000000000000001001000001000000000009900000000000909999000000010000000000009999
Anthracotherium	0000101000011000111011100011200000111000000010100100000000000000000000000000009011001000011100000000000000000000009999
Anthracothema	0001000001112000111011100111200000111100000010100100999000999009909999999999999999999999999999999999999999999999999999
Bothriodontinae	0010101010011000111010100011200101121000000010011100000000000000000000000100000011001000011100000000000000000000000000
Hippopotamidae	0001010101113101012010120001201000011100000010110100000000000001000000001000000111110000121010010000001010000000000000
D.pakistanensis	0000000000100000000000001000000010000000000000000000000001000010000000000000000100010009999999999999999999999999999999
Bunomerycidae	0000001010101000002000121101100111001000000010011100000000100000199000000000990910110000009109999099999999999999999999
Agriochoeridae	0010101010001000011100120111100111011001000010011111000001000001001000000000011011000010000101000000010000001000001100
Merycoidodontid	0010101010011200011200120111100111011100000010011111000001000001001000000000011011000010100101000000010000001000001110
Mixtotheriidae	0000101010001101011000110111110101011001000000011021000000000000000100001000011011100020999999999999999999999999999999
Cainotheriidae	0000101010011011012000120101100101011000000000001001000000001000010010000000011101000100011010000100011001001000001100
Anoplotheriinae	0010101010011011012000120100000101011000001000012022010010100101000000000000100100010100001010000100100000000001001110
Dacrytheriinae	0010101010011011012000120101100101011000000000012022010010100101000000000000100100010109999999999999100000999999999999
Xiphodontidae	0010101010011011011000120101100001011100020000012022010010100101000100000009100010001010121011100110010101000110011111
Protoceratidae	0000101010011200011200120001110001021100000000011021000001010001100000000000100110000101222111100000011010010110011101
Oromerycidae	0000000000121010111101201011000010111000001000120110000000100000010000099000200100010209999999100001100900101100110 0
Camelidae	0100000000022000112001201111100010211000000000110210000000100000010000000001200100010101200111201101110110101100111 1
Leptochoeridae	0000000000100000000000000100000000000010110000010000010999100099990000010001001101000009999990900109999919000099999
D.metsiacus	0000000000100000000000000100000000000000100000000000000000000000099000000099900999999999110001909091010000010000010 09
Microsus	0000000000000000000000001100001000011000090099010999999999999999999999999999909099999999999999999190000099999999 99
Antiacodon	0000000000003000012001011100101000111100000000010910999999999999990000000009090900999999999999999911900000900000 9999
Homacodon	0000000000000000012000111101101000111100000000000000000000000000990000000009009090090099999999999991990000010000 9999
Protodichobune	0000000000103000010000000100000000111100000099000999999999999999999999999999919199999999999999999999999999999999 99999
D.varleti	0000000000100000000000000100000001010110099999999999999999999999999999999999999999999999999999999999999999999999999
Messelobunodon	0000000000100000000000000100001001111110000000000000000000000001000010000000999999000000100010010000000001909009000001000
Dichobuninae	0000000000000000012000111101111000111100000000000000000000000001000010000000000990010000010101100000000019900000000099 99
Hyperdichobunin	0000101000001101000000101100011001111110001000101000001001010000909000000009000000001000999999999999999999999999999999
Amphimerycidae	0010101010011011001100120101121001031100020001010022010010100002990000000110000000000000999999999999100101901110999 9
Hypertragulidae	0000101000010101011100120111111001021100000010000100000001010002100000000000100000000011021010200011109101100000010 9
Tragulidae	0100000000010200011100120011121001031100000000001200010000000100022001000011111001000000112201000100011101011100101100
Leptomerycidae	0000101010011101011100120101121001031100000000101202210000100000200010000111010000000001112011001000110001110000011 00

Fig. 9.6. Data matrix 1—Artiodactyla. 9=missing state. Character marking is from left to right. 'Hyperdichobunin' in list of taxa =Hyperdichobuninae.

Data matrix 1

The computer produced 32 maximum parsimony trees, each with 488 steps and a consistency index of 0.285, reflecting much homoplasy and the unavoidable inclusion of missing character states. A consensus tree (Adams 1972) was obtained using the CONTREE program (Fig. 9.7). It shows a basal dichotomy corresponding approximately to the classical groupings of Bunodontia and Selenodontia. In the Bunodontia, Hippopotamidae belong to a trichotomy, nesting equally with *Achaenodon* and with Suoidea (Tayassuidea plus Suidae) plus Entelodontidae. This threesome is part of another trichotomy involving the Cebochoeridae and Anthracotheriidae, which in turn are nested with Raoellidae, and together form a trichotomy with *Helohyus* and *Gobiohyus*.

Within the Selenodontia, the 'Uintan homacodontines' (West 1984), here placed in a distinct family Bunomerycidae nov., are separated from *Homacodon* and form the sister-group to Merycoidodontoidea, Mixtotheriidae, Cainotheriidae, Anoplotheriidae, Xiphodontidae, Protoceratidae, Ruminantia (including Amphimerycidae and Hypertragulidae), Oromerycidae, and Camelidae. The remaining Dichobunidae (*s.l.*) are paraphyletic and successively nest the other groups, although Leptochoeridae plus *Diacodexis metsiacus* are the sister-group to all other Selenodontia. The Neoselenodontia of Webb and Taylor (1980), linking a restricted Tylopoda (comprising Camelidae, Oromerycidae, Protoceratidae, and Xiphodontidae) with Ruminantia, as distinct from other selenodonts, is largely supported except for exclusion of the Xiphodontidae and a closer linking of the Protoceratidae with the Ruminantia than with Camelidae plus Oromerycidae.

Study of this analysis shows that groupings depend on numerous reversals of character polarity. Three examples of the least likely ones are: redevelopment of $\overline{\mathrm{M}}$ paraconid after its fusion with the metaconid (28) in the independent origins of Choeropotamidae and Anoplotheriinae; redevelopment of $\underline{\mathrm{M}}$ postparaconule crista (18); and reformation of the lingual arm of the $\underline{\mathrm{M}_1}{}_{,2}$ postcristid (38) in deriving *Microsus*. There are many examples of $\overline{\mathrm{M}}$ paraconid loss in mammals, associated with a change in chewing mode and in diet towards herbivory; none of the reverse. Functionally, it is easier to understand the $\overline{\mathrm{M}}$ paraconid presence in anoplotheriines as a primitive retention rather than a regained structure. The $\underline{\mathrm{M}}$ postparaconule crista is a similar case (18), but here there often is a new structure formed in its place by distobuccal elongation of the paraconule itself into a loph (17). We feel, in general, that as structures transform, their interrelationships with other structures increase, thus reducing the chances of exact

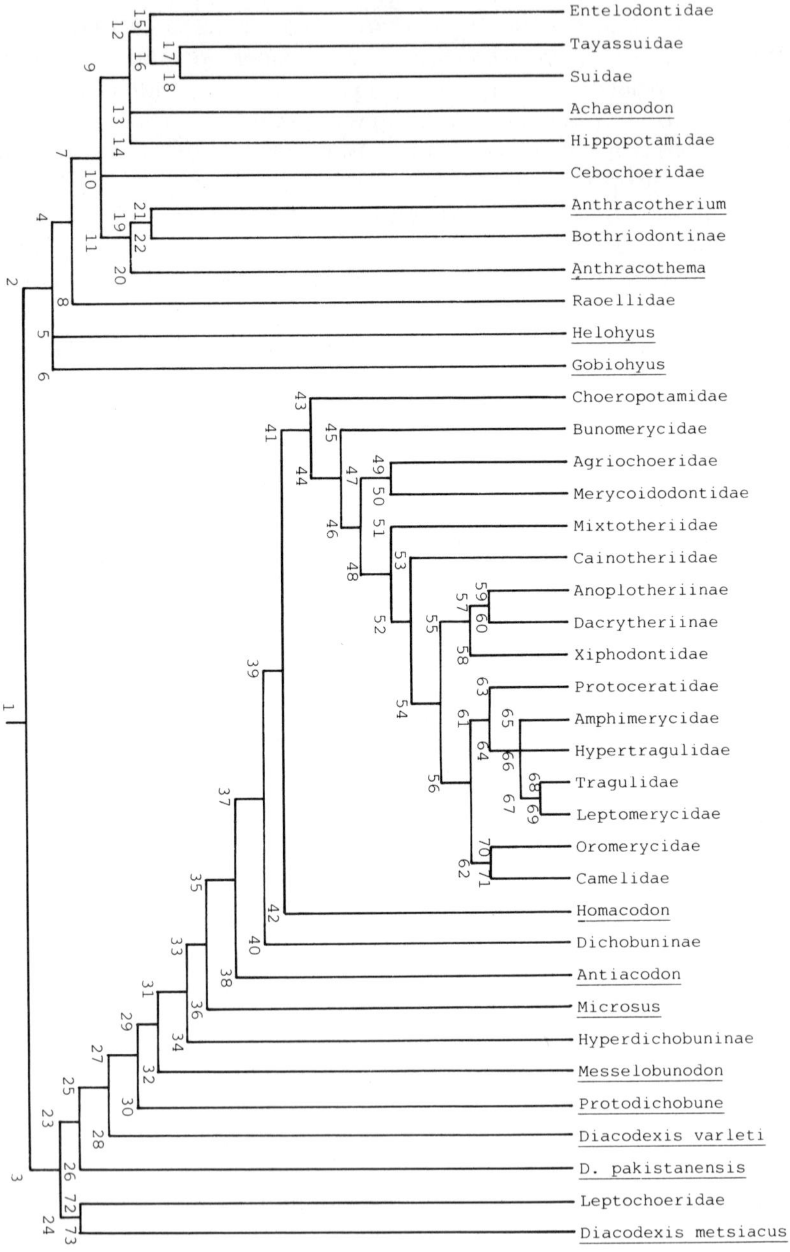

Entelodontidae
Tayassuidae
Suidae
Achaenodon
Hippopotamidae
Cebochoeridae
Anthracotherium
Bothriodontinae
Anthracothema
Raoellidae
Helohyus
Gobiohyus
Choeropotamidae
Bunomerycidae
Agriochoeridae
Merycoidodontidae
Mixtotheriidae
Cainotheriidae
Anoplotheriinae
Dacrytheriinae
Xiphodontidae
Protoceratidae
Amphimerycidae
Hypertragulidae
Tragulidae
Leptomerycidae
Oromerycidae
Camelidae
Homacodon
Dichobuninae
Antiacodon
Microsus
Hyperdichobuninae
Messelobunodon
Protodichobune
Diacodexis varleti
D. pakistanensis
Leptochoeridae
Diacodexis metsiacus

imperceptible reversal to the primitive state (Riedl's 1978 concept of burden and Dollo's Law). Distinction between characters likely to reverse or not, is not possible in the computer analysis. Thus, an assessment of a character reversal is felt to be as important as that of its initial polarity, and less reliance should be placed on its insertion simply as a device to find the shortest tree.

Data matrix 1 was therefore next assessed without the aid of a computer, by systematically searching for congruent characters and only inserting reversals where there appeared a good functional reason and/or because they were dependent on other characters. For instance, straightening out of the $\underline{M}$ ectoloph W (reversal of 3, 5, 7, and 9) is logical when the buccal and lingual tooth walls steepen with hypsodonty (2), as in Camelidae. Redevelopment of lost P_4 paraconid (47), by shift in the field along with the first development of P_4 metaconid (48) is also logical in Bothriodontinae, where strong trends in selenodonty affect the molars in a development analogous with that of ruminants. Loss of a whole series of premolar cusps and crests (reversal of 46, 48, 51A, 52B, A) in Hypertragulidae is the logical result of reduction at the front of the cheek tooth row associated with formation of diastemata.

The least-known taxa (*Microsus*, *Protodichobune*, and *Diacodexis varleti*) and character 88 were ignored to facilitate the exercise. Missing states were not relied upon for groupings, with the following exceptions: *Anthracothema* was assumed to have a downturned dentary angle like most other anthracotheres (83); absence of a dorsal hook on the dentary angle in such unmodified types as *Gobiohyus* or choeropotamids was judged as primitive; the presence of a hook (84) in primitive

---

Fig. 9.7. PAUP consensus cladogram 1—Artiodactyla. Characters at nodes below polychotomies are taken from one of the contributing cladograms. Node **1**: 11, 88; **2**: 23, 29A,B, 36A, 37, 47, 50; **3**: 26, 80, 101, 113, **4**: 35, 38, 45; **23**: 34, 59, 64A, 84; **24**: 42A, 99; **25**: 36A, 38; **27**: 35, 37; **28**: 39; **29**: 31, 80R, 91; **30**: 13A–C, 18, 34, 78; **31**: 11R, 25A, 48; **33**: 34R; **34**: 5, 7, 13A, 14A, 16, 23, 30A, 42A, 46, 54, 57; **35**: 18, 19A,B, 24A, 29A; **36**: 35R, 38R, 100; **37**: 23, 28, 48R, 84R; **38**: 13A–C, 22, 51A, 99, 100; **39**: 59R; **40**: 30A, 80, 86A, 90A; **41**: 7, 13A, 31R, 35R, 38R, 58; **42**: 64AR, 99, 107; **43**: 5, 12, 21, 24AR, 25AR, 26R, 28R, 29B, 55; **44**: 9, 24B, 32, 34, 48, 49A, 78, 80, 81; **45**: 11, 18R, 33, 36A, 45, 50, 89A; **46**: 5, 19BR, 25AR, 51A, 52A, 67, 77A, 86A, 107, 114; **47**: 3, 20A, 27, 33, 45, 50, 93; **48**: 16, 58R; **49**: 40, 88R; **50**: 12, 14A,B, 20B, 38, 115; **51**: 14A, 24BR, 27, 30A, 40, 51B, 64AR, 72, 82, 86B; **52**: 12, 59, 81R, 84, 89A; **53**: 15, 19B, 49AR, 52AR, 76, 86AR, 96, 100, 104; **54**: 51B, 67R, 78R, 93, 110; **55**: 3, 15, 49B, 52B, 54, 57, 62, 98, 107R, 115; **56**: 14A, 20A, 32R, 38, 92, 104, 109; **57**: 19B; **58**: 32R, 38, 42A,B, 68, 76, 77AR, 89B, 90A, 91R, 92, 94, 97, 103, 105, 109, 116; **59**: 28, 29AR, 43, 88R, 89AR; **61**: 30A, 36B, 84R, 90A,B; **62**: 5R, 7R, 9R, 13B, 64AR, 66, 77B, 95A, 100; **63**: 14B, 16R, 20B, 26R, 58, 59R, 60, 65A, 81, 89B, 94, 102, 116; **64**: 30B, 31, 36C, 64B, 74, 75, 80R, 86AR, 106; **70**: 21, 44, 49B, 51BR, 86B; **71**: 2, 12R, 14B, 16R, 20B, 27, 30A, 36B, 76, 89B, 91R, 94, 95B, 97, 98, 102, 105, 115, 116; **72**: 39, 41, 48, 54, 59, 72, 76, 79, 82, 106; **73**: 89A, 93, 107. States A, B, C indicate multiple states 1, 2, 3 in Fig. 9.6. R indicates a reversal from a state.

'dichobunids' like *Diacodexis pakistanensis* and *Messelobunodon* was taken as primitive for this group, and its absence in otherwise related groups (e.g. merycoidodonts, protoceratids) was assumed to be a secondary loss often associated with other evident jaw modifications (82, 83, 86).

The resultant cladogram may not be the most parsimonious in its number of steps (573), but we judge it to be more likely in terms of its pattern and development of character states than that produced by the PAUP program. Like the latter, it shows a primary division corresponding essentially to the Bunodontia and Selenodontia; splitting of the paraphyletic Palaeodonta and Ancodonta between Bunodontia and Selenodontia; and placement of the paraphyletic Dichobunidae within the Selenodontia. In broad contrast, however, the Choeropotamidae are nested with the Cebochoeridae in the Bunodontia; the Selenodontia trichotomise into firstly the Ruminantia plus most of the Dichobunidae, secondly the Camelidae, Oromerycidae, Protoceratidae, Xiphodontidae, Anoplotheriidae, Cainotheriidae, Mixtotheriidae, Merycoidodontoidea, 'Uintan homacodontines', and *Dia-*

Fig. 9.8. Cladogram 1 manual—Artiodactyla. Characters at nodes. Letters explained in Fig. 9.7. **1**: characters listed in Chapter 8 and in introduction to this chapter; **2**: 29A,B, 36A, 37, 90A; **3**: 11, 26, 84, 91, 101, 113; **4**: 1, 11, 18, 19A,B, 24A, 25A,B, 26, 38, 45, 72, 81, 89A, 90B, 92, 95A, 102; **5**: 18, 21, 23, 55, 58, 64A; **6**: 23, 27, 47, 50; **7**: 31, 39, 42A, 47, 50, 68, 69, 79, 80, 82, 83, 94, 95B, 97, 98, 101, 107, 109, 110; **8**: 23, 24B, 25C, 28, 35, 63, 70, 71, 104, 112, 113; **9**: 48, 49A, 50, 95B, 110, 111, 115; **10**: 61, 76, 80, 89B, 91, 99, 100; **11**: 11, 24A, 26, 35, 38, 47, 48, 50, 72, 77A, 78, 80, 81, 86A; **12**: 5, 7, 12, 13A, 19A,B, 101; **13**: 12, 13A, 17, 18, 19A, 21, 28, 45, 79, 80, 83, 89A; **14**: 29BR, 35, 38, 45, 48, 79; **15**: 1, 4, 10, 11, 14A, 27R, 28, 31, 35, 38, 45, 85; **16**: 11, 14A; **17**: 1, 11, 14A, 16, 18, 19A,B, 24A, 28, 35, 38, 45, 53, 56, 72, 79, 80, 82, 87; **18**: 12, 13A, 17, 18, 19A, 21, 28, 45, 79, 80, 83, 89A; **19**: 4, 10, 11, 13B, 38; **20**: 22; **21**: 3, 9, 32, 34, 36B, 47R, 48, 49A, 72; **22**: 22, 26, 35; **23**: 6, 8, 13C, 14A, 16, 17R, 19B, 24B, 27R, 31, 48, 63, 72, 81, 82, 89B, 92, 95A, 102, 104; **24**: 34, 59, 64A, 80; **25**: 42A, 99; **26**: 31, 35, 36A, 37, 38; **27**: 7, 9, 13A, 19A, 23, 24A, 37, 48; **29**: 5, 11R, 18, 36A, 51A, 113, 114; **30**: 19B, 24B, 25A, 28, 29A, 32, 33, 45, 49A, 50, 58, 59R, 78, 81, 84, 89A; **31**: 3, 49A, 52A, 86A, 93, 110, 12, 16, 24B; **32**: 12, 15, 16, 19B, 24B, 28, 29A, 32, 67, 76, 77A, 78, 89A, 96, 100, 104, 107; **33**: 28, 29A, 32, 49A, 52A, 59R, 67, 77A, 78, 81, 84, 86A; **34**: 3R, 14A, 20A, 28, 29A, 38, 77A, 89B, 92, 94, 104, 107, 109; **35**: 15, 49B, 51B, 52B, 54, 57, 62, 98, 115; **36**: 5R, 7R, 9R, 13B, 64AR, 66, 77B, 95A, 100; **37**: 14B, 26R, 30A, 36B, 51B, 58, 59R, 60, 65A, 81, 84, 90A,B, 102, 116; **38**: 2, 12R, 14B, 20B, 30A, 36B, 51B, 76, 91R, 95B, 97, 98, 102, 105, 115, 116; **39**: 20B, 21, 44, 49B, 86B; **40**: 28, 29A, 38, 42A,B, 68, 76, 89B, 90A, 91R, 92, 94, 97, 103, 105, 109, 116; **41**: 19B, 32, 77A; **42**: 43; **43**: 28, 29A; **44**: 14A, 16, 30A, 40, 51B, 64AR, 72, 82, 86B; **45**: 3, 20A, 24B, 33, 45, 50, 58, 93, 107; **46**: 12, 14A,B, 20B, 38, 115; **47**: 40; **48**: 39, 41, 48, 54, 59, 72, 76, 79, 80, 82, 84R, 106; **49**: 89A, 91R, 93, 107; **50**: 11R, 13A–C, 18, 19A,B, 22, 24A, 25A, 29A, 48, 51A, 99, 100; **51**: 11R, 18, 19A,B, 23, 24A, 25A, 28, 29A, 99, 107; **52**: 11R, 18, 19A,B, 23, 24A, 25A, 28, 29A, 30A, 59, 64A, 80, 84R, 86A, 90A; **53**: 34, 59, 64A; **54**: 5, 7, 11R, 16, 23, 30A, 34, 42A, 46, 48, 59, 64A; **55**: 13A, 14A, 25A, 54, 57; **56**: 3, 9, 12, 19A, 20A, 24A,B, 28, 29A, 35R, 36B, 42B, 51A,B, 52A,B, 64B, 84R, 90A,B, 106; **57**: 3R, 9R, 14A, 18, 42B, AR, 45, 46R, 48R, 50, 51B, AR, 52B, AR, 58, 59R, 60, 65A, 77A, 87, 93, 95A,B, 99, 100, 104, 107, 113R, 114; **58**: 30B, 36C, 74, 75; **59**: 13A, 15, 54, 57, 104, 109, 110, 111; **60**: 13A, 14A, 18, 42B, AR, 49A,B, 53, 59R, 68, 73, 77A, 87, 89A, 91R, 92, 96, 100, 107, 114; **61**: 2, 3R, 5R, 7R, 9R, 13AR, 14B, 26R, 46R; **62**: 3R, 58, 93, 105.

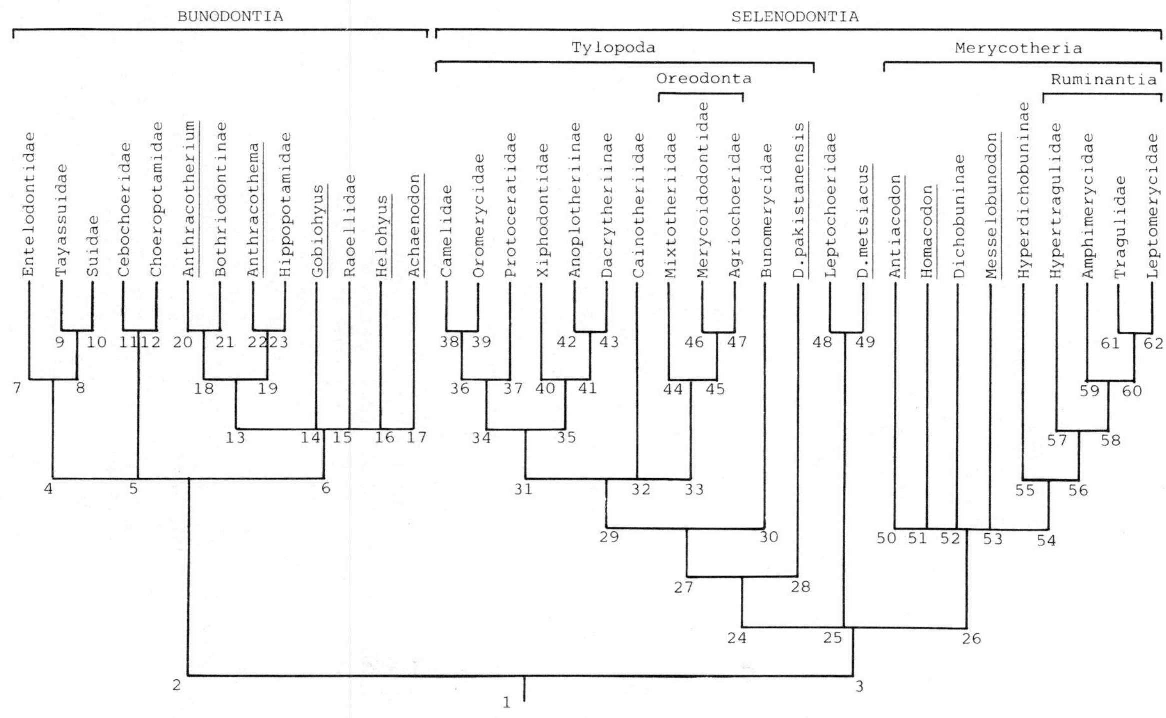

BUNODONTIA
SELENODONTIA
Tylopoda
Merycotheria
Oreodonta
Ruminantia
Entelodontidae
Tayassuidae
Suidae
Cebochoeridae
Choeropotamidae
Anthracotherium
Bothriodontinae
Anthracothema
Hippopotamidae
Gobiohyus
Raoellidae
Helohyus
Achaenodon
Camelidae
Oromerycidae
Protoceratidae
Xiphodontidae
Anoplotheriinae
Dacrytheriinae
Cainotheriidae
Mixtotheriidae
Merycoidodontidae
Agriochoeridae
Bunomerycidae
D.pakistanensis
Leptochoeridae
D.metsiacus
Antiacodon
Homacodon
Dichobuninae
Messelobunodon
Hyperdichobuninae
Hypertragulidae
Amphimerycidae
Tragulidae
Leptomerycidae
1
2
3
4
5
6
7
8
9
10
11
12
13
14
15
16
17
18
19
20
21
22
23
24
25
26
27
28
29
30
31
32
33
34
35
36
37
38
39
40
41
42
43
44
45
46
47
48
49
50
51
52
53
54
55
56
57
58
59
60
61
62

codexis pakistanensis, and thirdly the Leptochoeridae plus *Diacodexis metsiacus*.

The Bunodontia are defined mainly by broad notching of $\overline{M}$ paracristid (29B, Fig. 9.4) (unique, but needing a reversal in *Gobiohyus*), a mesial shift in $M_{1,2}$ entoconid to a position opposite the hypoconid, and distal displacement of the $M_{1,2}$ hypoconulid (the last two not present in the most primitive selenodonts). The strongest grouping in the Bunodontia is, not surprisingly, the Suidae plus Tayassuidae as Suoidea, and next their nesting with the Entelodontidae, which is based mainly on large $\underline{M}$ hypocone, high-ridged distal humerus intercondylar ridge, and reduction of cheek tooth cresting. The Cebochoeridae and Choeropotamidae are linked uniquely by a caniniform P^1, and within the Bunodontia also by caniniform P_1 and reduced lower canine. The Anthracotheriidae are supported uniquely by having the $\underline{M}$ paraconule extended distobuccally (17). Nesting the Hippopotamidae within them depends on reversal of this character, a high probability in view of the secondary loss of paraconule in advanced bothriodontines such as *Merycopotamus*. Anthracotheriidae and Hippopotamidae are also linked by a downturned dentary angle, various $\underline{M}$ mesostylar developments (the loop formed by the centrocrista occluding and disappearing to produce the isolated mesostyle of primitive hippopotamids: 13A–C), and obtuseness of the $\underline{M}$ distobuccal corner (reduced metastylar region). Hippopotamidae are specifically nested with *Anthracothema* by a lingual shift in the orientation of its $\underline{M}$ preparacrista and postmetacrista (paralleled by the Raoellidae). The Anthracotheriidae plus Hippopotamidae are linked in a pentachotomy with the largely primitive sister-taxa *Gobiohyus*, Raoellidae, *Helohyus*, and *Achaenodon* (Coombs and Coombs 1977*a*,*b*), by a molar enlargement gradient that especially affects $M^{2,3}$ (reversed in Raoellidae, as evidenced by retention of increased width versus distal contraction of M^3 outline), and loss of $P_{3,4}$ paraconids (paralleled by Cebochoeridae and Entelodontidae, and reversed in Bothriodontinae).

The Selenodontia are defined mainly by: dentary with dorsal hook, lateral border of distal humeral trochlea strongly ridged (both often lost in later groups), atlantoid facet extending half-way up neural arch (paralleled by Suoidea), and longitudinal ridge on distal trochlea of astragalus (paralleled by Entelodontidae and Choeropotamidae).

In the first selenodont clade, involving camels and relatives, authors' problems over relating xiphodonts to camels or to anoplotheres are reflected in the large amount of parallelism in this area. For instance, the strong linking of Camelidae to Oromerycidae by enlarged I^3, shape of the bulla (77B), distal fusion of radius and ulna, and proximal fusion of tibia and fibula, means that the neural canal exit position for the vertebrarterial canal evolved independently in camelids and protoceratids. It also evolved a third time in the Xiphodontidae because they and the Anoplotheriidae are linked by: premolariform upper canine, complex P_1^1, distally situated P_3 metaconid, and $P_{3,4}$ protocristid confluent with cristid obliqua. Likewise, reduction of the lateral metacarpals took place independently in Xiphodontidae and Camelidae. This mosaic of incongruent characters may nevertheless reflect a close relationship of Camelidae, Oromerycidae, and Protoceratidae (Cameloidea) with Xiphodontidae and Anoplotheriidae (Anoplotherioidea), but the link is not strong: incipient metatarsal reduction (110), incipient dorsal extension of hooked dentary angle, and a cylindrical distal humeral trochlea (shared with Merycoidodontoidea). Links with the Cainotheriidae are even weaker, based either on two characters shared with the Cameloidea plus Anoplotherioidea (paralleled in the Merycoidodontoidea), or a single more restricted character—large $\underline{M}$ paraconule shared with the Anoplotherioidea, but requiring parallelisms for three characters linking the other two groups. Clearly, resolution is low in this area and the preferred solution is a trichotomy. For an alternative scheme for cameloid/anoplotherioid relationships, see Prothero (in press).

Mixtotheriidae are supported as the sister-group to Merycoidodontoidea (Stehlin 1906–10; Thorpe 1937) by an $M^{2,3}$ enlargement gradient, incipient dorsal expansion of dentary angle, loss of hook on dentary angle (84R), reduction of coronoid process, and posterior deepening of the horizontal ramus, the last three shared with 'Uintan homacodontines' (Bunomerycidae). The number of characters linking Mixtotheriidae and Merycoidodontoidea with the latter nearly outweigh those linking them with Cameloidea, Anoplotherioidea, and Cainotheriidae. What is clear, however, is that the 'Uintan homacodontines' are linked with the foregoing groups rather than with *Homacodon*, reflecting West's (1984, p. 29) observation that 'typical *Homacodon* is markedly distinctive from its presumed Uintan derivatives'. This whole group is linked by a series of dental grade characters occurring elsewhere in the Artiodactyla, but not all together at one node. It is nested with *Diacodexis pakistanensis*, which shows only lingual displacement of $M_{1,2}$ hypoconulid, reduction of upper and lower canines and a flattening out of the ventral dentary concavity. Thus, support is given to Scott's (1940) and Romer's (1966) concepts of the Tylopoda to include Merycoidodontoidea, Anoplotherioidea, and Cainotheriidae, rather than to Webb and Taylor's (1980) Neoselenodontia, which excludes them from Cameloidea plus Ruminantia. The concept is extended slightly here to include the 'Uintan homacodontines' and, tentatively, *D. pakistanensis*.

Within the other major branch of selenodonts, the leptomerycids and tragulids (representing the basal stock of 'undoubted' ruminants) are nested with the Amphimerycidae on four unique characters (30B, 36C, 74, 75). The Hypertragulidae are the next closest group, having a variety of parallelisms with the Leptomerycidae and/or the Tragulidae. Together they are taken to constitute the Ruminantia, being linked mainly by: small spatulate lower canines resembling the incisors, the M entoconid slightly in front of hypoconid, and fused cuboid and navicular (paralleled by Leptochoeridae).

Of the paraphyletic Dichobunidae (including typical Homacodontinae but not 'Uintan homacodontines'), *Hyperdichobune* and *Mouillacitherium* form the sister-group to the Ruminantia, sharing with them especially a P_4 metaconid and a large P^3 parastyle (both reversed in Hypertragulidae—see above). The remaining dichobunids share with these two genera (Hyperdichobuninae nov.) plus Ruminantia a deeply notched M protocristid (31), a M entoconid that has moved mesially to opposite the hypoconid (36A), and loss of the M postcristid lingual arm (38). Reinsertion of *Microsus* with character 100 shared with *Antiacodon*, but lacking 24A, 35, and 38, and of *Diacodexis varleti* with character 34 shared with *Messelobunodon*, Hyperdichobuninae, and Ruminantia, but lacking 24A, 35, and 37, weakens the characters at node 26. It suggests instead the less parsimonious possibility of a relationship between *Antiacodon*, *Microsus*, *Homacodon*, *Diacodexis metsiacus*, and Leptochoeridae on characters 99 and 107 (although the latter's state is currently unknown in *Antiacodon* and Leptochoeridae). Otherwise, *D. metsiacus* and Leptochoeridae form the third unresolved branch of the Selenodontia trichotomy.

Data matrix 2

The PAUP analysis produced two maximally parsimonious cladograms with 86 steps for a selection of 16 ruminant taxa. The 41 characters were mostly long familiar in the taxonomy of the suborder. Both cladograms were largely identical and had a consistency index of 0.551.

The taxa above *Leptomeryx* plus *Bachitherium* (Fig. 9.10) constitute the infraorder Pecora. The first five of them are potential members of a superfamily Cervoidea, here paraphyletic, and they are succeeded by the three families or superfamilies Giraffoidea (*Climacoceras* plus Giraffidae), Bovidae, and Antilocapridae (Merycodontinae plus *Antilocapra*). Both cladograms selected as unique and important synapomorphies: loss of P_1 (19) above *Palaeomeryx* plus Dromomerycini, fusion of metatarsals III and IV (37) above *Archaeomeryx*, anteriorly complete distal keels on the metapodials (39) and fusion of metacarpals III and IV (40) above *Leptomeryx* plus *Bachitherium*, and enlargement of

Archaeomeryx	90000000901000000001001000000000100100000
Bachitherium	90000000999901010001110010111111290919001
Leptomeryx	90000000001001100001110010011110110111000
Tragulidae	00000000000102010100010000001101100011000
Gelocus	90000090999909019000000000001010011011100 0
Dremotherium	00000000009002019010101010111100990919111
Moschus	00000000101012010010201000100111220112111
Merycodontinae	91099001019012100010201100110190990911119
Antilocapra	11002001011012100010201100110110220112111
Palaeomeryx	01099010999919010000101011011100990112110
Dromomerycini	11099010009012010000101110011100990112110
Hydropotes	10000000111012010010201100110100220112111
Cervidae	11000101101002010010101011011100220112111
Climacoceras	91199001999919991010201100110100999111119
Giraffidae	01110000011012101010201110110100221111111
Bovidae	01011000211012100010201110110110220111111

Fig. 9.9. Data matrix 2—Ruminantia (except Hypertragulidae and Amphimerycidae). 9=missing state. Character marking is from left to right.

upper canine (16) above *Archaeomeryx*. The last, however, has to be followed by two reversals. Of the 86 steps on the cladograms, 56 (=66%) were parallels.

The second cladogram differed from the first in that Cervidae interchanged with *Dremotherium*. This was based on characters 9, 27, 33, and 34, but the last two were computer-assigned states for the unknown (to us) axis vertebra of *Dremotherium*. For character 9, size of foramen ovale, *Moschus*, *Hydropotes*, Cervidae, and Bovidae are all advanced on *Dremotherium*, Dromomerycini, Giraffidae, and Antilocapridae. We prefer to take the latter cluster as genuinely primitive and to favour cladogram 1, which places *Dremotherium* above Cervidae and has four independent evolutions of the larger foramen ovale. Character 27 has already been discussed. It is equally possible that *Dremotherium* is more advanced than *Palaeomeryx* and Cervidae, as on cladogram 1, or that it advanced in parallel to most other Pecora, as on cladogram 2. The remaining discussion will concern cladogram 1 (Fig. 9.10).

The placing of the five cervoids below Giraffoidea, Bovidae, and Antilocapridae means that the reduced upper canine of the latter three groups must be a reversal from a formerly enlarged state. Some support is given to this conception by Frick (1937, fig. 28B), who shows a *Meryceros* with a significantly-sized attenuated upper canine that could well be descended from an earlier enlarged condition. However, it is noteworthy that large upper canines have never been cited in association with finds of recognized early bovids and giraffoids; Gentry (1970, p. 249) noted that alveoli for a male Middle Miocene boselaphine antelope in early to middle maturity were already minute. A second and more unsatisfactory feature of the placing of cervoids on the cladogram is that the metatarsal gully (38) passes to a bridged condition below *Palaeomeryx*, reverses to open below giraffoids, and finally re-reverses to bridged in *Antilocapra*.

Bovidae are linked with Antilocapridae on the basis of horn cores with sheaths (5—assumed by the computer to be present in Merycodontinae) and the advanced selenodonty of character 31. This decision overrides characters 4 and 25, which would have linked bovids with giraffids but which now evolve in parallel, the latter after a reversal. Character 4 concerns horn core growth. In extant *Giraffa* and *Okapia* there is basal growth by an independent ossicone, which spreads over and fuses with the frontals and parietal of the skull roof. Geraads (1986) observes that basal growth could not have occurred in fossils like sivatheres with branched horns. It would be even less conceivable in *Climacoceras* (Hamilton 1978, fig. 21). In the early ontogeny of bovid horn cores, a tiny bone may exist independently of the upgrowth of the frontals (Bubenik 1983, fig. 12) but it is a rare occurrence (Duerst 1926, p. 23; Gadow 1902, p. 212), even in the minority of bovid species that have been investigated. Given these reservations, it is as well that character 4 has not been favoured as a bovid/giraffid synapomorphy. Growth of bovid horn cores takes place at their tips (Gadow 1902, p. 202), but of the sheaths at their bases.

The cladogram adds to the stock of autapomorphies within the Giraffoidea by using MacInnes' (1936, p. 525, fig. 3) observations of a thick cortical layer of compact bone in *Climacoceras* horn cores as a resemblance to later giraffids (3). This cladogram also invokes three independent evolutions of horns, once for the giraffoid–bovid–antilocaprid group, once in cervids, and once in *Palaeomeryx* plus dromomerycines.

The analysis of cervoid taxa is a little problematical. It is convenient and traditional to think of the Family Cervidae as having horns, called antlers, which are branched and seasonally shed. However, the living *Hydropotes* has long been judged to be an antlerless deer, and *Moschus*, the musk deer of Asia, is even more enigmatic. Flower (1875) painstakingly discussed *Moschus* and, a century later, Webb and Taylor (1980) commented on it anew and resurrected Moschidae as a fifth pecoran family to embrace some fossil forms as well. *Moschus* has no antlers (2) and does not possess the double lacrimal orifice of deer, dromomerycines (Leinders 1984, p. 41), and *Antilocapra* (1). On the other hand, its distal metatarsal gully is bridged (38), and there is a large upper canine in males, as in primitive antlered deer such as *Muntiacus* and *Elaphodus* (16). As to *Hydropotes*, Lister (1984, p. 207) has noted that if it is a cervid and its antlerless condition is primitive, then one or more among the earliest cervid species will also have to be antlerless. Early cervids without antlers had also been envisaged by Flower (1875, p. 183). The assertion or assumption of the cervid status of *Hydropotes* goes back to Garrod (1877*c*, p. 791), who had contrasted a new-born specimen with *Moschus* in:

(1) its fairly convoluted brain;

(2) quadruplicate psalterium (=omasum);

(3) no ilio-caecal gland;

(4) no gall bladder;

(5) fewer coils of the colon;

(6) few uterine cotyledons.

It may be noted that (2) is indeed like deer but of doubtful polarity; (3), (5) (Garrod 1877*b*, p. 288), and (6) are more specialized in *Moschus*; and (1) and (4) occur in other higher pecorans than deer. Thus, the cervid status of *Hydropotes* could be questioned.

Furthermore, a variety of horned or antlered, more or less cervid-like, fossils are found in the Eurasian Early and Middle Miocene, e.g. *Stephanocemas* and *Procervulus*, some with and others without deciduous antlers. Probably related to these is *Palaeomeryx*, of which a Chinese skull was described in outline by Qiu *et al.* (1985), and the North American Dromomerycini. *Hoplitomeryx* in the Hoplitomerycidae of Leinders (1984) will deserve further comment once its dentition and postcranial features have been documented. All these forms often, but not invariably, show two lacrimal orifices (1), forking of antlers (8), enlarged upper canines (16), and a bridged metatarsal gully (38). None has yet been alleged to possess a bi-lobed lower canine like giraffoids.

In earlier Miocene and Oligocene horizons are found antlerless cervoids like *Dremotherium* in Europe and the Blastomerycini in North America, all placed in the Moschidae by Webb and Taylor (1980).

This issue is too complicated to be decided here. The five taxa selected to represent this cervoid complex have come out adjacent on the cladogram. The paradoxical position of both *Moschus* and *Hydropotes* above Cervidae needs explanation. 'Cervidae' for this cladogram is taken as genera other than *Hydropotes*. Where the state of characters was known in earlier cervids such as the Miocene *Dicrocerus*, we have used them. Character states of *Moschus* and *Hydropotes*, on the other hand, were taken entirely from the extant species. Thus, hypsodonty of cheek teeth (13), further decline in saliency of hypoconids on P_3 and P_2 (21), and secondary disappearance of upper molar entostyles and of lower molar *Palaeomeryx*-folds (25, 29) have all advanced on these two lineages since the Miocene. On the cladogram, *Hydropotes* is the most advanced cervoid or, conceivably, the most primitive of the pecorans above cervoids. It is linked with the succeeding pecorans by the bovid-like condition of the tympanohyal pit (10) and by loss of the lingual cingulum on its upper molars (24).

Palaeomeryx and Dromomerycini are associated on the cladogram by the possession of horns (2), an occipital horn (7), and hypsodont cheek teeth (13). Together with higher taxa, they are set off by one of the better concentrations of advanced characters on the cladogram: final disappearance of upper incisors (14), appearance of a transverse entoconid ridge on P_4 (23), bridging of the metatarsal gully (38), metapodial distal keels complete anteriorly (39), and fusion of metacarpals III and IV (40). These characters are here acting to define something close to the old infraorder Pecora, and 39 and 40 are two of the rare unique synapomorphies.

Among living ruminants, the primitive sister-group of Pecora is certainly the Tragulidae (chevrotains). They share with Pecora, and contrast with camels in, for example, the loss of upper incisors, an incisiform lower canine, fused magnumtrapezoid in the carpus, and fused naviculocuboid in the tarsus. They are primitive in shorter legs, lack of omasum, and cancellous auditory bulla (but hollow in *Tragulus meminna*). However, the Leptomerycidae are an alternative fossil sister-group and, in fact, the cladogram has favoured them above Tragulidae. They share many character states with the tragulids but their teeth look more fully selenodont, the bulla is hollowed, P_1 is caniniform, and they lack the enlarged male upper canine of the tragulids. On the cladogram

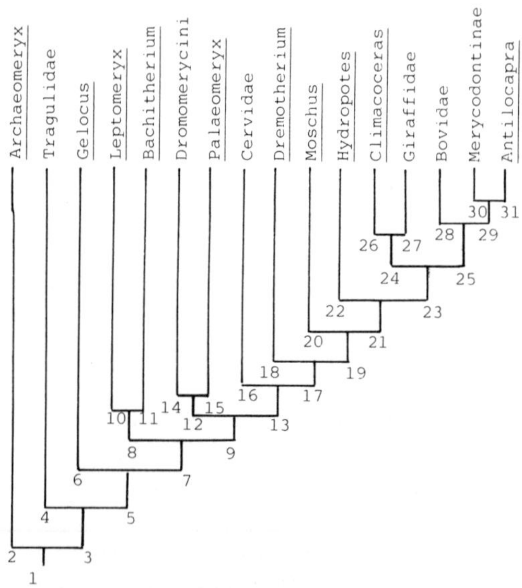

Fig. 9.10. PAUP cladogram for ruminants. Nodes 1–31 are listed here with their character state changes. **1**: 33A; **2**: 11, 20, 23, 36; **3**: 14A, 16, 30, 37, 38A; **4**: 12, 14B, 18, 22, 29, 32; **5**: 28, 34A, 36; **6**: –; **7**: 11, 21A, 25, 29; **8**: 20, 22, 31; **9**: 14A, 23, 38B, 39, 40; **10**: 15, 16R; **11**: 27, 32, 33B, 41; **12**: 2, 7, 13; **13**: 19, 33B, 34B, 41; **14**: 1, 24; **15**: 26; **16**: 1, 2, 6, 8, 9, 26; **17**: 27; **18**: –; **19**: 13, 21B, 25R, 29R; **20**: 9, 28R, 31, 32; **21**: 10, 24; **22**: 1, 9; **23**: 2, 15, 16R, 38A; **24**: 3, 17; **25**: 5A, 31; **26**: 8; **27**: 4, 25, 35; **28**: 4, 9B, 25; **29**: 8; **30**: –; **31**: 1, 5B, 38B.

they have been selected on the strength of the hollow bulla (11), a less salient hypoconid on $P_{2,3}$ (21), entostyles on upper molars (25), and a *Palaeomeryx*-fold on the lower molars (29).

Archaeomeryx is well separated from Leptomerycidae on the cladogram, although this necessitates parallel evolution of hollowed bullae (11), caniniform P_1 (20), parallel proximal and distal axes of the astragalus (36), and perhaps a transverse entoconid ridge on the P_4 (23). The outstandingly primitive backward position of the metaconid on its lower molars (30) and the separate metatarsals III and IV (37) are responsible for this outcome.

Gelocus occupies a position below Leptomerycidae, to which it is primitive in the continued separation of metacarpals III and IV (40), in tooth characters 21, 25, and 29, and by the computer's assignment of a non-hollowed interior to its bulla in the interests of parsimony (11). However, it is still held to be advanced over tragulids by the metastylids on the lower molars (28), the state of its odontoid process (34), and the straight astragalus (36). Others have viewed it as the most primitive pecoran (Webb and Taylor 1980, fig. 10), but on our cladogram, Leptomerycidae would have to be advanced to this status before *Gelocus* could be admitted as well.

As an alternative to the most-parsimonious cladogram and its attendant problems, we would prefer one in which the Cervoidea (perhaps excluding *Hydropotes*) was monophyletic. *Gelocus* would represent the most primitive pecorans preceding the split between the cervoids and the other higher pecoran families. Below *Gelocus* would come a clade for *Archaeomeryx*, *Leptomeryx*, and *Bachitherium*, and then Tragulidae as a still lower clade, much as in Webb and Taylor (1980). Such a cladogram would require even more parallel evolution: perhaps 84 parallels out of a minimum of 102 steps. In the fossil record as known at present, merycodontines in North America, and possibly giraffoids in Africa, occur as early as the first antlered/horned cervoids in Europe, suggesting great scope for parallel evolution from a primitive pecoran group of low diversity. Another version of ruminant phylogeny is given in Chapter 10 of this volume.

Taxonomic modifications

Changes have been kept to a minimum and restricted to the more robust groupings. The Bunodontia and Selenodontia are resurrected as unranked groups between order and suborder. Infraorder Suiformes is modified for Suoidea plus Entelodontidae. Suborder Tylopoda is extended to include Anoplotherioidea, Cainotheriidae, Mixtotheriidae,

Merycoidodontoidea, Bunomerycidae nov., and *Diacodexis pakistanensis*. Infraorder Oreodonta is extended to encompass Mixtotheriidae. Superfamily Hippopotamoidea is extended to include 'Anthracotheriidae', and superfamily Choeropotamoidea (new rank) is extended to include Cebochoeridae. Oromerycidae and Protoceratidae are included in the Cameloidea, whilst the Xiphodontidae are included in the Anoplotherioidea. It is probably useful to retain the paraphyletic families 'Dichobunidae' and 'Anthracotheriidae', at least provisionally. New taxa are defined as follows, and references to authors of taxa in this section can be found in Gazin (1955) and Hooker (1986).

Family 'Dichobunidae' Gill 1872
Subfamily **Hyperdichobuninae** nov.

Type genus: *Hyperdichobune* Stehlin 1910.

Included genus: *Mouillacitherium* Filhol 1882.

Range: M–U Eocene, Europe.

Diagnosis: Artiodactyls with characters 5, 7, 13A, 14A, 16, 23, 25A, 26, 30A, 31, 34, 35, 36A, 37, 38, 42A, 46, 48, 54, 57, 59, 84 of data matrix 1.

Note that the nominate subfamily Dichobuninae is restricted here to the genera *Dichobune* Cuvier 1822, *Meniscodon* Rütimeyer 1888 and *Metriotherium* Filhol 1882.

Family **Bunomerycidae** nov.

Type genus: *Bunomeryx* Wortman 1898.

Included genera: *Hylomeryx* Peterson 1919, *Mesomeryx* Peterson 1919, *Mytonomeryx* Gazin 1955, and *Pentacemylus* Peterson 1931.

Range: M–U Eocene, North America.

Diagnosis: Artiodactyls with characters 7, 9, 11, 13A, 19B, 23, 24B, 25A, 26, 28, 29A, 32, 33, 34, 37, 45, 48, 49A, 50, 58, 64A, 78, 80, 81, 89A of data matrix 1.

Suborder **Merycotheria** nov.

Included taxa: infraorder Ruminantia plus 'Dichobunidae' *s.s.* (i.e. minus Bunomerycidae, *Diacodexis pakistanensis*, and probably other *Diacodexis* spp.).

Range: L. Eocene–Recent; worldwide today, at first restricted to Europe and North America.

Diagnosis: Artiodactyls with characters 31, 36, 37, 38 of data matrix 1. Primitively, the state of character 36 is A and character 35 is present. Many taxa have more advanced states of 36 and have reversed 35.

Acknowledgements

We thank N. Goldman, C. J. Humphries, and A. B. Smith for much help with computer work, and P. M. Butler and D. R. Prothero for helpful discussion.

References

Adams, E. N. (1972). Consensus techniques and the comparison of taxonomic trees. *Syst. Zool.* **21**, 390–7.

Bergounioux, F.-M. and Crouzel, F. (1967). Suoides de Sansan. *Ann. Paléontol. (Vertébr.)* **53**, 1–24.

Black, C. C. (1978). Paleontology and geology of the Badwater Creek Area, central Wyoming. Part 14: The artiodactyls. *Ann. Carnegie Mus.* **47**, 223–59.

Bouvrain, G. and Geraads, D. (1985). Un squelette complet de *Bachitherium* (Artiodactyla, Mammalia) de l'Oligocène de Céreste (Alpes de Haute-Provence). Remarques sur la systématique des Ruminants primitifs. *C. R. Acad. Sci., Sér. D.* **300**(2), 75–8.

Bubenik, A. B. (1983). Taxonomy of Pecora in relation to morphophysiology of their cranial appendices. In *Antler Development in Cervidae* (ed. R. D. Brown), pp. 163–85. Caesar Kleberg Wildlife Research Institute, Kingsville, Texas.

Butler, P. M. (1951). The milk-molars of Perissodactyla, with remarks on molar occlusion. *Proc. Zool. Soc. London* **121**, 777–817.

Colbert, E. H. (1935). Distributional and phylogenetic studies on Indian fossil mammals. IV. The phylogeny of the Indian Suidae and the origin of the Hippopotamidae. *Am. Mus. Novit.* **799**, 1–24.

Colbert, E. H. (1938). Fossil mammals from Burma in the American Museum of Natural History. *Bull. Am. Mus. Nat. Hist.* **74**, 255–436.

Colbert, E. H. (1941). The osteology and relationships of *Archaeomeryx*, an ancestral ruminant. *Am. Mus. Novit.* **1135**, 1–24.

Coombs, M. C. and Coombs, W. P. (1977*a*). The dentition of *Gobiohyus* and a re-evaluation of the Helohyidae. *J. Mammal.* **58**, 291–308.

Coombs, W. P. and Coombs, M. C. (1977*b*). The origin of anthracotheres. *Neues Jahrb. Geol. Paläontol. Monatsh.*, **1977**, 584–99.

Coombs, M. C. and Coombs, W. P. (1982). Anatomy of the ear region of four Eocene artiodactyls: *Gobiohyus, ?Helohyus, Diacodexis* and *Homacodon*. *J. Vertebr. Paleontol.* **2**, 219–36.

Cuvier, G. (1822). *Recherches sur les ossemens fossiles* (new edn.), Vol. 2, pp. 648; Vol. 3, pp. 412. G. Dufour et E. d'Ocagne, Paris.

de Blainville, H. M. D. (1839–64). *Ostéographie, ou description iconographique composée du squelette et du système dentaire de mammifères récents et fossiles pour servir de base à la zoologie ou à la géologie*, 8 vols. Baillière, Paris.

Dechaseaux, C. (1965). Artiodactyles des Phosphorites du Quercy, 1. Étude sur le genre *Dichodon*. *Ann. Paleóntol. (Vertébr.)* **51**, 91–108.

Dechaseaux, C. (1967). Artiodactyles des Phosphorites du Quercy, 2. Étude sur le genre *Xiphodon*. *Ann. Paleóntol. (Vertébr.)* **53**, 27–47.
Dechaseaux, C. (1974). Artiodactyles primitifs des Phosphorites du Quercy. *Ann. Paleóntol. (Vertébr.)* **60**, 58–100.
Depéret, C. (1917). Monographie de la faune de mammifères fossiles du Ludien inférieur d'Euzet-les-Bains (Gard.). *Ann. Univ. Lyon* **40**(1), 1–290.
Domning, D. P., Ray, C. E., and McKenna, M. C. (1986). Two new Oligocene desmostylians and a discussion of tethytherian systematics. *Smithson. Contrib. Paleobiol.* **59**, 1–56.
Duerst, J. U. (1926). Das Horn der Cavicornia. *Denkschr. Schweiz. Naturforsch. Ges.* **63**(1), 1–180.
Edwards, P. (1976). The subfamily Leptochoerinae (Artiodactyla, Dichobunidae) of North America (Oligocene). *Univ. Wyo. Contrib. Geol.* **14**, 99–113.
Filhol, H. (1877). Recherches sur les phosphorites du Quercy, Part 2. *Ann. Sci. Géol. Paris* **8**(1), 1–340.
Filhol, H. (1881). Étude des mammifères fossiles de Saint-Gérand le Puy (Allier), Part 2. *Ann. Sci. Géol. Paris* **11**(1), 1–86.
Filhol, H. (1882). Étude des mammifères fossiles de Ronzon (Haute-Loire). *Ann. Sci. Géol. Paris* **12**(3), 1–270.
Flower, W. H. (1875). On the structure and affinities of the musk deer (*Moschus moschiferus* Linn.). *Proc. Zool. Soc. London* **1875**, 159–90.
Flower, W. H. and Lydekker, R. (1891). *An Introduction to the Study of Mammals, Living and Extinct.* A. & C. Black, London.
Franzen, J.-L. (1981). Das erste Skelett eines Dichobuniden (Mammalia Artiodactyla) geborgen aus mitteleozänen Ölschiefern der 'Grube Messel' bei Darmstadt (Deutschland, S-Hessen). *Senckenbergiana Lethaea* **61**, 299–353.
Franzen, J.-L. (1983). Ein zweites Skelett von *Messelobunodon* (Mammalia Artiodactyla, Dichobunidae) aus der 'Grube Messel' bei Darmstadt (Deutschland, S-Hessen). *Senckenbergiana Lethaea* **64**, 403–45.
Franzen, J.-L. and Krumbiegel, G. (1980). *Messelobunodon ceciliensis* n.sp (Mammalia, Artiodactyla)—ein neuer Dichobunide aus der mitteleozäner Fauna des Geiseltales bei Halle (DDR). *Z. Geol. Wiss.* **8**, 1585–92.
Frick, C. (1937). Horned ruminants of North America. *Bull. Am. Mus. Nat. Hist.* **69**, 1–669.
Furlong, E. L. (1927). The occurrence and phylogenetic status of *Merycodus* from the Mohave desert Tertiary. *Univ. Calif. Publ. Geol. Sci.* **17**, 145–76.
Gadow, H. (1902). The evolution of horns and antlers. *Proc. Zool. Soc. London* **1902**, 206–22.
Garrod, A. H. (1877*a*). Notes on the visceral anatomy and osteology of the ruminants. *Proc. Zool. Soc. London* **1877**, 2–18.
Garrod, A. H. (1877*b*). Notes on the anatomy of the musk-deer (*Moschus moschiferus*). *Proc. Zool. Soc. London* **1877**, 287–92.
Garrod, A. H. (1877*c*). Notes on the anatomy of the Chinese water-deer (*Hydropotes inermis*). *Proc. Zool. Soc. London* **1877**, 789–93.

Gazin, C. L. (1955). A review of the Upper Eocene Artiodactyla of North America. *Smithson. Misc. Collect.* **128**, 1–96.

Gentry, A. W. (1970). The Bovidae (Mammalia) of the Fort Ternan fossil fauna. In *Fossil Vertebrates of Africa* (eds. L. S. B. Leakey and R. J. G. Savage), Vol. 2, pp. 243–323. Academic Press, London.

Geraads, D. (1986). Remarques sur la systématique et la phylogénie des Giraffidae (Artiodactyla, Mammalia). *Géobios* **19**, 465–77.

Golz, D. J. (1976). Eocene Artiodactyla of southern California. *Sci. Bull. Nat. Hist. Mus. Los Angeles County* **26**, 1–85.

Hamilton, W. R. (1978). Fossil giraffes from the Miocene of Africa and a revision of the phylogeny of the Giraffoidea. *Philos. Trans. R. Soc. London, Ser. B* **283**, 165–229.

Heintz, E. (1963). Les caractères distinctifs entre metatarses de Cervidae et Bovidae actuels et fossiles. *Mammalia* **27**, 200–9.

Heller, F. (1934). *Anthracobunodon weigelti* n.gen. et n.sp., ein Artiodactyle aus dem Mitteleozän des Geiseltales bei Halle a.S. *Paläontol. Z.* **16**, 247–63.

Hendey, Q. B. (1978). Preliminary report on the Miocene vertebrates from Arrisdrift, South West Africa. *Ann. S. Afr. Mus.* **76**, 1–41.

Hooker, J. J. (1986). Mammals from the Bartonian (middle/late Eocene) of the Hampshire Basin, southern England. *Bull. Br. Mus. (Nat. Hist.)*, London (Geol.) **39**, 191–478.

Hürzeler, J. (1936). Osteologie und Odontologie der Cainotheriden. *Abh. Schweiz. Paläontol. Ges.* **58**, **59**, 1–112.

Kermack, K. A., Lee, A. J., Lees, P. M. and Mussett, F. (1987). A new docodont from the Forest Marble. *Zool. J. Linn. Soc.* **89**, 1–39.

Kowalevsky, W. (1873). On the osteology of the Hyopotamidae. *Philos. Trans. R. Soc. London* **163**, 19–94.

Kowalevsky, W. (1874). Monographie der Gattung *Anthracotherium* Cuv. und Versuch einer naturlichen Classification der fossilen Huftiere (2). *Palaeontographica* **2**, 211–90.

Krishtalka, L. and Stucky, R. K. (1985). Revision of the Wind River faunas, Early Eocene of central Wyoming. Part 7. Revision of *Diacodexis* (Mammalia, Artiodactyla). *Ann. Carnegie Mus.* **54**(14), 413–86.

Leinders, J. (1984). Hoplitomerycidae fam.nov. (Ruminantia, Mammalia) from Neogene fissure fillings in Gargano (Italy). Part 1: The cranial osteology of *Hoplitomeryx* gen.nov. and a discussion on the classification of pecoran families. *Scr. geol.* **70**, 1–68.

Lister, A. M. (1984). Evolutionary and ecological origins of British deer. *Proc. R. Soc. Edinburgh, B* **82**, 205–29.

Macdonald, J. R. (1955). The Leptochoeridae. *J. Paleontol.* **29**, 439–59.

MacInnes, D. (1936). A new genus of fossil deer from the Miocene of Africa. *J. Linn. Soc., Zool.* **39**, 521–30.

Matthew, W. D. (1908). Osteology of *Blastomeryx* and phylogeny of the American Cervidae. *Bull. Am. Mus. Nat. Hist.* **24**, 535–62.

Matthew, W. D. (1929). Reclassification of the artiodactyl families. *Bull. Geol. Soc. Am.* **40**, 403–8.

Mayet, L. (1908). Étude des mammifères Miocènes des Sables de l'Orleanais et des faluns de la Touraine. *Annl. Univ. Lyon* **24**, 1–336.

Owen, R. (1848). Description of teeth and portions of jaws of two extinct Anthracotherioid quadrupeds (*Hyopotamus vectianus* and *Hyop. bovinus*) discovered by the Marchioness of Hastings in the Eocene deposits on the N.W. coast of the Isle of Wight: with an attempt to develope Cuvier's idea of the classification of pachyderms by the number of their toes. *Q. J. Geol. Soc. London* **4**, 103–41.

Patton, T. H. and Taylor, B. E. (1971). The Synthetoceratinae (Mammalia, Tylopoda, Protoceratidae). *Bull. Am. Mus. Nat. Hist.* **145**, 119–218.

Patton, T. H. and Taylor, B. E. (1973). The Protoceratinae (Mammalia, Tylopoda, Protoceratidae) and the systematics of the Protoceratidae. *Bull. Am. Mus. Nat. Hist.* **150**, 347–414.

Pearson, H. S. (1927). On the skulls of early Tertiary Suidae, together with an account of the otic region in some other primitive Artiodactyla. *Philos. Trans. R. Soc. London, B* **215**, 389–460.

Peterson, O. A. (1919). Report upon the material discovered in the Upper Eocene of the Uinta Basin by Earl Douglas in the years 1908–1909, and by O. A. Peterson in 1912. *Ann. Carnegie Mus.* **12**, 40–168.

Pickford, M. (1983). On the origins of Hippopotamidae together with descriptions of two new species, a new genus and a new subfamily from the Miocene of Kenya. *Géobios* **16**, 193–217.

Pickford, M. (1984). A revision of the Sanitheriidae, a new family of Suiformes (Mammalia). *Géobios* **17**, 133–54.

Pilgrim, G. E. (1928). The Artiodactyla of the Eocene of Burma. *Mem. Geol. Surv. India, Palaeontol. Indica* **13**, 1–39.

Pilgrim, G. E. (1941). The dispersal of the Artiodactyla. *Biol. Rev.* **16**, 134–63.

Pilgrim, G. E. and Cotter, G. de P. (1916). Some newly discovered Eocene mammals from Burma. *Rec. Geol. Surv. India* **47**, 42–77.

Prothero, D. R. (In press). In *The Tertiary Mammals of North America* (eds. C. Janis, L. Jacobs, and K. M. Scott). Cambridge University Press, New York.

Qiu, Z., Yan, D., Jia, H., and Sun, B. (1985). Preliminary observations on the newly found skeletons of *Palaeomoeryx* from Shanwang, Shandong. *Vertebr. Palasiat.* **23**, 173–95.

Riedl, R. (1978). *Order in Living Organisms: A Systems Analysis of Evolution*, translated by R. P. S. Jefferies. Wiley, Chichester.

Romer, A. S. (1966). *Vertebrate Paleontology*, 3rd edn. University of Chicago Press, Chicago.

Rose, K. D. (1982). Skeleton of *Diacodexis*, oldest known artiodactyl. *Science N.Y.* **216**, 621–3.

Rose, K. D. (1985). Comparative osteology of north American dichobunid artiodactyls. *J. Paleontol.* **59**, 1203–26.

Rose, K. D. (1987). Climbing adaptations in the Early Eocene mammal *Chriacus* and the origin of Artiodactyla. *Science N.Y.* **236**, 314–16.

Russell, D. E., Thewissen, J. G. M., and Sigogneau-Russell, D. (1983). A new dichobunid artiodactyl (Mammalia) from the Eocene of north-west Pakistan. Part 2: cranial osteology. *Proc. K. ned. Akad. Wet* **86B**, 285–300.

Sahni, A., Bhatia, S. B., Hartenberger, J.-L., Jaeger, J.-J., Kumar, K., Sudre, J. and Vianey-Liaud, M. (1981). Vertebrates from the Subathu formation and comments on the biogeography of Indian subcontinent during the early Paleogene. *Bull. Soc. Géol. Fr. Sér. 7*, **23**, 689–95.

Schaeffer, B. (1948). The origin of a mammalian ordinal character. *Evolution* **2**, 164–75.

Schlosser, M. (1886). Beiträge zur Kenntniss der Stammesgeschichte der Hufthiere und Versuch einer Systematik der Paar- und Unpaarhufer. *Morphol. Jahrb.* **12**, 1–136.

Schlosser, M. (1902). Beiträge zur Kenntniss der Säugethierreste aus den Süddeutschen Bohnerzen. *Geol. Paläontol. Abh.* **5**, 117–258.

Scott, W. B. (1940). The mammalian fauna of the White River Oligocene. Part 4. Artiodactyla. *Trans. Am. Philos. Soc.* **28**, 363–746.

Sigogneau, D. (1968). Le genre *Dremotherium* (Cervoidea). *Ann. Paleóntol. (Vertébr.)* **54**, 39–100.

Simpson, G. G. (1945). The principles of classification and a classification of mammals. *Bull. Am. Mus. Nat. Hist.* **85**, 1–350.

Sinclair, W. J. (1914). A revision of the bunodont Artiodactyla of the Middle and Lower Eocene of north America. *Bull. Am. Mus. Nat. Hist.* **33**, 267–95.

Stehlin, H. G. (1899–1900). Ueber die Geschichte des Suiden-Gebisses. *Abh. Schweiz. Paläontol. Ges.* **26**, **27**, 1–527.

Stehlin, H. G. (1906–10). Die Säugetiere des schweizerischen Eocäns. Critischer Catalog der Materialen 4, 5 & 6. *Abh. Schweiz. Paläontol. Ges.* **33**, **35**, **36**, 597–1164.

Stehlin, H. G. (1937). Bemerkungen über die miocänen Hirschgenera *Stephanocemas* und *Lagomeryx*. *Verh. Naturforsch. Ges. Basel* **48**, 193–214.

Stehlin, H. G. (1939). *Dicroceros elegans* Lartet und sein Geweihwechsel. *Eclogae Geol. Helv.* **32**, 162–79.

Stock, C. (1934). A hypertragulid from the Sespe uppermost Eocene, California. *Proc. Natl. Acad. Sci. U.S.A.* **20**, 625–9.

Swofford, D. L. (1984). *PAUP—Phylogenetic analysis using parsimony*, Program and user's manual. Illinois Natural History Survey, Champaign.

Sudre, J. (1977). L'évolution du genre *Robiacina* Sudre 1969, et l'origine des Cainotheriidae. *Géobios, Mém. Spéc.* **1**, 213–31.

Sudre, J. (1978). Les artiodactyles de l'Éocène moyen et supérieur d'Europe occidentale. *Mém. Trav. Inst. Montpellier* **7**, 1–229.

Sudre, J. (1984). *Cryptomeryx* Schlosser 1886, tragulidé de l'Oligocène d'Europe; relations du genre et considérations sur l'origine des ruminants. *Palaeovertebrata* **14**, 1–31.

Sudre, J., Russell, D. E., Louis, P. and Savage, D. E. (1983). Les artiodactyles de l'Éocène inférieur d'Europe, I & II. *Bull. Mus. Natl. Hist. Nat. Paris Sér. 4*, **5C**, 281–333; 339–65.

Thewissen, J. G. M., Russell, D. E., Gingerich, P. D., and Hussain, S. T. (1983). A new dichobunid artiodactyl (Mammalia) from the Eocene of north-west Pakistan. *Proc. K. Ned. Akad. Wet.* **86B**, 153–80.

Thorpe, M. H. (1937). The Merycoidodontidae. An extinct group of ruminant mammals. *Mem. Peabody Mus. Nat. Hist.* **3**, 1–428.

Tobien, H. (1985). Zur Osteologie von *Masillabune* (Mammalia, Artiodactyla, Haplobunodontidae) aus dem Mitteleozän der Fossilfundstatte Messel bei Darmstadt (S-Hessen, Bundesrepublik Deutschland). *Geol. Jahrb. Hessen* **113**, 5–58.

Van Valen, L. (1971). Toward the origin of artiodactyls. *Evolution* **25**, 523–9.

Viret, J. (1961). Artiodactyla. In *Traité de Paléontologie* (ed. J. Piveteau), Vol. 6, pp. 887–1084. Masson, Paris.

Webb, S. D. and Taylor, B. (1980). The phylogeny of hornless ruminants and a description of the cranium of *Archaeomeryx*. *Bull. Am. Mus. Nat. Hist.* **167**, 117–58.

West, R. M. (1971). Deciduous dentition of the early Tertiary Phenacodontidae (Condylarthra, Mammalia). *Am. Mus. Novit.* **2461**, 1–37.

West, R. M. (1972). Upper deciduous dentition of the Oligocene insectivore *Leptictis (=Ictops) acutidens*. *Ann. Carnegie Mus.* **44**, 25–32.

West, R. M. (1980). Middle Eocene large mammal assemblage with Tethyan affinities, Ganda Kas region, Pakistan. *J. Paleontol.* **54**, 508–33.

West, R.M. (1984). Paleontology and geology of the Bridger Formation, southern Green River basin, southwestern Wyoming. Part 7. Survey of Bridgerian Artiodactyla, including description of a skull and partial skeleton of *Antiacodon pygmaeus*. *Contrib. Biol. Geol. Milwaukee Publ. Mus.* **56**, 1–47.

Whitworth, T. (1958). Miocene ruminants of east Africa. *Fossil Mammal. Afr.* **15**, 1–50.

Wortman, J. L. (1898). The extinct Camelidae of North America and some associated forms. *Bull. Am. Mus. Nat. Hist.* **10**, 93–142.

Zittel, K. A. (1925). *Textbook of Palaeontology*, Vol. 3, *Mammalia*, translated by C. R. Eastman and A. S. Woodward. Macmillan, London.

10. The phylogeny of the Ruminantia (Artiodactyla, Mammalia)

CHRISTINE M. JANIS
Division of Biology and Medicine, Brown University, Providence, Rhode Island, USA

KATHLEEN M. SCOTT
Department of Biological Sciences, Rutgers University, Piscataway, New Jersey, USA

Abstract

The main new aspects of this classification of the Ruminantia are as follows: *Bachitherium* and *Lophiomeryx* are traguloids that possess certain synapomorphies with the Pecora, and are placed in their own families within the Tragulina; the pecoran family 'Gelocidae' is a polyphyletic assemblage, rather than the sister-taxon to the other Pecora, and the derivation of the higher pecoran families can be found within the 'gelocid' assemblage; pecoran cranial appendages are non-homologous, and there is no basis for placing the Moschidae as the sister-taxon to the horned families (moschids in fact possess cervoid synapomorphies); the Giraffoidea is the most primitive of the living pecoran families, as synapomorphies can be found to unite the Bovidae and the Cervoidea; the family Palaeomerycidae is re-established, uniting *Palaeomeryx*, *Amphitragulus*, and the Dromomerycinae by a synapomorphic suite of dental characters, and all palaeomerycids possess derived characters typical of higher cervoids.

We recognize outstanding problems as follows: the affinities of eight genera placed as *incertae sedis* on the legend for Fig. 10.1; the taxonomic position of the genera *Hydropotes*, *Prolibytherium*, and *Propalaeoryx*; the taxonomic position of the Antilocapridae, and the definition and unity

The Phylogeny and Classification of the Tetrapods, Volume 2: Mammals (ed. M. J. Benton), Systematics Association Special Volume No. 35B, pp. 273–82. Clarendon Press, Oxford, 1988.

of this family; and the definition and interrelationships of the moschid subfamily Blastomerycinae.

Introduction

The suborder Ruminantia has long been a problematical group for taxonomists, despite the fact that ruminant artiodactyls represent the majority of large living mammals. The reason may be that their evolutionary history has been beset by multiple parallel radiations during the Tertiary, from forested habitats into more open areas. Thus, many ruminant families have evolved in parallel those derived characteristics typical of open-habitat mammals: longer legs (including fusion of the metapodials, reduction or loss of the lateral digits, and development of complete distal metapodial keels); more high-crowned (hypsodont) cheek teeth (involving the simplification of molar form, and the molarization of the premolars, which may obscure many of the original diagnostic morphological features); and skulls that are more dolichocephalic, with lengthening of the diastema. Similar changes occurred in the perissodactyl family Equidae during the later Tertiary, so it is clear that these characteristics cannot be used in a simplistic fashion as synapomorphies to unite ruminant families (e.g. as commonly used to unite the Antilocapridae with the Bovidae).

The suborder Ruminantia can be divided into the infraorders Tragulina and Pecora (Webb and Taylor 1980), which broadly separates the smaller, more primitive forms from the generally larger and more open-habitat forms possessing the aforementioned derived features (although certain traguloids also acquired some of these in parallel). Pecorans may also possess cranial appendages. The Tragulina is the paraphyletic assemblage of primitive families that, with the exception of the living Tragulidae, did not survive past the Middle Miocene, while the Pecora is defined by a suite of synapomorphies (Fig. 10.1). On the basis of these characters, Webb and Taylor (1980) transferred the hornless family Gelocidae from the Tragulina to the Pecora. However, many problems remain. The common Eurasian traguloids *Lophiomeryx* and *Bachitherium*, variously ascribed to the Leptomerycidae (Tragulina) or Gelocidae (Pecora), lack pecoran synapomorphies, yet cannot be assigned to any of the existing traguloid families. Cranial appendages cannot simply be used to unite the 'higher' pecoran families as a 'Eupecoran' clade (Webb and Taylor 1980), as these structures are clearly non-homologous between the living families (see later discussion).

In addition, the fossil record abounds with problematic taxa of pecoran status. These include both hornless genera, such as *Amphimos-*

chus, Amphitragulus, Blastomeryx, Dremotherium, Micromeryx, Propalaeoryx, and *Walangania,* and horned genera such as *Climacoceras, Lagomeryx, Palaeomeryx, Prolibytherium, Triceromeryx,* and *Zarafa.* The North American horned pecoran family Dromomerycidae is still of dubious taxonomic affinities, variously allied with the Giraffidae (e.g. Crusafont 1961) or the Cervidae (e.g. Leinders 1983). Many of these taxa have been placed in the 'Palaeomerycidae', but Hamilton (1978*a,b*) clearly showed that this family, as then defined, was a 'waste-basket' category that was cladistically inviable. Hamilton (1978*a*) showed that *Climacoceras* and *Zarafa* (=*Canthumeryx*) could be included in the Giraffoidea by the synapomorphy of a bi-lobed lower canine, and *Triceromeryx* could be included with the giraffids on the basis of the P_4 morphology; although Janis and Lister (1985) showed that these premolar characters are not unique to giraffids, and suggested that *Triceromeryx* should remain as Pecora *incertae sedis.* Webb and Taylor (1980) grouped the European genera *Amphitragulus* and *Dremotherium,* and the North American Blastomerycinae, with the living genus *Moschus* in the family Moschidae on the basis of ear region characters. *Amphimoschus* and *Walangania,* along with certain small Oligocene taxa known primarily from teeth alone (see legend for Fig. 10.1), have been considered bovids by various authors, because they possess fairly hypsodont cheek teeth, but this is hardly a valid bovid autapomorphy. However, these revisions still leave a large number of taxa as Pecora *incertae sedis.*

This paper represents a condensed abstract of our previous work (Janis 1987; Scott and Janis 1987; Janis and Scott 1987), and the reader is referred to these papers for a more detailed account of our phylogenetic reasoning. We have reinvestigated the character states of the living and fossil ruminant taxa, examining collections in Europe, North America, and Africa, and casts of Chinese material. We have been influenced strongly by certain recent work on ruminant taxonomy, but have also discovered a number of new characters that led to this new synthesis. This analysis has been carried out independently of that by Gentry and Hooker (Chapter 9), who have come to different conclusions on some points.

Discussion

Figure 10.1 depicts our cladogram of the Ruminantia, and the legend for the figure details the relevant characters at the nodes, and the taxa in the various families. The definition of the Ruminantia, and the interrelationships and definitions of the families Hypertragulidae, Tragulidae, and Leptomerycidae are taken from Webb and Taylor (1980). Our only contribution to this scheme is the recognition that the

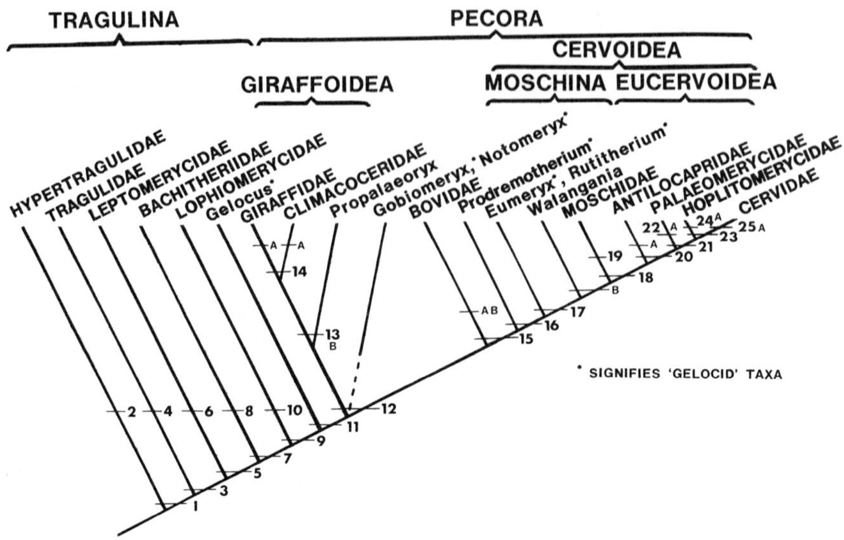

Fig. 10.1. Characters used in cladogram. **1**: RUMINANTIA. Fusion of cuboid and navicular in tarsus; upper incisors reduced or lost; lower canine incisiform. **2**: HYPERTRAGULIDAE. Postglenoid foramen laterally enclosed; radius fused to complete ulna; complete fibula fused proximally and distally with tibia. **3**: Amastoid condition of skull; jugular foramen confluent with posterior lacerate foramen; loss of trapezium, and fusion of trapezoid and magnum in manus; metacarpal I lost; P^1 lost; mesostyles in upper molars and ectostylids in lower molars; '*Dorcatherium* fold' in lower molars; **4**: TRAGULIDAE. Complete postorbital bar formed from jugal; tensor tympani chamber posteriorly restricted; postglenoid process reduced and overridden by expanded auditory bulla; palatine bones extended posteriorly; fusion of cubonavicular with ectocuneiform in tarsus; malleolar articulation of calcaneum convex. **5**: Postorbital bar formed from frontal; axis with odontoid process possessing high dorsal crest with high anterior articulatory surfaces; fibular facet of calcaneum with large proximal concavity and small distal convexity. **6**: LEPTOMERYCIDAE. Large mastoid fissure; I_1 enlarged and tusk-like; P_1 smaller and conical. **7**: Anterior cingulum on lower molars; P^3 with posteriorly situated and posteriorly directed protocone. **8**: BACHITHERIIDAE. Pinched M_3 hypoconulid. **9**: P_1 premolariform, separated from P_2 by a small diastema. **10**: LOPHIOMERYCIDAE. 'Figure eight' configuration of lingual border of lower molars; posterior fossette on lower molars open lingually. **11**: PECORA. Compact, parallel-sided astragalus; stapedial artery lost; enlarged fossa for stapedial muscle on petrosal, and subarcuate fossa reduced; broadened basioccipital with strong flexion stops on condyles; laterally enclosed postglenoid foramen; fore- and hindlimbs almost equal in length; four-chambered stomach with omasum. **12**: Metastylids in lower molars. **13**: GIRAFFOIDEA (including *Propalaeoryx*). Bifurcation of posterior wing of metaconule in upper molars. **14**: GIRAFFOIDEA (restricted). Bilobed lower canine; flat and broad cubonavicular facet on proximal matatarsus. **15**: Entostyle on upper molars; more ventral entrance of oesophagus to rumen. **16**: Closed metatarsal gully. **17**: CERVOIDEA. '*Palaeomeryx* fold' in lower molars (brachydont taxa). **18**: Sabre-like upper canines. **19**: MOSCHIDAE. Laterally enclosed, sub-central tympanohyal vagina on auditory bulla. **20**: EUCERVOIDEA. Double lacrimal foramen. **21**: Posterior tuberosity on metatarsus. **22**: PALAEOMERYCIDAE. Double posterior lobe on M_3, closed posteriorly; reduced metastyle in upper molars; reduced metacone on P^4. **23**: Bifurcation of posterior wing of protocone in upper molars. **24**: HOPLITOMERYCIDAE. Double posterior lobe on M_3, open posteriorly. **25**: CERVIDAE. Lacrimal fossa; 'cervid' features of soft anatomy, notably

'*Dorcatherium* fold' on the lower molars is not a unique tragulid character (*contra* Sudre 1984), but is rather symplesiomorphic for ruminants above the level of the Hypertragulidae. This feature, or remnants of it, can be seen in various traguloids, and occasionally in pecorans (Janis 1987). *Lophiomeryx* and *Bachitherium* do not possess any of the derived features of the other traguloid families, and share certain derived dental features with the Pecora (see Fig. 10.1). *Bachitherium* appears to be a monotypic genus in its own family, the Bachitheriidae (Janis 1987). *Lophiomeryx* can be distinguished by unique dental features, which also characterize *Cryptomeryx* and *Iberomeryx*. These genera are now united in the Lophiomerycidae, forming the sister-taxon to the Pecora (Janis 1987).

We disagree with the division of the Pecora by Webb and Taylor (1980) into the hornless 'Moschina' (families Gelocidae and Moschidae) and the horned 'Eupecora'. They defined the Gelocidae on the basis of a posteriorly situated and directed protocone on P^3, but this character is also seen in *Lophiomeryx* and *Bachitherium* (Janis 1987), and as a more weakly defined retention in certain higher pecorans (Janis and Scott 1987). This dental character is evidently symplesiomorphic for ruminants above the level of the Leptomerycidae, and the 'Gelocidae' is then a polyphyletic assemblage of primitive pecoran taxa. Although other pecorans can be distinguished from 'gelocids' by complete distal metapodial keels, this character is probably

---

absence of gall bladder. **Pecoran characteristics that have evolved in parallel in different families**; **A**: Cranial appendages. **B**: Complete distal metapodial keels. **Taxa included in families listed in cladogram** (*signifies extant taxon): ANTILOCAPRIDAE* (pronghorns): Antilocaprinae* and Merycodontinae. BACHITHERIIDAE: *Bachitherium*. BOVIDAE* (cattle and antelope): living and fossil genera, lacking a sabre-like upper canine and with non-deciduous, unbranched keratin-covered horn cores. CERVIDAE* (deer): *Hydropotes** and living and fossil genera with deciduous, branched antlers, including *Dicrocerus*, *Heterocemas*, *Lagomeryx*, and *Procervulus*. CLIMACOCERIDAE: *Climacoceras*, *Canthumeryx*, *Injanatherium*, and *Nyanzameryx*. GIRAFFIDAE* (giraffe and okapi): Giraffinae* and Sivatheriinae. HOPLITOMERYCIDAE: *Amphimoschus* and *Hoplitomeryx*. HYPERTRAGULIDAE: *Andegameryx*, *Hypertragulus*, *Hypisodus*, *Nanotragulus*, *Parvitragulus*, and *Simimeryx*. LEPTOMERYCIDAE: *Archaeomeryx*, *Hendryomeryx*, *Leptomeryx*, *Miomeryx*, *Pronodens*, *Pseudoparablastomeryx*, and *Xinjiangmeryx*. LOPHIOMERYCIDAE: *Cryptomeryx*, *Lophiomeryx*, and *Iberomeryx*. MOSCHIDAE* (musk 'deer'): Blastomerycinae (*Blastomeryx*, *Longirostromeyx*, *Machaeomeryx*, *Parablastomeryx*, and *Problastomeryx*), *Dremotherium*, *Hispanomeryx*, *Micromeryx*, and *Moschus**. PALAEOMERYCIDAE: Dromomerycinae (*Aletomeryx*, *Barbouromeryx*, *Bouromeryx*, *Cranioceras*, *Drepanomeryx*, *Dromomeryx*, *Rakomeryx*, *Pediomeryx*, *Procranioceras*, and *Sinclairomeryx*), *Amphitragulus*, *Palaeomeryx* (and possibly also *Prolibytherium*). TRAGULIDAE* (mouse 'deer'): *Dorcabune*, *Dorcatherium*, *Hyemoschus**, and *Tragulus**. **Problematical taxa**: *Indomeryx*, *Pseudoceras*, and *Pseudomeryx* (variously grouped as 'Hypertragulids' or 'Gelocids') remain as Ruminantia *incertae sedis*. *Gobiocerus*, *Hypsodontus*, *Kubanotragus*, and *Palaeohypsodontus* (termed Oligocene and Miocene Asian 'bovids') remain as Pecora *incertae sedis*. *Triceromeryx* remains as Cervoidea *incertae sedis*.

homoplasic, as it is functionally associated with increased cursoriality, and also evolved in parallel in the Equidae.

Cranial appendages have been used not only to define the 'Eupecora' but apparent similarities in these appendages have also been used as synapomorphies to unite horned pecoran families, such as bovids with giraffids (Hamilton 1978*a*) or with antilocaprids (O'Gara and Matson 1975). However, ruminant-like cranial appendages have evolved more than once in mammals (see Janis 1982), and their mode of development is non-homologous in living families. Bovid horns are outgrowths of the frontal bone, induced by dermal tissues, and growth is from the tip outwards by intermembranous ossification (Dove 1935). Cervid antlers are also outgrowths of the frontal bone but are induced by the periosteum, and although growth is also from the tip, it is by endochondral ossification (Goss 1983). Giraffid ossicones are preformed in cartilage as a separate ossification, and growth is at the base by endochondral ossification (Lankester 1907). The branched cranial appendages of *Climacoceras* cannot have developed in the mode typical of living giraffids, and thus cranial appendages must have arisen at least twice within the Giraffoidea (see also Geraads 1986). We suggest that cranial appendages arose independently on many occasions within the Pecora (Fig. 10.1), and are invalid as taxonomic characters at the family level of classification (Scott and Janis 1987; Janis and Scott 1987). Thus, there is no basis for placing the Moschidae as the sister-taxon to the horned Pecora (*contra* Webb and Taylor 1980).

The most primitive pecoran taxon is *Gelocus communis*, as it lacks metastylids in the lower dentition (Janis 1987). (Other species ascribed to this genus, such as *Gelocus whitworthi*, are probably not congeneric with *Gelocus communis*; Janis and Scott 1987.) Bovids, cervoids, and *Prodremotherium* all share the synapomorphy of entostyles on the upper molars (although this character does occasionally appear in certain traguloids and in occasional individuals of *Giraffa*; Janis and Scott 1987). Living cervids and bovids may also be united by the more ventral entrance of the oesophagus into the rumen (Hofmann 1973), although the validity of this character needs further investigation. Todd (1975) also defined the Giraffidae as primitive, on the basis of chromosomal characters, although we do not agree with his reasoning (Scott and Janis 1987). This leaves an unresolved trichotomy between the Giraffoidea (*sensu* Hamilton 1978*a*), the African genus *Propalaeoryx*, and Asian 'gelocids' such as *Notomeryx* and *Gobiomeryx*. Gentry (1978) suggested that *Propalaeoryx* might be the sister-taxon to the Giraffoidea, and a weak synapomorphy exists in the bifurcation of the metaconule in the upper molars (Janis and Scott 1987). Other giraffoids are united by the bi-lobed lower canine (Hamilton 1978*a*), and the flat and broad cubonavicular facet on the proximal metatarsus (Janis and Scott 1987).

The Giraffidae are distinguished from the Climacoceridae primarily by the possession of unbranched ossicones (the dental synapomorphies of giraffids cited by Hamilton (1978*a*) also occur in other pecoran taxa; Janis and Scott 1987), and the Climacoceridae remain as a paraphyletic assemblage of primitive giraffoids.

No autopomorphic character can be found to unite the Bovidae, other than the possession of horns (cranial appendages that are unbranched, non-deciduous, and covered with a permanent keratin sheath). However, similar kinds of appendages were seen in the cervoid *Hoplitomeryx* (Leinders 1983). The 'gelocid' *Prodremotherium* is placed as the sister-taxon to the cervoids, on the basis of a closed metatarsal gully (Leinders 1979). Other ruminants have either unfused metapodials, or have fused the metatarsals with a resultant open gully (e.g. bovids, giraffoids, *Leptomeryx*, *Bachitherium*, and *Gelocus*). However, an open metatarsal gully is probably merely the 'normal' method of fusion of the metapodials, rather than the symplesiomorphic condition. The closed metatarsal gully apparently has no functional significance, and elsewhere is seen only in living tragulids (where it is formed by a different developmental process) and in the problematical, possibly 'gelocid' genus *Pseudoceras*. Although this character state is variable in the blastomerycine moschids and the merycodontine antilocaprids, fossil record sequences clearly show that the polarity of this character is from an initially closed gully to a secondarily open gully (Janis and Scott 1987).

We have redefined the 'Cervoidea' by the possession of a '*Palaeomeryx* fold' in the lower molars (Janis and Scott 1987), thereby including the 'gelocids' *Eumeryx* and *Rutitherium*. The '*Palaeomeryx* fold' is clearly a derived pecoran character, not symplesiomorphic (*contra* Stirton 1944). However, while it is present in early brachydont moschids, dromomerycids, and cervids, it is lost in later, more hypsodont forms. Since the earliest antilocaprids and hoplitomerycids were already hypsodont, its apparent absence in these families cannot be held as diagnostic. *Walangania* is more derived than these 'gelocids' in the possession of complete distal metapodial keels, but it lacks the sabre-like upper canine (which is distinct from the traguloid shorter type of canine) characteristic of other hornless cervoids such as the moschids, *Hydropotes*, and certain cervoids with small cranial appendages (e.g. *Aletomeryx*, *Barbouromeryx*, *Hoplitomeryx*, and *Palaeomeryx*). The shorter canine of the muntiacine cervids represents a reversal to the traguloid-like condition, probably in association with their particular mode of intraspecific fighting (see Janis 1982).

We agree with Webb and Taylor's (1980) definition of the Family Moschidae, but note that it is clearly of cervoid affinities (Fig. 10.1; Leinders and Heintz 1980), and we have removed *Amphitragulus*, which lacks moschid synapomorphies but does possess certain characters typical of higher cervoids (see later discussion). Cervoids above this level are united into the Eucervoidea by the double lacrimal orifice (Janis and Scott 1987), although this character is problematical since it also occurs in certain bovids, and is individually variable among antilocaprids (Leinders and Heintz 1980). No apomorphic feature can be found to define the Merycodontinae or the Antilocaprinae, nor to unite them into the Antilocapridae. We have retained this problematical grouping primarily on the basis of historical precedence, general similarity (although the cranial appendages may not be homologous), and biogeographical occurrence in North America. Other cervoids can be united by the presence of the posterior tuberosity on the metatarsus, although we note that this character evolved in parallel within the Moschidae, seen in *Moschus, Parablastomeryx*, and *Micromeryx* (Janis and Scott 1987).

We have redefined the 'Palaeomerycidae', by the possession of a derived suite of dental characters (Fig. 10.1), to include *Palaeomeryx, Amphitragulus*, and the Dromomerycidae (now demoted to the Dromomerycinae) (Janis and Scott 1987). These taxa all possess derived cervoid features, although the condition of the lacrimal orifice in *Palaeomeryx* is uncertain. Qiu *et al.* (1985) claim it to be single, and ally *Palaeomeryx* with the Giraffidae, following Ginsberg and Heintz (1966). *Palaeomeryx* is clearly a cervoid (Leinders 1983), and we regard the condition of the lacrimal orifice to be as yet undetermined, based on the material discussed by Qiu *et al.* (1985). The problematical African genus *Prolibytherium* shares these dental characters, together with a probably double lacrimal orifice, and may belong in the Palaeomerycidae, although more material is needed for confirmation. *Triceromeryx*, although possessing a '*Palaeomeryx* fold' and cranial appendages (including a single occipital appendage) similar to those in *Palaeomeryx* and the craniocerative dromomerycids, does not possess these dental characters, and we leave it for the time being as Cervoidea *incertae sedis* (Janis and Scott 1987).

We unite the Hoplitomerycidae (Leinders 1983) with the Cervidae on the admittedly rather weak synapomorphy of a bifurcation of the posterior wing of the protocone in the upper molars (Janis and Scott, in press). The apparently hornless *Amphimoschus* can be allied with *Hoplitomeryx* on dental and cranial characters (Leinders 1983). We have allied the living antlerless *Hydropotes* with the Cervidae on the basis of soft anatomy characters that are absent in *Moschus* and *Antilocapra* (Fig. 10.1), but this condition is of course unknown for the extinct cervoids. The presence of a lacrimal fossa, although found in other pecoran taxa (e.g. certain bovids and dromomerycids), may possibly be used at this level to unite *Hydropotes* with the cervids. The problematical 'lagomerycids', such as *Lagomeryx*, *Procervulus*, and *Heterocemas*, previously considered to be giraffoids (e.g. Pilgrim 1941) or palaeomerycids (e.g. Stirton 1944), have been shown to be muntiacine cervids on the basis of newly discovered material (Chow and Shih 1978).

Acknowledgements

Helpful comments on manuscripts at various stages were provided by A. Lister, A. Gentry, R. Tedford, and D. Webb, and we also thank M. Brunet, D. Geraads, Y. Jehenne, J. Leinders, E. Manning, and D. Savage for discussion and correspondence. This work was funded by a Brown University Biomedical Research Grant (RR0785-19) and an N.S.F. grant (BR-8418148) to the senior author, and a Rutgers Junior Faculty Fellowship and an N.S.F. grant (BSR 84-18349) to the junior author.

References

Chow, B. and M. Shih. (1978). A skull of *Lagomeryx* from the middle Miocene of Linchu, Shantung. *Vertebr. palasiat.* **16**, 111–22.

Crusafont, M. (1961). Giraffoidea. In *Traité de Paléontologie*, Vol. 6 (ed. J. Piveteau), pp. 1022–37. Masson et Cie, Paris.

Dove, W. F. (1935). The physiology of horn growth: a study of the morphogenesis, interaction of tissues, and the evolutionary process of a Mendelian recessive character by means of transplantation of tissues. *J. Exp. Biol.* **69**, 347–405.

Gentry, A. W. (1978). Quoted *pers. comm.* in Hendey, Q. B. (1978). Preliminary report on the Miocene vertebrates from Arrisdrift, South West Africa. *Ann. S. Afr. Mus.* **76**, 1–41.

Geraads, D. (1986). Remarques sur la systématique et la phylogénie des Giraffidae (Artiodactyla, Mammalia). *Géobios* **19**, 465–77.

Ginsburg, L. and Heintz, E. (1966). Sur les affinités du genre *Palaeomeryx* (ruminant du Miocène Européen). *C.R. Acad. Sci. Paris, Ser. D* **262**, 979–82.

Goss, R. J. (1983). *Deer Antlers: Regeneration, Function and Evolution.* Academic Press, New York.

Hamilton, W. R. (1978*a*). Fossil giraffes from the Miocene of Africa, and a revision of the phylogeny of the Giraffoidea. *Philos. Trans. R. Soc. London, B* **282**, 165–229.

Hamilton, W. R. (1978*b*). Cervidae and Palaeomerycidae. In *Evolution of African Mammals* (eds. V. J. Maglio and H. B. S. Cooke), pp. 496–508. Harvard University Press, Cambridge, Mass.

Hofmann, R. R. (1978). *The Ruminant Stomach.* East African Monographs in Biology (2), (ed. T. R. Odhiambo), 349 pp. East African Literature Bureau.

Janis, C. M. (1982). Evolution of horns in ungulates: ecology and palaeoecology. *Biol. Rev.* **57**, 261–318.

Janis, C. M. (1987). Grades and clades in hornless ruminant evolution: the reality of the Gelocidae and the systematic position of *Lophiomeryx* and *Bachitherium*. *J. Vertebr. Paleontol.* **7**, 200–16.

Janis, C. M. and Lister, A. (1985). The morphology of the lower fourth premolar as a taxonomic character in the Ruminantia (Mammalia, Artiodactyla), and the systematic position of *Triceromeryx*. *J. Paleontol.* **59**, 405–10.

Janis, C. M. and Scott, K. M. (1987). The interrelationships of higher ruminant families, with special emphasis on the members of the Cervoidea. *Am. Mus. Novit.* **2893**, 1–85.

Lankester, E. R. (1907). The origin of the lateral horns of the giraffe in foetal life on the areas of the parietal bone. *Proc. Zool. Soc. London* **1907**, 24–36.

Leinders, J. J. M. (1979). On the osteology and function of the digits in some ruminants and their bearing on taxonomy. *Z. Säugetierkd.* **44**, 302–18.

Leinders, J. J. M. (1983). Hoplitomerycidae fam. nov. (Ruminantia, Mammalia), from Neogene fissure fillings in Gargano, Italy. *Scr. Geol.* **70**, 1–51.

Leinders, J. J. M. and Heintz, E. (1980). The configuration of the lacrimal orifice in pecorans and tragulids (Artiodactyla; Mammalia) and its significance for the distinction between Bovidae and Cervidae. *Beaufortia* **30**, 155–60.

O'Gara, B. W. and Matson, G. (1975). Growth and casting of horns by pronghorns and exfoliation of horns by bovids. *J. Mammal.* **56**, 829–46.

Pilgrim, G. (1941). The dispersal of the Artiodactyla. *Biol. Rev.* **16**, 155–8.

Qiu Z., Yan, D., Jia, H., and Sun, B. (1985). Preliminary observations on the newly found skeleton of a *Palaeomeryx* from Shanwag, Shandong. *Vertebr. Palasiat.* **23**, 173–95.

Scott, K. M. and Janis, C. M. (1987). The phylogenetic position of the Cervidae, and the case for a superfamily Cervoidea. In *The Biology and Management of the Cervidae* (ed. C. Wemmer), pp. 3–20. Research Symposia of the National Zoological Park. Smithsonian Institution Press, Washington, DC.

Stirton, R. A. (1944). Comments on the relationships of the Palaeomerycidae. *Am. J. Sci.* **242**, 633–55.

Sudre, J. (1984). *Cryptomeryx* (Schlosser, 1886) tragulidé de l'Oligocène d'Europe: relations du genre et considérations sur l'origine des ruminants. *Palaeovertebrata* **14**, 1–31.

Todd, N. B. (1975). Chromosomal mechanisms in the evolution of artiodactyls. *Paleobiology* **1**, 175–88.

Webb, S. D. and Taylor, B. E. (1980). The phylogeny of higher ruminants and a description of the cranium of *Archaeomeryx*. *Bull. Am. Mus. Nat. Hist.* **167**, 117–58.

11. The Tethytheria: elephants and their relatives

PASCAL TASSY
Laboratoire de Paléontologie des Vertébrés et de Paléontologie Humaine, Université Pierre et Marie Curie, Paris, Cedex 05, France; and INSERM, Unité 155, Château de Longchamp, Bois de Boulogne, 75016 Paris, France

JEHESKEL SHOSHANI
Department of Biology, Wayne State University, Detroit, Michigan, USA

Abstract

The monophyly of the Tethytheria *sensu* McKenna (1975), including extant orders Proboscidea and Sirenia and extinct order Desmostylia, is supported on the basis of soft anatomy and osteological traits. A detailed cladogram of the Proboscidea is given, with a discussion on the relationships of the primitive genera *Moeritherium*, *Barytherium*, and *Numidotherium*. The genus *Anthracobune* and allied forms—the so-called 'Asian moeritheres'—could either be classified as desmostylians (the opinion defended here) or as proboscideans (but not moeritheres), i.e. Tethytheria *incertae sedis*. Finally, the controversy about the relationships of Proboscidea and Sirenia to other mammalian orders, namely extant Hyracoidea, is examined: some characters support a sister-group relationship between tethytheres and hyracoids, though striking similarities exist between hyracoids and perissodactyls.

Introduction

The Tethytheria of McKenna (1975), which united only elephants (Order Proboscidea Illiger) and sirenians (Order Sirenia Illiger)

The Phylogeny and Classification of the Tetrapods, Volume 2: Mammals (ed. M. J. Benton), Systematics Association Special Volume No. 35B, pp. 283–315. Clarendon Press, Oxford, 1988.

among extant mammals, was a challenge to Simpson's (1945) Paenungulata. The latter embraced another extant member, namely the hyracoids (Order Hyracoidea Huxley). The Tethytheria or 'beasts of the Tethys' (because early members were thought to have inhabited the shores of the ancient sea of Tethys) included also an extinct order, the Desmostylia Reinhart. The Paenungulata of Simpson, on the other hand, included all the above-mentioned taxa plus four extinct orders, Pantodonta Cope, Dinocerata Marsh, Pyrotheria Ameghino, and Embrithopoda Andrews.

Among the Tethytheria, the two extant orders (Proboscidea, Sirenia) and the extinct Desmostylia have sharply contrasting external appearances. Perhaps even more contrasting is the morphology of putative relatives, such as as the extant Hyracoidea and extinct Embrithopoda (other fossil taxa are not considered in this analysis). Elephants (*Loxodonta africana* and *Elephas maximus*) are the largest terrestrial mammals. Manatees and dugongs or sea-cows (Sirenia) are the size of, and are somewhat reminiscent of, a dolphin and are strictly aquatic. Desmostylians (e.g. *Paleoparadoxia*) are an 'old paradox' that approximated to a hippopotamus in size, and were amphibious mammals that became extinct by the end of the Miocene; hyraxes or conies (e.g. *Procavia capensis*) are rabbit-like in size but with small external ears. The last group that will be discussed, the embrithopods (e.g. *Arsinoitherium zitteli*), were the most mysterious mammals found in the Fayum of Egypt; they approximated to an elephant in size and possessed a gigantic pair of horns on the front of their skulls and a much smaller pair of horns behind them. Clearly, from any superficial examination of this bizarre assemblage of mammals, there would be no reason to suspect any particular relationships among them. Even more detailed examination shows discrepancies in the phylogenetic hypotheses.

The relationships of elephants and sirenians to other orders of living mammals is also still very much under debate. As will be discussed later, evidence for the monophyly of Tethytheria plus Hyracoidea comes from molecular data. Morphological sources give contradictory results: some data support the monophyly of Tethytheria plus Hyracoidea (Shoshani 1986*a*; Novacek 1986*a*; Novacek and Wyss 1986; Chapter 3 of this volume), while others support a closer relationship of Hyracoidea to Perissodactyla (Fischer 1986; Chapter 8 of this volume).

This paper is a tentative synthesis based partly on recent work by the authors (Tassy 1985; Shoshani 1986*a*) on proboscidean, tethythere, and paenungulate relationships, though the question of sister-group relationships of the Tethytheria to other mammalian orders remains open. We shall begin with the relationships of extant tethytheres, followed by discussions on the monophyly of Proboscidea and of

Sirenia, and lead into the relationships within Tethytheria, emphasizing the phylogenetic positions of Desmostylia and *Anthracobune*. We shall end with the controversies over Hyracoidea and Paenungulata.

The extant sister-group of elephants: the concept of Tethytheria

The hypothesis of sister-group relationships for elephants and sirenians can be traced back to Blainville (in Gervais 1836), who classified within the 'Monodelphes' his group 'les Gravigrades', including *only* elephants and sirenians (Blainville's 'lamantins'). The term 'gravigrades' means 'heavy gait' and was coined in 1816 by Blainville for elephants only. This term has since been applied also to fossil sloths. Earlier, Linnaeus' order Bruta included not only elephants and sirenians but also other mammals (pangolins, sloths, and anteaters) on the basis of the lack of front teeth. It was only in 1975 that McKenna, in his phylogenetic classification of the Mammalia, coined the name Tethytheria for his mirorder (a new category)—which can be applied more simply as a superorder (Wiley 1979; Tassy 1981)—for both extant elephants and sirenians. In the interval, no sister-group relationships were hypothesized for the two groups. In their review of tethytherian systematics, Domning *et al.* (1986, p. 38) emphasized the fact that 'most [late-19th-century] taxonomic treatises made no special attempt to relate the two (proboscideans and sirenians) or in fact argued against the relationship and maintained the cetacean connection [for sirenians]' (see also Simpson 1932, pp. 492–4). After Andrews' early discoveries of proboscideans and sirenians in the Fayum of Egypt, and his monograph (Andrews 1906), hypotheses on the close relationship of proboscideans and sirenians could be found in the literature (e.g. Winge as early as 1906 (Winge 1942); Abel 1919) mainly because of overall similarity in the shape of the skull of early proboscideans (*Moeritherium*) and early sirenians (*Eotheroides*).

Nevertheless Simpson's (1945) grouping (named Paenungulata) for extant proboscideans, sirenians, hyracoids, and various fossil groups was very successful and has recently been re-evaluated on the basis of molecular and morphological data (e.g. Shoshani *et al.* 1977; de Jong 1982; Novacek 1986*a*; Shoshani 1986*a*; Chapter 3 of this volume). Conversely, the close relationship of Hyracoidea to Perissodactyla, following McKenna's (1975) concept of Phenacodonta, was argued by Fischer (1986) and by Prothero *et al.* (Chapter 8).

Moreover, after the publication of some papers supporting the monophyly of the Tethytheria *sensu* McKenna (1975) (see Tassy 1979, 1981; Novacek 1982, 1986*a*; Domning *et al.* 1986; Shoshani 1986*b*), Novacek (1986*b*) questioned the hypothesis of sister-group relationships between elephants and sirenians. In addition, McKenna (*pers. comm.* to J. S. in Shoshani 1986*a*, appendix D) left Sirenia as Epitheria *incertae sedis*, leaving only the Proboscidea and the extinct Desmostylia in the Tethytheria. In this paper, we support the monophyly of Tethytheria *sensu* McKenna (1975).

An early presentation of characters supporting close relationships between elephants and sirenians can be found in Blainville (1844, pp. 37–40), a list that influenced many (see, for instance, Andrews 1906; Gregory 1910; Eales 1929). Blainville cited the shape of molars (lophodonty), the horizontal tooth replacement, the wide external nares and shape and contact of surrounding bones, the loss of a lacrimal foramen, the pectoral mammae, the internal testes, the thick skin, the expanded distal end of the ulna, the long intestinal tract, and other characters. Likewise, the list given by Gregory (1910, p. 407) mixed primitive characters (such as internal testes, long intestinal tract) with derived, 'gradistic', or convergent characters such as the well-known 'horizontal tooth replacement'. This latter character appeared twice, with different modes, during the evolution of the proboscideans (in post-Fayum elephantoids) and of sirenians (in manatees but not in dugongs), not to mention a third time (in *Desmostylus* among Desmostylia).

Only more recently have characters been analysed from a phylogenetic viewpoint. Yet few characters remain as synapomorphies for elephants and sirenians, and fewer still are not to be found in any other mammalian groups (Fig. 11.1)

What could be a derived feature not observed in other mammalian taxa is the bifid apex of the heart, each ventricle being separated from the other at the tip. This character, well known for both groups, had probably been overlooked as a synapomorphy because it is associated with a primitive character and interpreted as such: the occurrence of two anterior venae cavae (see in particular Eales (1929, p. 238) and Shoshani *et al.* (1982, p. 54), who support the primitiveness of this character). Yet, according to Watson (1872, p. 87) and others (see discussion in Shoshani *et al.* 1982), some large cetaceans (*Balaenoptera, Physeter*) have a somewhat similar condition associated with a broadening of the heart, which is due to the peculiar shape of the thorax (Slijper 1979, p. 159). The mapping of this character within cetaceans still needs further examination. Among osteological characters, the lateral expansion of the *processus zygomaticus* of the squamosal is also a rare character. The pair of pectoral mammae is also seen among other mammalian groups but not in taxa more or less related to elephants. Other osteological characters can be found in other taxa, such as the orbits placed forward, with the infraorbital foramen under the orbit (in a few rodents, for example, the infraorbital canal is under the orbit);

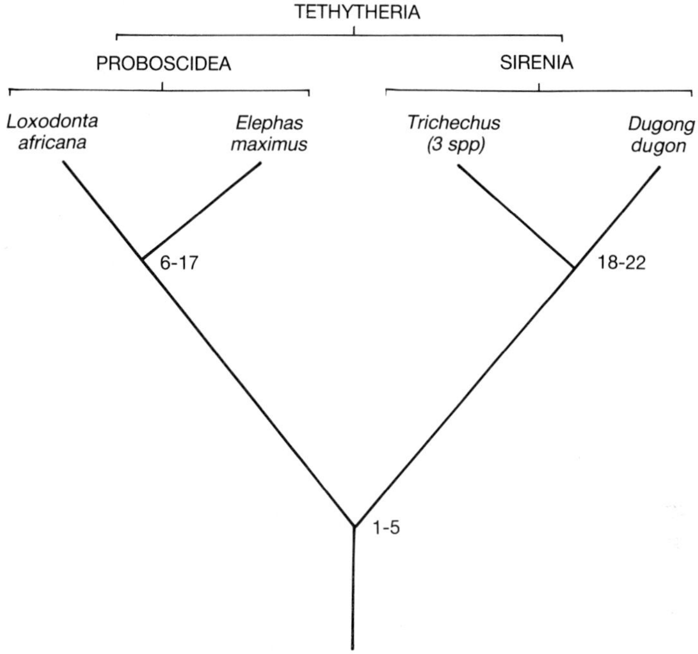

Fig. 11.1. Relationships among extant tethytheres. Characters 1a, 5 (modified), 10, 11, from Andrews (1906); 1b, 5 (modified), 19, from Blainville (1844); 2, 3, 6, 9, from Tassy (1979); 3 (modified), from Osborn (1909); 4, from Gregory (1910); 7, 8 (modified), from Matsumoto (1923); 12, from Tassy (1981); 13–17, from Tassy (1982, 1985); 20, from Shoshani (1986*b*); 21 (modified), from Savage (1976). TETHYTHERIA: 1(A) bifid apex of the heart; (B) a pair of pectoral mammae; (2) processus zygomaticus of the squamosal expanded laterally; (3) orbit forward, above anterior premolars, with infraorbital canal under orbit; (4) amastoidy: reduction of the mastoid process; (5) incipiently bilophodont bunolophodont dentition. PROBOSCIDEA: (6) no processus ascendens of palatine in the orbitotemporal fossa; (7) jugal reduced anteriorly: orbit opens in the maxilla; (8) jugal lengthened posteriorly to the rear border of the glenoid fossa of the squamosal; (9) cerebral part of the squamosal inflated dorsally; (10) external auditory meatus closed ventrally by post-tympanic and postglenoid parts of the squamosal; (11) loss of the condyloid foramen; (12) optic foramen (canalis opticus) and fissura sphenorbitale on the same line under the crista orbitotemporalis of the orbitotemporal fossa; (13) metacone on P^4; (14) upper molars with paracone and metacone transversally thickened and subdivided; paraconule touches the precingulum, protocone with a postprotocone crest; (15) $M^{1,2}$ with a distocrista; (16) lower molars with metaconid and entoconid transversely thickened; hypolophid as high as the protolophid; hypoconulid developed labially and postentoconulid outlined; (17) astragalus with enlarged lateral calcaneal facet and protruding tuberculum mediale. SIRENIA: (18) pachyostosis and osteosclerosis; (19) wide external nares and long processus nasalis of premaxilla which is in contact with the frontal; (20) loss of the ethmoid foramen; (21) inflated petrosum; well ossified drop-like ectotympanic ring with anterior and posterior attachments; (22) reduced hind limbs.

mastoid process reduced (also in various ungulate groups). Novacek (1986*a*, table 3, p. 98) adds the ventral expansion of the frontal in the orbital wall, a character also found in edentates and pholidotes as he notes.

Our concept of the incipient bilophodont bunolophodont condition (character 5 of Fig. 11.1) must be explained here. In our view, the morphology of the molars of the ancestral morphotype of the Proboscidea is not changed in the genus *Moeritherium* (Fig. 11.2). The bunolophodont condition could be the primitive state for the fully lophodont condition seen in manatees and some proboscideans, and so characterizes the ancestral morphotype of the Tethytheria. The morphology of the molars of primitive desmostylians (see later) and of the Eocene Asian genus *Anthracobune* limits the pattern of the incipient bilophodonty/bunolophodonty to the following characters: upper molars with metaconule close to hypocone (the condition needed for a transverse metaloph); lower molars with the four main cusps arranged opposite each other transversely. It may be noted that the former character is hypothesized by McKenna and Manning (1977) in the common ancestor of Tethytheria + Hyracoidea + Perissodactyla + Embrithopoda (but in the Tethytheria only the metaconule is clearly on the line between hypocone and metacone), while the latter is hypothesized in the ancestor of Tethytheria + Hyracoidea + Perissodactyla (see discussion in the last part of this paper).

In giving the diagnostic features of Tethytheria in Fig. 11.1, palaeontological information has been taken into account, i.e. the characters of the most primitive members of each group. Derived characters, found in extant elephants and sirenians, and sometimes cited as synapomorphies for the group (e.g. retracted nares with contact between premaxilla and frontal, lacrimal foramen indistinct externally (Novacek 1986*a*, p. 98; table 3, pp. 82–4)), should in fact be treated as convergent characters (Tassy 1981, pp. 111 and 116, figs. 3, 5 and 9). The primitive Eo-Oligocene proboscidean genus *Moeritherium* shows the plesiomorphic condition of the nasal area (Fig. 11.3): the premaxilla does not touch the frontal (the nasal bones are small but it is because the whole snout is short due to the position of the orbits). Likewise, in primitive elephantoids (*Phiomia*, Mammutidae; Tassy 1981, p. 111; 1982, p. 241) and primitive sirenians (*Prorastomus, Eotheroides*; Savage 1976, p. 346) a distinct lacrimal foramen is seen on mature skulls. Moreover, the interpretation of the position of the orbits as a shared derived feature for both sirenians and elephants (character 3 of Fig. 11.1) is based on the condition seen in primitive proboscideans such as *Moeritherium* and deinotheres, because a double reversal occurred in the history of Proboscidea (Fig. 11.4). The condition seen in *Moeritherium* was interpreted by Osborn (1909) as an aquatic adaptation and, as such, as a

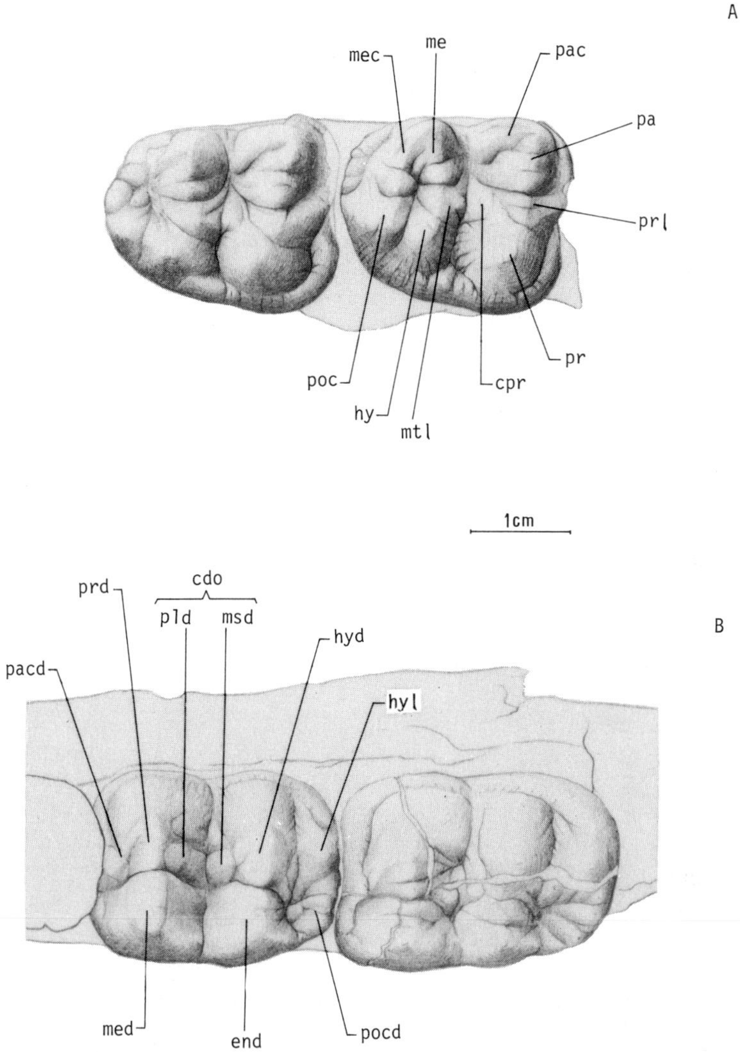

Fig. 11.2. Right M^{2,3} (A) and right M$_{2,3}$ (B) of *Moeritherium*, Fayum, Gebel el Qatrani (A: 40014, B: 13092; SMN, Stuttgart). cdo: cristid obliqua; cpr: postprotocone crest (outlined posterior central conule of the first pretrite half-loph); end: entoconid; hy: hypocone; hyd: hypoconid; hyl: hypoconulid; me: metacone; mec: postmetacrista; med: metaconid; msd: mesoconid (posterior part of the cristid obliqua); mtl: metaconule (with outlined anterior central conule of the second pretrite half-loph); pa: paracone; pac: postparacrista; pacd: paracristid; pld: protoconulid (= postprotoconid crest: anterior part of the cristid obliqua); poc: postentoconule; pocd: postentoconulid; pr: protocone: prd: protoconid; prl: paraconule (From Tassy 1985).

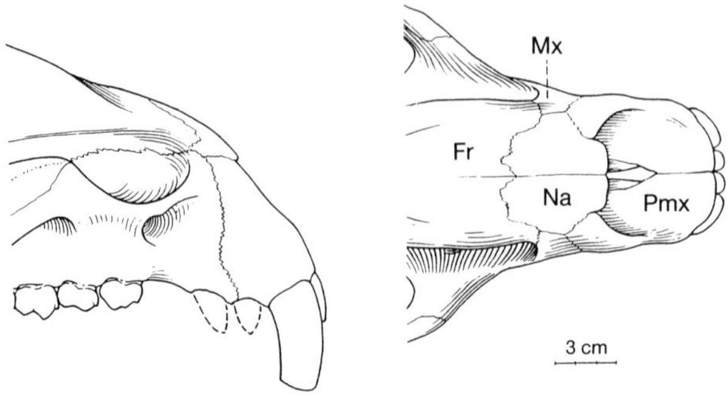

Fig. 11.3. Snout of *Moeritherium* showing the plesiomorphic condition of the connections of the nasal bone (Na): the short processus nasalis of the premaxilla (Pmx) is not in contact with the frontal (Fr) in left lateral (left) and dorsal (right) views (from Tassy 1981).

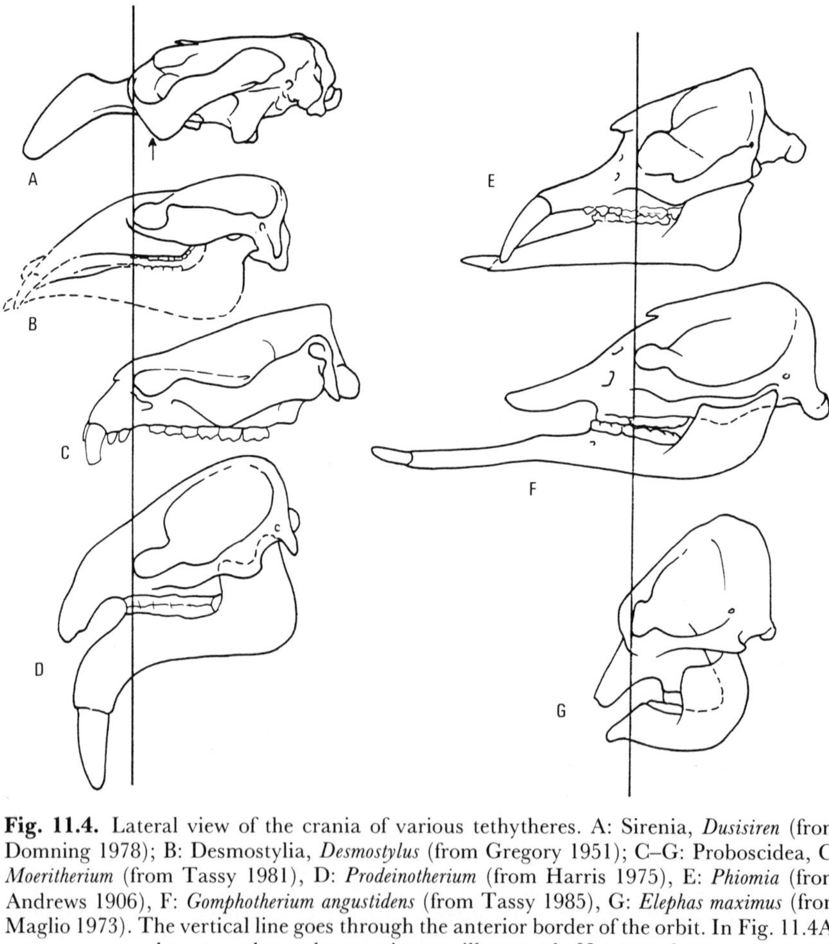

Fig. 11.4. Lateral view of the crania of various tethytheres. A: Sirenia, *Dusisiren* (from Domning 1978); B: Desmostylia, *Desmostylus* (from Gregory 1951); C–G: Proboscidea, C: *Moeritherium* (from Tassy 1981), D: *Prodeinotherium* (from Harris 1975), E: *Phiomia* (from Andrews 1906), F: *Gomphotherium angustidens* (from Tassy 1985), G: *Elephas maximus* (from Maglio 1973). The vertical line goes through the anterior border of the orbit. In Fig. 11.4A, the arrow shows the anterior maxillary tooth. Not to scale.

sirenian character. The reversals are: retracted orbits in correlation with the retraction of external nares (trunk) and development of tusks in *Palaeomastodon* and *Phiomia*; and subsequently a forward shifting of the orbit in elephants due to the extreme forshortening of the skull (strong vertical component in the growth of the skull, elevated basicranium, and anteriorly displaced face). Sometimes simple characters such as the position of the orbits have a complex history.

Monophyly of Elephantidae and the extent of Proboscidea

The diagnosis of Elephantidae Gray, a family-group taxon, within Proboscidea Illiger, a taxon of ordinal rank, is only meaningful if fossils are included, giving the Proboscidea a wider definition.

The characters that make the extant elephants so distinctive appeared in a sequence, the first fossil records of which go back to the Eocene (Algeria, Egypt, Libya). Recent proboscidean systematics produced synapomorphic schemes for the order, based on 140 dental and non-dental, mainly cranial, characters (Tassy 1982, 1985: fig. 19, table 19, p. 63) and on 58 non-dental, mainly postcranial, characters (Shoshani 1986*a*, table 54, p. 399).

Here, we will summarize only the main transformation series within Elephantiformes (*sensu* Tassy 1988: i.e. from the Fayum genera *Palaeomastodon* and *Phiomia* up to the extant elephants) and detail the first dichotomies, including the origin of the Proboscidea (a problem tied to the recognition of the fossil sister-group of Proboscidea).

Elephants and elephant-like mammals

Figure 11.5 summarizes the state of knowledge on elephantiform interrelationships (detailed autapomorphies for terminal taxa are found in Tassy 1982, 1985; Shoshani 1986*a*). The Mammutidae include *Eozygodon*, *Zygolophodon*, *Miomastodon*, *Pliomastodon*, and *Mammut*. The Amebelodontidae include *Archaeobelodon*, *Serbelodon*, *Protanancus*, *Amebelodon*, *Platybelodon*, and perhaps *Gnathabelodon*. The trilophodont gomphotheres include *Gomphotherium*, *Rhynchotherium*, *Eubelodon*, *Cuvieronius*, *Haplomastodon*, *Notiomastodon*, and *Stegomastodon*; the tetralophodont gomphotheres include *Tetralophodon*, *Anancus*, and *Paratetralophodon*. The Stegodontidae include *Stegolophodon* and *Stegodon*. Figure 11.5 shows unresolved multifurcations (multifurcations E and F/G corresponding to the 'gomphotheres', the stem group of both Elephantidae and Stegodontidae). The monophyly of what we call here Elephantoidea (dichotomy D) is argued, and challenges the hypothesis of polyphyly for that group supported by Matsumoto (1924), Tobien (1976), and Madden (1980). These authors favour close relationships

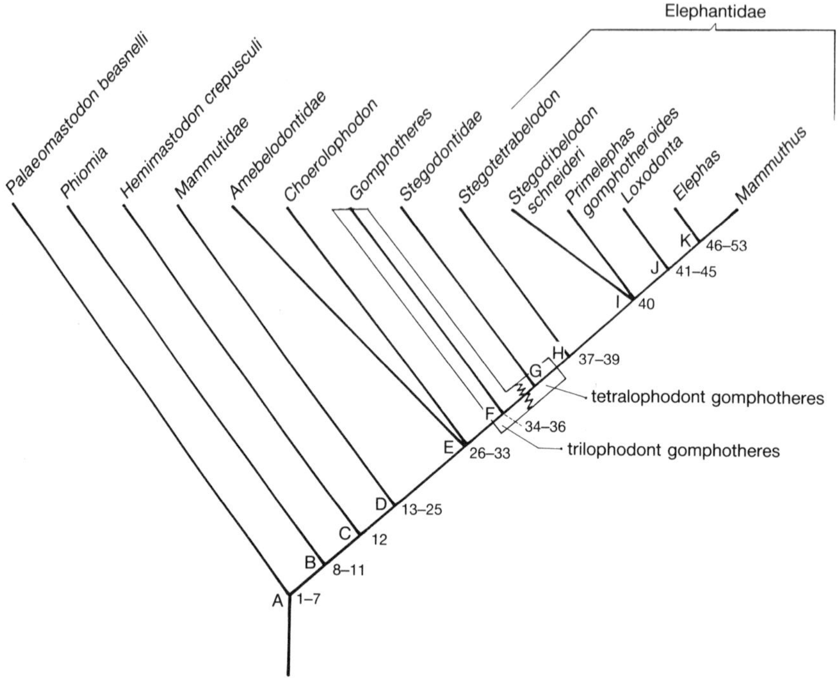

Fig. 11.5. Interrelationships of Elephantiformes. The white area that embodies the terminal taxon 'gomphotheres' and dichotomies F and G, shows the topographic extension of the gomphotheres as a stem group. Characters 1–30, 34, 36–41, 45, 53, after Tassy (1982, 1985); characters 31–33, 35, 42–44, 46–52 after Shoshani (1986*a*). A. ELEPHANTIFORMES: (1) tusks (enlarged dI^2/dI_1); (2) orbits retracted above M^1; (3) palatine with spina nasalis posterior above choanae; (4) bilophodont P^4 with interloph slightly opened labially; (5) lingual cusps of upper molars and labial cusps of lower molars enlarged transversely; (6) M^3 with enlarged postentoconule; (7) enlarged protoconulid (M_{1-3}) and posthypoconid central conule ($M_{1,2}$). B. UN-NAMED: (8) premaxilla slightly expanded dorsally at the base of the external nasal opening (see character 16 and Fig. 11.6); (9) foramen supraorbitale in the maxilla (conditions 8 and 9 not seen in *Palaeomastodon*); (10) M^3 with enlarged posterolabial cusp; (11) fully trilophodont $M^{1,2}$ and M_3. C. UN-NAMED: (12) fully trilophodont M^3 (i.e. subdivided postentoconule and large post-trite cusps on the third loph), postcingulum connected to postentoconule. D. ELEPHANTOIDEA: (13) elevation of the skull; wider cerebral part and dissociated lineae temporale (i.e. no sagittal crest); (14) premaxilla with lengthened fossa incisiva; (15) in the nasal fossa, shortening of the insertion of the ethmoid cartilage on the premaxilla; (16) premaxilla strongly expanded dorsally at the base of the external nares (see Fig. 11.6); (17) bulla (ectotympanic area) slightly enlarged posterolaterally; (18) loss of the canalis temporalis (= squamosal–sinus canals); (19) foramen ovale and foramen lacerum medium confluent; (20) horizontal functional succession of maxillary teeth in association with a peculiar growth pattern and remodelling of adjacent bones; (21) P^2 reduced; (22) reduced protocone on P^3; (23) reduced protoconid on P_3; (24) fully molarized P^4 (i.e. interloph entirely opened labially); (25) M_3 tetralophodont with a central conule on the posterior side of the third pretrite half-lophid. E. UN-NAMED: (26) lacrimal foramen generally not opened in the orbit at adult growth-stage; (27) palatine with reduced spina nasalis posterior; (28) external nasal opening enlarged laterally; (29) enlarged tritolophid on $M_{1,2}$; (30) cement on crowns of teeth (parallelism in advanced mammutids); (31) deep bicipital groove of humerus and higher

between *Palaeomastodon beadnelli* and Mammutidae on the basis of dental characters (a less parsimonious hypothesis: see discussion in Tassy 1982). Figure 11.5 also shows both Elephantidae and Stegodontidae (two genera *Stegolophodon* and *Stegodon*) as derivatives of tetralophodont gomphotheres, though the Stegodontidae had been regarded as specialized mammutids (Maglio and Hendey 1970; Maglio 1973; Tobien 1976, 1978*b*) on the basis of dental characters. Newly available data on *Stegolophodon* fit best with a bunodont (=gomphothere) origin (Tassy 1983). Nevertheless, the sequence of early events relative to the differentiation of Elephantidae and Stegodontidae is obscured by the paraphyletic nature of the gomphotheres. One primitive trilophodont grade distinct from the more derived type species *Gomphotherium angustidens* from Western Europe has been recognized in the Early Miocene of the Old World and named the '*G. annectens* group' by Tassy (1985). North American trilophodont gomphotheres probably derived from that early stock, but this needs a systematic revision to elucidate the phylogeny more precisely. To complete this discussion on extant elephants, it may be noted that a sister-group relationship between *Elephas* and *Mammuthus* is better supported by osteological criteria than by biochemical data (Shoshani *et al.* 1985*a*,*b*; Shoshani 1986*a*).

Early proboscideans and the extent of Proboscidea

The classification of deinotheres (two genera *Prodeinotherium* and *Deinotherium*, known from Miocene to Early Pleistocene), of the Eocene and Oligocene genus *Moeritherium*, and of the Eocene genus *Barytherium*

---

trochanter; (32) carpus and tarsus less 'cursorial' (parallelism in advanced mammutids); (33) lunar with enlarged articular surface for ulna. F. UN-NAMED: (34) cranium with enlarged cerebral part; the tympanic participates in the construction of the internal side of the base of the skull (condition not seen in previous taxa); (35) presence of a posterior ramus on the stylohyoid (probably present at trifurcation E but condition not seen in *Choerolophodon*); (36) lower tusks with a piriform transverse section. H. ELEPHANTIDAE: (37) elevation of the basicranium; (38) loss of the enamel band on the upper tusks; (39) molars with five loph(id)s or more; loph(id)s become laminae. I. UN-NAMED: (40) reduced mandibular symphysis and reduced lower tusks. J. UN-NAMED: (41) molar crowns as high as wide; (42) reduced shelf at the base of the inferior ramus of stylohyoid (condition not seen in other Elephantidae: dichotomy H–I); (43) Olecranon process more tilted medially and extends proximally beyond the trochlear notch; (44) pelvis with well-delineated distal obturator notch; (45) loss of supraorbital foramen (probably already at dichotomy H). K. UN-NAMED: (46) large dorsal parietal bulges and fronto-parietal concavity; (47) rim of the orbit with angular corners; (48) maxillary ridges prominent and close to each other and to the midline; (49) process at the anterior border of the mandibular foramen; (50) stylohyoid more gracile (Fig. 11.7); (51) 18 and 19 thoracic vertebrae (20 and 21 in *Loxodonta*), though 20 in *E. m. sumatrensis*; (52) reduced lateral bulging of the magnum articular facet of the scaphoid; (53) molars with thinner and more numerous laminae.

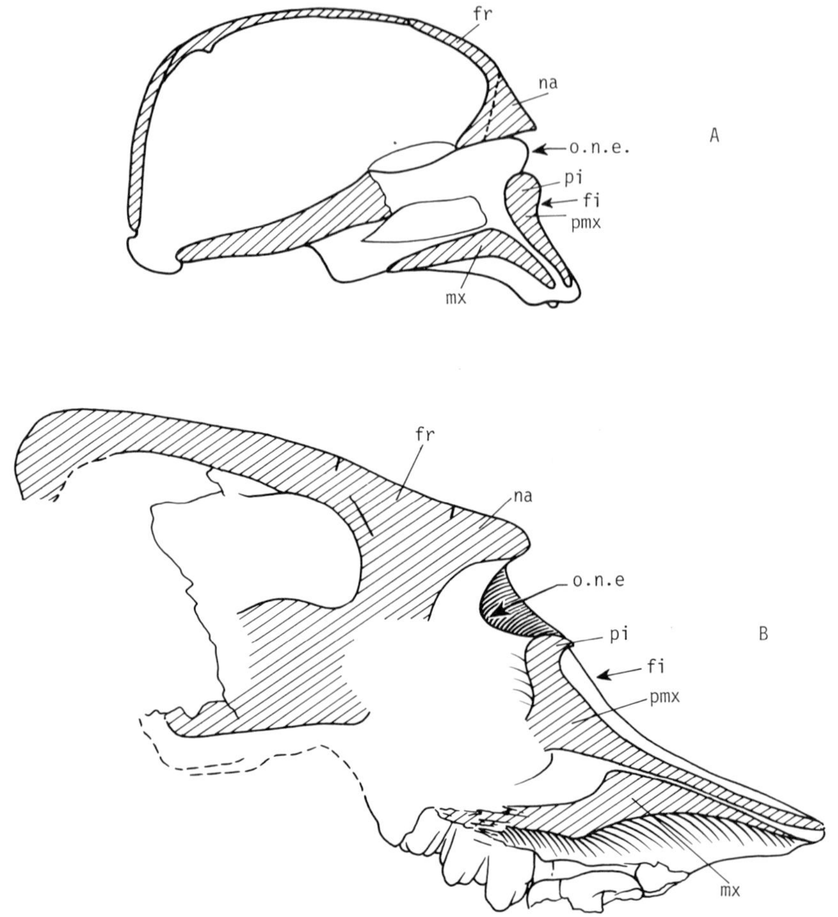

Fig. 11.6. Sagittal section of crania of *Loxodonta africana* (A) and *Gomphotherium angustidens* (B) from the Miocene of Gers, France (228; MHN, Toulouse). A: foetus (from Eales, 1926), B: juvenile with $Dp^{2,3}$ in function and erupting Dp^{4} (from Tassy, 1985). fi: fossa incisiva; fr: frontal; mx: maxilla; na: nasal; o.n.e.: external nares; pi: posterodorsal processus of the fossa incisiva (synapomorphy of elephantiforms); pmx: premaxilla. Not to scale.

has long been discussed. The problem recently has been renewed with the discovery of a new species from the late Early Eocene of Algeria, *Numidotherium koholense*, by Mahboubi *et al.* (1984, 1986).

Maglio (1973) questioned the inclusion of the deinotheres, *Moeritherium*, and *Barytherium* in the Proboscidea, and treated these taxa as separate orders. In parallel, Tobien (1971, 1976, 1978*a*) excluded *Moeritherium* from the proboscideans to classify it more generally among Paenungulata. In the book '*Evolution of African Mammals*', Moeritherioidea, Deinotherioidea, and Barytherioidea were treated separately

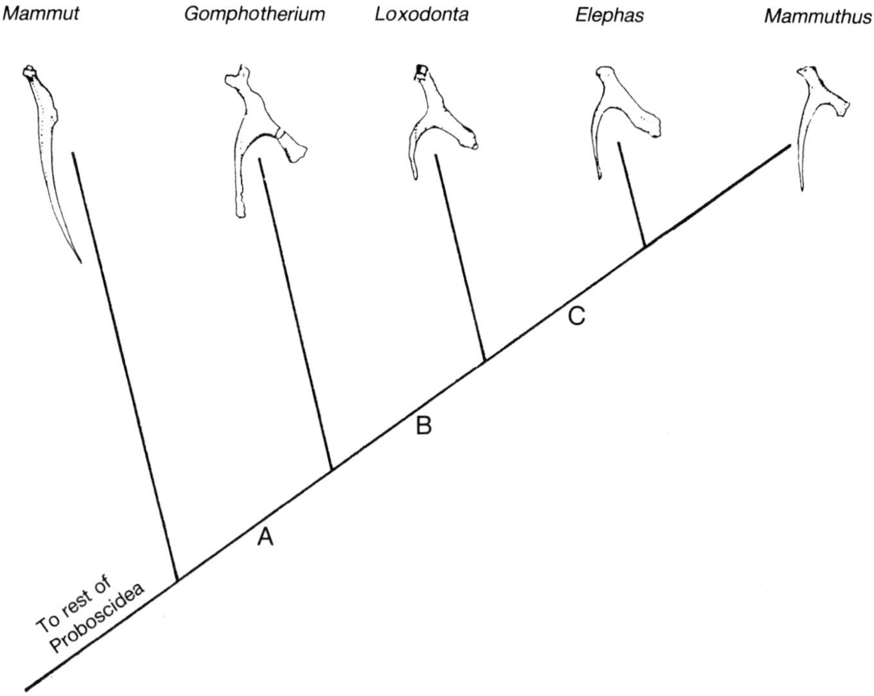

Fig. 11.7. Relationship among five elephantoid genera based on characters of the stylohyoid bone. A: presence of a posterior ramus, B: a reduced shelf at the base of the inferior ramus, C: delicate bone, flat and oblong in cross-section (from Shoshani 1986*a*).

from Proboscidea (respectively by Coppens and Beden 1978; Harris 1978; and Coppens *et al.* 1978).

The discussion of the definition and diagnosis of Proboscidea is endless. One approach is to include in Proboscidea every taxon that is closer to living proboscideans than to their extant sister-group, the Sirenia (following Hennig's (1969) concept of '*group'). Hence, the diagnostic characters of Proboscidea become few if early known fossil forms are poorly represented. Close relationships of deinotheres and *Moeritherium* to elephants were hypothesized by Tassy (1979, 1981, 1982) and confirmed by Domning *et al.* (1986). The proboscidean characters listed in the caption to Fig. 11.1 are based on the analysis of *Moeritherium*, but very few characters are left if *Barytherium* and *Numidotherium* are included in the Proboscidea (Fig. 11.8). The reinsertion of *Barytherium* in the Proboscidea has been argued by Tassy (1982, 1985), Shoshani (1986*a*), and Domning *et al.* (1986) with different opinions on its exact position. Tassy (1982, 1985) used only one character (the morphology of the astragalus with a salient *tuberculum mediale*) to include the genus in the Proboscidea, and further added two

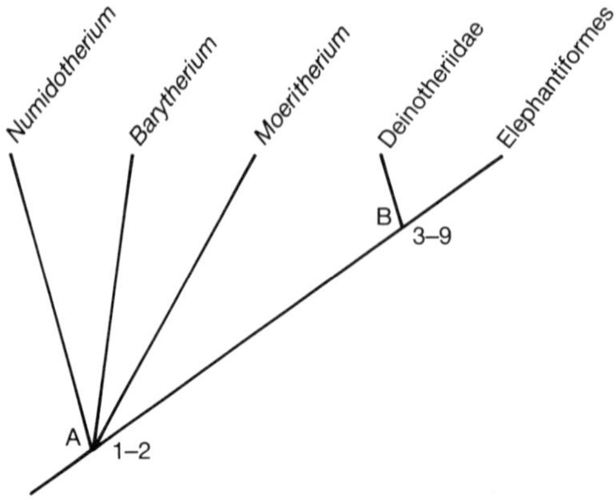

Fig. 11.8. The multifurcation at the origin of the Proboscidea summarizes the conflicting evidence (see Fig. 11.9) and the sister-group relationships between deinotheres and elephantiforms. A. PROBOSCIDEA: (1) lower molars with cristid obliqua entirely labial; (2) astragalus with prominent tuberculum mediale. B. DEINOTHERIIDAE + ELEPHANTIFORMES: (3) ethmoid foramen situated backward in the orbitotemporal fossa; (4) cranial pneumatization (also in *Barytherium* and *Numidotherium*); (5) nasal fossa retracted with contact between frontal and premaxilla; (6) strong ligamentum nuchae: insertion on the occipital through a deep nuchal fossa with a median crest and passage through a deep groove on top of the neural apophysis of the axis; (7) hypocone on $P^{3,4}$, P_4 bilophodont, with a posthypoconid crest, M^1 and M_1 with a third loph(id), distocrista on M^3 (a third crest not separated from the second by an interloph); (8) loss of P_2 (though some isolated premolars had been described as P_2 of deinotheres; see discussion in Tassy 1982, p. 231; (9) elephant-like skeleton (though the body of deinotheres—as for primitive elephantiforms—is longer than the body of extant elephants, the postcraniae as a whole are very much alike; derived features of the astragalus and calcaneum seen in the primitive condition in the other three proboscidean genera are the reduced fibular facet of the astragalus and the laterally-protruding fibular facet of the calcaneum).

dental characters: loss of P^1 and P_1 (already listed by Harris 1978) and the entirely labial cristid obliqua. Hence, *Barytherium* appeared to be less close to elephants plus deinotheres than was *Moeritherium*. Domning *et al.* (1986) placed *Barytherium* closer to deinotheres and elephants on the basis of loss of I^3, C^1, I_1, the enlarged I_2, the enlarged coracoid process of the scapula (but perhaps due only to larger size than *Moeritherium*), graviportal stance, and 'size larger (and still larger).' (The latter character is weak since earliest elephantiforms were smaller than *Barytherium*.) The manus and pes of *Barytherium* remain poorly known. A detailed study of that genus is awaited.

The 'Algerian form' (Mahboubi *et al.* 1984), already included in the Proboscidea by Shoshani (1986*a*) and Domning *et al.* (1986), was recently described by Jaeger (in Mahboubi *et al.* 1986) as *Numidotherium koholense*. He provided a list of characters to support either a sister-group relationship of *Numidotherium* with deinotheres + elephants, or direct ancestor-descendent relationship between *Numidotherium* and deinotheres. In both alternatives (where *Barytherium* is lacking) *Moeritherium* is depicted as a sister-group to the rest of the proboscidean taxa (see fig. 16 *in* Mahboubi *et al.* 1986). In Fig. 11.9 we summarize the conflicting characters and the different parsimonious solutions.

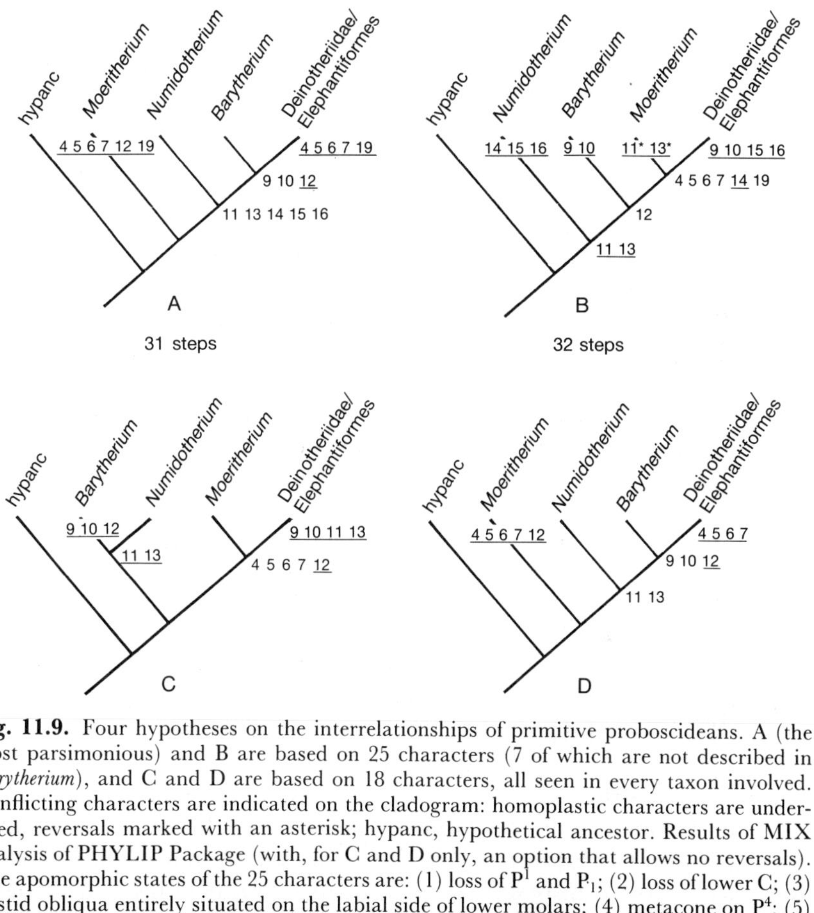

Fig. 11.9. Four hypotheses on the interrelationships of primitive proboscideans. A (the most parsimonious) and B are based on 25 characters (7 of which are not described in *Barytherium*), and C and D are based on 18 characters, all seen in every taxon involved. Conflicting characters are indicated on the cladogram: homoplastic characters are underlined, reversals marked with an asterisk; hypanc, hypothetical ancestor. Results of MIX analysis of PHYLIP Package (with, for C and D only, an option that allows no reversals). The apomorphic states of the 25 characters are: (1) loss of P^1 and P_1; (2) loss of lower C; (3) cristid obliqua entirely situated on the labial side of lower molars; (4) metacone on P^4; (5) lengthening of the lower diastema: P_4 behind the hind border of symphysis; (6) postentoconulid on M_3, outlined on $M_{1,2}$; (7) lower molars with posterior lophid as high as the anterior one; (8) P_4 bilophodont; (9) loss of an upper I (probably I^1); (10) loss of upper C; (11) enlarged infraspinous fossa of scapula; (12) loss of entepicondylar foramen of humerus; (13) high supracondylar crest on humerus; (14) enlarged proximal epiphysis on ulna; (15) massive olecranon on ulna; (16) femur longer than tibia; (17) enlarged lateral facet on the astragalus for calcaneum; (18) enlarged medial facet on the astragalus for calcaneum; (19) external auditory meatus closed ventrally by squamosal; (20) retracted nares: contact between premaxilla and frontal; (21) occipital with deep nuchal fossa, with a median crest for the insertion of the ligamentum nuchae; (22) pneumatization of skull; (23) retracted ethmoid foramen; (24) loss of P_2; (25) on the astragalus, shortening of the fibular facet.

Not all characters checked in *Moeritherium* and used as a diagnosis of Proboscidea (Fig. 11.1) are observed in *Barytherium* and *Numidotherium* (Fig. 11.8). The latter genus shows a remarkable association of primitive and derived features; some of them are autapomorphic features such as the extreme elevation of the basicranium. The postcranial skeleton shows derived characters shared also by deinotheres and elephantiforms, but not by *Moeritherium* (Fig. 11.9, characters 11 and 13–16). Some are known also in *Barytherium* (11, 13: scapula with large infraspinous fossa and great development of supracondylar crest of humerus); others (14–15) are not fully described in *Barytherium* but are considered as derived in a parsimony analysis based on the combination of other characters (Fig. 11.9). On the other hand, a feature such as the entepicondylar foramen of the humerus (12 of Fig. 11.9), already lost in *Moeritherium* and *Barytherium*, is a surprising retention of a primitive trait for proboscideans (see Jaeger in Mahboubi *et al.* 1986, pp. 34–35, fig. 10c: anterior view of a right humerus *contra* the caption). Among cranial characters of *Numidotherium*, another primitive feature described by Jaeger (in Mahboubi *et al.* 1986, pp. 27, 42) is the lack of a ventral closure of the external auditory meatus by the post-tympanic and postglenoid parts of the squamosal. This character is known as the derived condition in *Moeritherium*. Based on these last characters only, *Numidotherium* might be considered as the most primitive proboscidean.

Finally, one dental character must be discussed here (character 1, Fig. 11.5): the hypothesis of homology of the tusks in all proboscideans defended by Shoshani (1986*a*), Domning *et al.* (1986), and Jaeger (in Mahboubi *et al.* 1986) and interpreted as I^2 and I_2. The inflated upper incisor belongs to the second locus in all known proboscideans (in *Numidotherium*, the alveolus of I^1 is small). The sequence of incisor eruption is not known in every taxon. Tassy (1981, pp. 112–15; 1986, pp. 82–4) supports the hypothesis that elephant tusks are milk incisors, and that the enlarged I^2 of *Moeritherium* is an incisor of the second dentition (the same, probably, for *Numidotherium*). In the mandible, the inflated incisor is not the same in *Moeritherium* (lateral one) and in *Numidotherium* (medial one), which implies a parallelism in the loss of the third incisor, or a convergence in the enlargement of one incisor. Moreover, an undescribed symphysis of *Barytherium grave* (MNHN, coll. Arambourg) shows an unerupted incisor between the two erupted ones. If the three incisors belong to the same dentition, and if this trait is confirmed in other specimens, this species would be the only proboscidean with three lower incisors.

Other dental traits, such as the loss of an upper I and of an upper C, shared by *Barytherium*, deinotheres, and elephantiforms, are treated as synapomorphies in parsimony analyses (Fig. 11.9A). On the basis of 25 characters (7 of which are not fully described in *Barytherium*), the shortest tree takes 31 steps (Fig. 11.9A, which agrees with Domning *et al.*'s (1986) hypothesis). It takes only one more step to get *Moeritherium* closer to deinotheres and elephantiforms (Fig. 11.9B). If the analysis is restricted to the 18 characters observed in each taxon, the most-parsimonious tree contains 22 steps (with *Moeritherium* as a sister-group of deinotheres plus elephantiforms, and *Barytherium* the sister-group of both). This hypothesis contains some unlikely reversals (a common problem with parsimony analyses), so that if a 'no reversal' option is introduced, two trees with 23 steps are produced (Fig. 11.9C, D). These results show the extreme flexibility of the constructions. Too many characters remain not fully described. New detailed analyses, particularly of *Barytherium*, will surely provide new tree(s).

The sirenians

Elsewhere in this chapter, we discuss in detail the characters influencing the phylogenetic position of Sirenia within Tethytheria. In this section, we elaborate on two specific issues, namely: the monophyly of sirenians, and do the sirenians belong within Tethytheria?

The monophyly of Sirenia

Sirenians show clear-cut characters that are undoubtedly evidence of the monophyletic nature of the group. Nevertheless, the condition of two characters listed in Fig. 11.1 is not known yet in *Prorastomus*, the most primitive sirenian genus from the Eocene of Jamaica (Owen 1875; Savage 1976): the loss of the ethmoid foramen (character 20) and the reduced hind limbs (character 22). The loss of the ethmoid foramen (Shoshani 1986*b*) is also a derived condition for desmostylians (see later). The auditory region of *Prorastomus* is primitive compared to other sirenians (Savage 1976), but the inflated petrosum and the drop-like ectotympanic can be used as early derived conditions of the morphocline of the auditory region of sirenians. We must emphasize that the premaxillo–frontal contact (19 of Fig. 11.1) seen in all sirenians including *Prorastomus* is surely a synapomorphy for Sirenia, not for Sirenia plus Proboscidea since this contact occurred among the Proboscidea but not at the origin of the group (Fig. 11.3).

Do the sirenians belong within Tethytheria?

Figure 11.1 explicitly answers the question. But this dilemma originated with McKenna's (1975) concept of Preptotheria for Deltatheridia plus Tokotheria, a monophyletic group defined by the loss of one postcanine locus (last molar), i.e. with seven postcanine loci. The existence of eight postcanine loci in primitive sirenians like *Prorastomus*, *Protosiren*, *Eotheroides*, and *Prototherium* (Owen 1875; Sickenberg 1934; Savage 1976; Domning 1978; Domning *et al.* 1982) would exclude the Sirenia from the Tokotheria. The same observation and conclusion had been reached for other mammalian groups: primates, erinaceid insectivores, and dermopterans (Krishtalka 1978; Krishtalka and Schwartz 1978; Schwartz and Krishtalka 1976) but was refuted for the primates (Luckett and Maier 1982). Nevertheless, the existence of eight postcanine loci in Sirenia is not equivocal and apparently falsifies the taxon Tokotheria (Domning *et al.* 1982; Tassy 1985). Another position is held by Novacek (1986*b*), McKenna (in Shoshani 1986*a*, appendix D), and Domning *et al.* (1986, p. 46): the taxon Tokotheria is not falsified but sirenians are not tokotheres. Hence, Sirenia would not be tethytheres any longer. Then, many of the morphoclines reviewed here should be withdrawn. Consequently, numerous parallelisms should be invoked for the derived characters shared by sirenians on the one hand and proboscideans plus desmostylians, and even hyracoids, embrithopods, and perissodactyls, on the other. The interpretation of the loss of one postcanine locus as a unique evolutionary novelty for Tokotheria (sirenians excluded), with no homoplasy, remains to be argued against contradictory evidence.

The origin of Proboscidea in relation to desmostylian relationships, and the status of *Anthracobune*

Desmostylians are amphibious animals known from marine deposits in the norther circum-Pacific region (Japan, North America) of Oligocene to Late Miocene age. They were included in Tethytheria by McKenna (1975), followed by Tassy (1981), who considered a sister-group relationship between Desmostylia and Proboscidea plus Sirenia, Shoshani (1986*a*, p. 293) accepts sister-group relationships between Desmostylia and Sirenia only, and Domning *et al.* (1986), after a description of a new Oligocene primitive desmostylian genus, *Behemotops*, concluded that Desmostylia are the sister-group of Proboscidea (see history of desmostylian systematics in Domning *et al.* (1986)). Figure 11.10 summarizes the contradictory evidence on tethytherian interrelationships.

The reduction of the mastoid in Sirenia and Proboscidea only was used by Tassy (1981) as evidence of a sister-group relationship between the two groups. Recently, this morphocline has been refuted (Novacek and Wyss 1987) with the demonstration that the mastoid previously described by VanderHoof (1937) in the genus *Desmostylus* was a

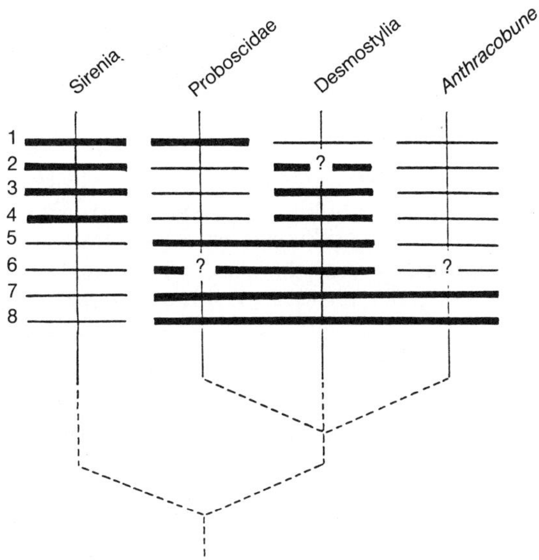

Fig. 11.10. Contradictory evidence on tethytherian interrelationships (see text). Black bars indicate derived states of characters: (1) true bunolophodonty (narrow median sulcus); (2) pachyostosis and osteosclerosis; (3) loss of ethmoid foramen; (4) dorsal extension of zygomatic part of squamosal; (5) high external auditory meatus; (6) external auditory meatus closed ventrally by the post-tympanic and postglenoid parts of squamosal; (7) seven postcanine loci; (8) inflated postcingulum with outlined postentoconulid on lower molars.

misinterpretation. Yet, no suture is seen on the *Paleoparadoxia tabatai* specimen, a more primitive species. Novacek and Wyss concluded convincingly that the amastoid condition was present in the ancestor of tethytheres: the sirenians could have developed secondarily an inflated, externally apparent mastoid in connection with the opening of an occipital fenestra (it was not an intermediate state before complete reduction). Amastoidy was the basis of sister-group relationships between Sirenia and Proboscidea. Since it occurs also in Desmostylia, we are left (Fig. 11.10) with the hypothesis of sister-group relationships between Proboscidea and Desmostylia (a hypothesis favoured by Domning *et al.* 1986) on the basis of the high position of the external auditory meatus. It is less certain that the ventral closure of the external auditory meatus is a synapomorphy, since *Numidotherium* lacks the derived condition and a reversal of this character is unlikely (in Fig. 11.10 this trait is coded '?' for Proboscidea). Domning *et al.* (1986) added two characters: on lower molars, a small entoconid II (= postentoconulid = pcld in Fig 11.2) adjacent to the hypoconulid (character 8 of Fig. 11.10), and former last molars lost (character 7 of Fig. 11.10). Character 8 is tied to the enlargement of the postcingulum and, as we will discuss later, some parallelism must occur for this trait within Tethytheria. Character 7 is relevant to the questions of the monophyly of Tethytheria and the relationships to other mammalian orders, and is discussed elsewhere.

Figure 11.10 shows parallel development of the dorsal extension of the zygomatic part of squamosal (character 4) between Sirenia and Desmostylia. Indeed, this extension is weaker in *Prorastomus* than in other sirenians, and even weaker in primitive desmostylians than in *Desmostylus*. There is no clear evidence of true pachyostosis and osteosclerosis (character 2) in desmostylians, though an increased proportion of dense bone is reported in a rib assigned to *Paleoparadoxia* (Mitchell 1963) and in a horizontal ramus of a *Desmostylus* by VanderHoof (1937, p. 188). In desmostylians, osteosclerosis is not as extensive as in sirenians. Ribs of *Desmostylus* are partly osteosclerotic, like ribs of some mysticete whales (Domning, *pers. comm.*, 1987); hence, osteosclerosis not truly present at the origin of Desmostylia evolved independently of Sirenia. Ribs of *Paleoparadoxia* and *Desmostylus* do not show the typical pachyostosis (increased volume of bones) seen in sirenians (Domning, *pers. comm.*, 1987). We prefer to consider these characters united in character 2 in Fig. 11.10 as a synapomorphy of Sirenia (Savage 1976; Tassy 1981; Domning *et al.* 1986).

Among the tethytherian characters (Fig. 11.1) found in Desmostylia are the anterior position of the orbits and the laterally expanded squamosals. The condition of these characters is not known in *Behemotops*.

The Asian genus *Anthracobune* from the Eocene of Pakistan is not included here within the proboscideans. This genus, and allied forms such as *Jozaria* described by West (1980, 1983) and Wells and Gingerich (1983), have already been classified among moeritheres or more generally among Proboscidea or Tethytheria, as the Family Anthracobunidae. The family was classified among Proboscidea by Wells and Gingerich, who concluded also that the Anthracobunidae were likely to be '"basal tethytheres" *sensu* McKenna, 1975', a statement that emphasizes uncertainties surrounding those Asian animals. No derived characters shared only by *Anthracobune* and Proboscidea had then been described. Indeed, the recent character analysis by Domning *et al.* (1986) retains only one character as a synapomorphy for these two groups: two cusps that form the posterior part of the lower molars, the posterolingual cusp being the evolutionary novelty (entoconid II = postentoconulid = plcd of Fig. 11.2). This cusp is already outlined in the primitive desmostylian genus *Behemotops* and, following Domning *et al.* (1986), in the ancestral morphotype of Desmostylia plus Proboscidea. Moreover, the cusp in *Anthracobune* is much more enlarged than in *Moeritherium* (Fig. 11.2), where it is sometimes only an inflated wrinkled part of the postcingulum. In primitive lophodont proboscideans (*Numidotherium* and *Barytherium*), the posterolingual area of the cingulum does not form a distinct cusp, and on the M_3 only one inflated hypoconulid is seen. If the condition seen in *Anthracobune* is present in the ancestral morphotype of Proboscidea, a reversal would be needed for primitive proboscideans, followed by a secondary enlargement of the cusp in elephantiforms. Alternatively, a parallelism could have occurred in *Anthracobune* on the one hand and among proboscideans on the other, which is likely. As was shown by Domning *et al.* (1986, p. 44), the tendency to form a posterolingual cusp on lower molars is already outlined in the species *Minchenella grandis* from the Upper Palaeocene of Nanxiong, China, a species previously described as a phenacolophid (Zhang 1978) and recognized by Gingerich and Russell (1981, p. 237) as related to *Moeritherium* and *Anthracobune*. Clearly, the big posterolingual cusp seen on the M_3 of *Anthracobune* could easily be an autapomorphic feature, or a desmostylian tendency, since this cusp in primitive proboscideans is smaller than the hypoconulid and never became higher in more evolved forms, a condition seen in desmostylians such as *Paleoparadoxia* (but not in *Behemotops*): see Domning *et al.* (1986, fig. 12, p. 16). This homoplasy precludes any confident allocation of *Anthracobune* to Desmostylia, which is suspected mainly because of the overall similarity displayed by the dentitions of *Anthracobune* and *Behemotops*.

Other dental traits are affected by homoplasy. The derived condition of the cristid obliqua (Fig. 11.2)—held as a synapomorphy of Proboscidea—is not seen in *Anthracobune* and allied forms. But it is nearly indicated in *Behemotops* (Domning *et al.* fig. 7, pp. 9 and 18) where the anterior part of the cristid obliqua forms a cuspule at the base of the posterolabial slope of the metaconid: i.e. it is labially displaced. The same observation can be made on a M_3 of *Lammidhania wardi* described by West (1980, p. 529); see Tassy 1985, p. 192). Certainly, *Behemotops* is not a proboscidean; subsequently in other desmostylians, the cuspules of the cristid obliqua do not form true cusps. Another parallelism between *Anthracobune* on the one hand, and *Moeritherium* and primitive proboscideans on the other, affected the more molarized premolars of the first genus (West 1980). However, the bunolophodont structure is not realized in *Anthracobune* and allied forms (plesiomorphic condition), where the median sulcus is still very wide: there is no high cuspule on each side of the sulcus to unite the labial and lingual main cusps. Here, again, we are left with extreme difficulty in identifying firm synapomorphic traits for dental structures.

In conclusion, *Anthracobune* could either be a desmostylian (our opinion) or a proboscidean genus, or even be classified as Tethytheria *incertae sedis*. D. P. Domning (*pers. comm.* to J. S., 1987), however, noted that the dentition of a new specimen of *Behemotops* shows close resemblance, but not necessarily close cladistic relationships, to the dentition of *Anthracobune*. These uncertainties perhaps explain the phylogram given by Domning *et al.* (1986, fig. 23, p. 45), which shows as trifurcation between *Anthracobune*, Proboscidea, and Desmostylia, as in Fig. 11.5. Interestingly, the ancestral position for all tethytheres of *Anthracobune* (= *Pilgrimella*) corresponds to E. Manning's original idea (*pers. comm.* to P. T., 1979).

Relationships of Tethytheria to other mammalian orders: the paenungulate controversy

The molecular evidence

Weitz (1953) studied *Elephas maximus* and *Procavia capensis*, and found them to be closely related. The first study to indicate that Proboscidea, Sirenia, and Hyracoidea (hereafter P, S, H) were closely related from the molecular standpoint was presented by Shoshani *et al.* (1977, 1981). They also stated that *Orycteropus afer* (Tubulidentata) was the most closely related extant order to P, S, H (this hypothesis was restated by Shoshani 1986*a*,*b*).

Subsequent molecular studies that support the relationships among P, S, H and between P, S, H and Tubulidentata include: de Jong *et al.* (1981), Rainey *et al.* (1984), Kleinschmidt *et al.* (1986), and Miyamoto and Goodman (1986). Other molecular studies that augment our knowledge of paenungulate relationships are those of Prager *et al.* (1980) and Benveniste (*pers. comm.* to J. S.); these latter researchers, however, investigated only two of the three (P, S, H) orders. None of the eleven mentioned works provided evidence for close relationships between Hyracoidea and Perissodactyla, the original hypothesis of Owen (1848), which was revived more recently by McKenna (1975) and Fischer (1986) and is discussed below.

Because of the lack of sufficient sequence data, it is not possible to delineate which of the three named orders are closer to each other. Nonetheless, from the molecular viewpoint, the evidence is strong that P, S, H are a monophyletic group that would require seven nucleotide replacements to be 'broken'.

The morphological evidence

Recent examinations of paenungulate interrelationships have produced contradictory results. On the one hand, the monophyly of a group composed of Tethytheria plus Hyracoidea, and sister-group relationships between this group and extinct Embrithopoda, were supported by Shoshani (1986*b*). On the other hand, Fischer (1986) and Prothero *et al.* (Chapter 8 of this volume) supported the monophyly of hyracoids plus perissodactyls (perissodactyls here being called Mesaxonia by Fischer)—a concept related to the Phenacodonta *sensu* McKenna (1975)—and also a sister-group relationship between this group and tethytheres. Shoshani's approach revived the Paenungulata concept, which is also the position supported by Novacek (1982, 1986*a*), Novacek and Wyss (1986), and Novacek *et al.* (Chapter 3 of this volume). On the contrary, Fischer's examination of hyracoid anatomy discarded paenungulate characters. Obviously, convergent evolution, as ever, is the main cause of these discrepancies.

The hypothesis of close relationships between elephants and hyraxes has long been present in the literature and has never completely disappeared (see, for instance, Hanks 1979, p. 103), but most of the characters usually used to assess such relations are clearly plesiomorphic traits (e.g. the abdominal testes), or convergence (e.g. the enlargement of a pair of incisors). Also, the interpretation of the placenta and foetal membranes as synapomorphies of Hyracoidea plus Tethytheria by Novacek and Wyss (1986), after Wislocki and Westhuysen (1940), has been criticized by Perry (1974), Shoshani (1986*a*), and Fischer (1986); early development in *Loxodonta*, *Trichechus*, and *Procavia* shows numerous differences, among which (Amoroso and Perry 1964, *fide* Fischer 1986) the placenta of the elephant is of the endotheliochorial type, while those of *Trichechus* and *Procavia* are of the haemochorial type.

Figure 11.11 shows Shoshani's (1986*b*) and Fischer's (1986) hypotheses. Among characters shared by Tethytheria and hyracoids (1–11, Fig. 11.11), some are also seen in Embrithopoda (1–5) and interpreted by Shoshani (1986*b*) as evidence of close relationships between Embrithopoda and Hyracoidea plus Tethytheria. The problem is that none of these characters is uniquely derived for the groups involved. The reduced mastoid (1)—which is cited by Novacek and Wyss (1986, 1987) as a paenungulate trait—is seen among other ungulates such as suoids. Homoplasy affects the exclusion of the supraoccipital in the formation of the foramen magnum (2) within the tethytheres (reversal in derived dugongids). The polarity of the coronoid canal character (3) is not clear. It is also seen in edentates (sloths: McKenna, *pers. comm.* to J. S.), lagomorphs, and artiodactyls and perissodactyls where it is small. Surely the occurrence of the canal is derived for the Eutheria relative to other mammals, but it could be either primitive within the Eutheria (and its loss would then be derived for some eutherian mammals), or its presence could be derived within the Eutheria. We are left with multiple losses or multiple appearances. Homoplasy affects the reduction of the acromion process of the scapula (4), used as

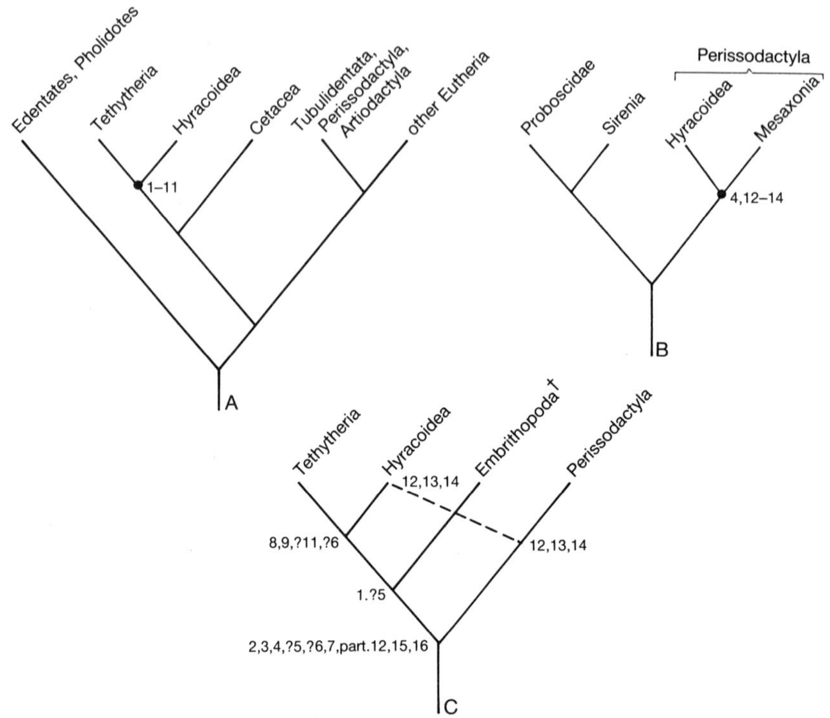

Fig. 11.11. Three hypotheses of the relationships of Tethytheria to Hyracoidea and other mammalian orders: A, the paenungulate hypothesis (from Shoshani 1986*b*); B, the perissodactyl and tethytherian sister-group relationships (from Fischer 1986); C, tentative synthesis of Paenungulata relationships. Character 1, after Novacek and Wyss (1986); 2, 3, 7, after Shoshani (1986*a*); 4, after Shoshani (1986*a*); and Fischer (1986); 5, 6, after Novacek (1982); 8, 12–16, after Fischer (1986); 9, 10, after Novacek (1986*a*). The characters are: (1) amastoidy (mastoid not seen on lateral side of cranium), see Fig. 11.1, character 4; (2) foramen magnum formed by basioccipital and exoccipitals; (3) coronoid canal present; (4) acromion process of scapula reduced; (5) carpal bones serially arranged: lunar–unciform contact very weak or absent; (6) tarsal bones serially arranged: astragalus–cuboid contact absent; (7) coronoid process of dentary: axis is perpendicular to or forms an acute angle with occlusal plane; (8) distal doubling of the M. styloglossus; (9) fenestra rotunda with vertical orientation, partly shielded ventrally by expansion of promontorium cochleae; (10) lambdoid crest weak or occiput expanded; (11) post-tympanic ridge of squamosal very weak or absent; (12) same cursorial pattern of locomotory organs (morphology of the capitulum radii, occurrence and course of M. sternoscapularis, flattened talo-navicular facet, complete reduction of the 1st and 5th toes, additional caudal origin of the M. semimembranosus, morphology of the terminal phalanx, and the formation of the organon digital as a hoof); (13) development on both sides of a voluminous diverticulum from the membranous part of the eustachian tube and derived course of the arteria carotis interna through the foramen lacerum medium; (14) persistence of the tuber maxillare in the adult; (15) presence of a lacrimal tubercle; (16) fossa glandis at the tip of the penis.

a derived trait for Paenungulata by Shoshani (1986*b*) and as a derived trait for hyracoids plus perissodactyls by Fischer (1986). However, this character, contrary to Fischer's (1986, p. 95) comment, could well be present in the ancestor of the tethytheres even if the scapula of elephants shows a strong acromion: this condition is probably derived, since the scapula of *Moeritherium* shows a reduced acromion; the same argument is also validly suggested to explain the reduced acromion of primitive sirenians. Yet, this trait could also be a derived feature for *Moeritherium* and early sirenians (parallelism). Nevertheless, it is possible that the two authors have missed the point. Since the reduced acromion occurred in Perissodactyla, it could well be a synapomorphy for a more inclusive group (tethytheres, hyracoids, perissodactyls, and embrithopods as well).

The problem of the serial manus and pes (5–6, Fig. 11.11) is still puzzling, though a few important contributions have appeared recently on this topic. This could be the best character for paenungulate monophyly and, together with the molecular data, it is used by Fischer (1986, fig. 39, p. 102) to falsify his monophyletic group Perissodactyla (hyracoids plus his 'Mesaxonia'). Tassy (1985) used the serial arrangement for Proboscidea only (which implies that the serial arrangement would not be homologous for the different taxa where it is described). Nevertheless, the question should be divided into two parts, relating to the structure of the manus and of the pes. The serial arrangement of the hand is also known in sirenians, embrithopods (*Arsinoitherium zitteli*), and two 'condylarths' (*Phenacodus* and *Meniscotherium*). Recent reinterpretations of the manus of the Desmostylia (Shoshani 1986*a*, p. 303; Novacek and Wyss 1986, 1987) led to the conclusion that the arrangement could be of the serial type. Novacek and Wyss (1987) propose a different reconstruction of the carpal bones from Shikama (1966), with a contact between the distally expanded lunar and the McIII, and a smaller contact between lunar and unciform. They conclude that this 'unique system' could have been derived from a more basic serial carpal arrangement. Then, the same explanation (a highly specialized arrangement) would be necessary to explain the tarsal arrangement in desmostylians, with a wide contact between astragalus and cuboid (Shikama 1966). If the Sirenia are *hors de combat* for the analysis of the foot, no direct argument can be used to assume such a morphocline, since the desmostylian astragalus is not reminiscent of the proboscidean astragalus. Nonetheless, if we accept a reversal in desmostylians due to their own specialization, the serial pattern of the hand could be derived from a more inclusive group, i.e. hyracoids, *Arsinoitherium*, as well as *Meniscotherium* and *Phenacodus* and, *via* the last genus, Perissodactyla (the alternate pattern seen in this group and in embrithopods being secondarily derived). Moreover, there is a clear-cut discrepancy between the arrangement of the carpals and tarsals. In *Arsinoitherium*, the manus of which shows a serial arrangement, the astragalus and cuboid are in contact, a condition also probably present in *Phenacolophus* (see McKenna and Manning 1977, p. 70). Our conclusion is that direct observation does not give definitive evidence on the carpal and tarsal arrangement of Desmostylia, hence of the ancestral morphotype of Tethytheria. The strength of this character for sister-group relationships between hyracoids and tethytheres is then somewhat weakened.

Character 8 (Fig. 11.11) was used by Fischer (1986, p. 87) to falsify his hypothesis of a hyracoid–perissodactyl relationship. Among characters 9–11 (Fig. 11.11), used by Novacek (1986*a*, p. 98) to support sister-group relationships between hyracoids and tethytheres, character 9 is considered by the present authors as the best evidence for the monophyly of the Paenungulata (Novacek 1986*a*, fig. 31, p. 91): the morphology and orientation of the fenestra rotunda. The fenestra rotunda of *Prorastomus* is more ventrally situated (plesiomorphic condition) than in other sirenians (*pers. obs.*, P. T.), so that some homoplasy could have affected the evolution of this trait among paenungulates. Character 10 does not apply to fossil Hyracoidea, as Fischer (1986, p. 96) already remarked, and must be withdrawn. The distribution of character 11 is not conclusive since it shows homoplasy: its condition is primitive in the Sirenia (Novacek 1986*a*, p. 86).

The extension of the jugal to the posterior border of the glenoid fossa of the squamosal is considered in this paper as a proboscidean synapomorphy (character 8 of Fig. 11.1), and not—as Novacek and Wyss (1986, p. 279) argue—a paenungulate synapomorphy. A posteriorly lengthened jugal is surely a plesiomorphic feature for Eutheria, being already seen in fossils such as *Asioryctes* (Kielan-Jaworowska 1981, p. 31). But, as Novacek and Wyss stress, the backward extension of the jugal to the rear border of the glenoid fossa (proboscideans), or even further (hyracoids), is derived. As the jugal is shorter in sirenians and desmostylians, the lengthening is interpreted as a parallelism between proboscideans and hyracoids; maybe also the variability of this character has been overlooked.

Fischer's (1986) arguments for his concept of Perissodactyla, including Mesaxonia and Hyracoidea, are based on a detailed study of the anatomy of extant hyracoids. He retained three 'character complexes' (12, 13 and 14, Fig. 11.11). The first one is the same cursorial pattern of the locomotory organs: a combination of several traits (12, Fig. 11.11), of which we have withdrawn the reduced scapular acromion (see above), and consequently the acromial reduction of the M. deltoideus. Not only detailed anatomical studies of adults but also the ontogeny of the limbs give good evidence that hyracoids had acquired a secondary plantigrady (Fischer's conclusion) rather than evolved cursorial traits from a plantigrade condition, as Novacek and Wyss (1986, p. 267) suppose. One element of this complex, the specialized hoof, is similar in hyracoids and perissodactyls, but not in elephants and other ungulates (Fischer 1986, p. 37). The other two character complexes (13 and 14, Fig. 11.11) are undoubtedly derived characters for both hyracoids and perissodactyls. The diverticulum of the eustachian tube (13) is a remarkably unique feature. Though Novacek and Wyss (1986, p. 267) remark that the persistence of a tuber maxillare in the adult (14) is seen in elephants and manatees, due to the enlarged maxillary capsule (associated with the 'horizontal' functional tooth succession), it is clearly a parallelism. The plesiomorphic condition is seen in primitive proboscidean genera with no 'horizontal' tooth succession (*Moeritherium*, *Numidotherium*, *Phiomia*, and deinotheres).

One solution that integrates the contradictory conclusions reached by Shoshani (1986*a*,*b*) and Novacek and Wyss (1986) on the one hand, and Fischer (1986) on the other, is to consider several characters as defining a more inclusive group (Fig. 11.11C): Tethytheria plus Hyracoidea plus Perissodactyla (characters 2, 3, 4, 7, part of 12, and perhaps 5 and 6 in Fig. 11.11). Considering the state of knowledge, a parsimony analysis would not be enough to solve the problem. Though such an analysis tends to favour the Paenungulata solution (in agreement with the molecular approach), it must be noticed that Fischer's (1986) character complex 1 (character 12 of Fig. 11.11) is composed of seven characters, showing a close correspondence between perissodactyls and hyracoids. Though some digitigrade characters (part of 12) could have been present in the ancestor of tethytheres, hyracoids, and perissodactyls, the occurrence and course of M. sternoscapularis and the specialized hoof are striking features only of hyracoids and perissodactyls. However, seeing parsimony merely as a kind of enumeration is rather naive. On what grounds should we give the same weight, say, to the amastoid condition of hyracoids and tethytheres seen in some other ungulates, and to the specialized hoof (a trait included in character 12), or to the expanded diverticulum from the tuba eustachii (character 13 of Fig. 11.11)? The last character is uniquely derived in hyracoids and perissodactyls (tapirs and horses; no information is available for rhinoceroses). Figure 11.11C shows that, as several 'paenungulate' characters may be hypothesized to have been present in the common ancestor of the tethytheres plus hyracoids plus perissodactyls, paenungulate synapomorphies are reduced; they are even more reduced if we accept that the serial manus and pes (characters 5 and 6) is a primitive condition inherited from *Phenacodus* and *Meniscotherium*.

The hypothesis of close relationships between Tethytheria, Hyracoidea, and Perissodactyla would contradict the sister-group relationships favoured by Shoshani (1986*b*) for Perissodactyla and Artiodactyla (Fig. 11.11A). We are left with a concept close to Szalay's (1977) Mesaxonia, which includes extant Perissodactyla, Hyracoidea, Proboscidea, and Sirenia. A closer relationship of Perissodactyla to hyracoids and tethytheres (and also cetaceans) is also considered with prudency by Novacek and Wyss (1986, p. 267). In this view, Fischer (1986, pp. 102–103) used two characters to unite tethytheres, hyracoids, and perissodactyls (Fischer's Pantomesaxonia): the morphology of the tip of the penis (a fossa glandis) (character 16 in Fig. 11.11C) and the presence of a lacrimal tubercle (character 15 in Fig. 11.11C). Again, some homoplasy must have occurred in the evolution of this latter character, which is also considered as derived by Novacek (1986*a*, p. 84), but is present in leptictids, edentates, and pholidotans. This lacrimal tubercle is present in primitive sirenians (a smooth bump in *Prorastomus*), and is missing in evolved sirenians; it disappeared with the entire lacrimal bone in *Moeritherium*.

Conclusions

Often in the literature, concluding remarks are general statements in support of a hypothesis proposed by the author(s). Although it appeared to us initially that our hypothesis on the monophyly of the Tethytheria was based on solid ground, upon detailed scrutiny, and cross-examination, we acknowledge that there is much more homoplasy (parallelism, convergence, and reversals) than we might have wished. Furthermore, homoplasy has no boundaries, and it extends much farther than the limits of Tethytheria. Thus, the same statement is true for the hypothesis of paenungulate monophyly, seriously disputed by Fischer's original approach. An attempt has been made in this paper to understand better the basic units of evolution: the evolution of the characters.

Acknowledgements

D. P. Domning, M. Fischer, N. Inuzuka, M. C. McKenna, R. J. G. Savage, S. Lee Shoshani, and R. M. West, are acknowledged for making critical comments and/or sharing their expertise with us. Some specimens figured here were studied courtesy of E. P. J. Heizmann (Stuttgart) and Cl. Sudre (Toulouse). Illustrations were done by D. Visset. Funds to P. T. were provided by INSERM (Unité 155).

References

Abel, O. (1919). *Die Stämme der Wirbeltiere.* Walter de Gruyter, Berlin and Leipzig.

Andrews, C. W. (1906). *A Descriptive Catalogue of the Tertiary Vertebrata of the Fayûm, Egypt.* British Museum (Nat. Hist.), London.

Blainville, H. de (1844). *Ostéographie, fasc. 15: Des lamantins (Buffon), (Manatus, Scopoli), ou gravigrades aquatiques*, pp. 1–140. Fain et Thunot, Paris.

Coppens, Y and Beden, M. (1978). Moerithericoidea. In *Evolution of African Mammals* (eds. V. J. Maglio and H. B. S. Cooke), pp. 333–5. Harvard University Press, Cambridge, Massachusetts.

Coppens, Y., Maglio, V. J., Madden, C. T., and Beden, M. (1978). Proboscidea. In *Evolution of African Mammals* (eds. V. J. Maglio and H. B. S. Cooke), 336–67. Harvard University Press, Cambridge, Massachusetts.

de Jong, W. W. (1982). Eye lens proteins and vertebrate phylogeny. In *Macromolecular Sequences in Systematic and Evolutionary Biology* (ed. M. Goodman) pp. 75–114. Plenum Press, New York.

de Jong, W. W., Zweers, A. and Goodman, M. (1981). Relationship of aardvark to elephants, hyraxes and sea cows from α-crystallin sequences. *Nature (London)* **292**, 538–40.

Domning, D. P. (1978). Sirenian evolution in the North Pacific Ocean. *Univ. Calif. Publ. Geol. Sci.* **118**, 1–176.

Domning, D. P., Morgan, G. S., and Ray, C. E. (1982). North American Eocene sea cows (Mammalia: Sirenia). *Smithson. Contrib. Paleobiol.* **52**, 1–69.

Domning D. P., Ray, C. and McKenna, M. C. (1986). Two new Oligocene desmostylians and a discussion of tethytherian systematics. *Smithson. Contrib. Paleobiol.* **59**, 1–56.

Eales, N. B. (1926). The anatomy of the head of a foetal African elephant, *Elephas africanus (Loxodonta africana). Trans. R. Soc. Edinburgh* **54**, 491–546.

Eales, N. B. (1929). The anatomy of a foetal African elephant. Section 7—The skeleton. *Trans. R. Soc. Edinburgh* **56**, 220–46.

Fischer, M. (1986). Die Stellung der Schliefer (Hydracoidea) im phylogenetischen System der Eutheria. *Cour. Forsch. -Inst. Senckenberg* **84**, 1–132.

Gervais, P. (1836). Mammalogie. In *Dictionnaire Pittoresque d'Histoire Naturelle et des Phénomènes de la Nature*, (ed. F. E. Guérin), pp. 614–40. Cosson, Paris.

Gingerich, P. D. and Russell, D. E. (1981). *Pakicetus inachus*, a new archaeocete (Mammalia, Cetacea) from the Early–Middle Eocene Kuldana Formation of Kohat (Pakistan). *Contrib. Mus. Paleontol. Univ. Mich.* **25**, 235–46.

Gregory, W. K. (1910). The orders of mammals. *Bull. Am. Mus. Nat. Hist.* **27**, 1–524.

Gregory, W. K. (1951). *Evolution Emerging.* McMillan, New York.

Hanks, J. (1979). *A Struggle for Survival. The Elephant Problem*, Country Life Books. Hamlyn, London.

Harris, J. M. (1975). Evolution of feeding mechanisms in the family Deinotheriidae (Mammalia: Proboscidea). *Zool. J. Linn. Soc.* **56**, 331–62.

Harris, J. M. (1978). Deinotherioidea and Barytherioidea. In *Evolution of African Mammals* (eds. V. J. Maglio and H. B. S. Cooke) pp. 315–32. Harvard University Press, Cambridge, Massachusetts.
Hennig, W. (1969). *Die Stammesgeschichte der Insekten.* Senckenberg Buch Vol. 49. W. Kramer, Frankfurt.
Kielan-Jaworowska, Z. (1981). Evolution of the therian mammals in the late Cretaceous of Asia. Part IV. Skull structure in *Kennalestes* and *Asioryctes*. *Palaeontol. Pol.* **42**, 25–78.
Kleinschmidt, T., Czelusniak, J., Goodman, M., and Braunitzer, G. (1986). Paenungulata: a comparison of the hemoglobin sequences from elephant, hyrax, and manatee. *Mol. Biol. Evol.* **3**, 427–35.
Krishtalka, L. (1978). Paleontology and geology of the Badwater Creek area, Central Wyoming. Part 15. Review of the Late Eocene Primates from Wyoming and Utah, and the plesitarsiiforms. *Ann. Carnegie Mus.* **47**, 335–60.
Krishtalka, L. and Schwartz, J. H. (1978). Phylogenetic relationships of plesiadapiform–tarsiiform Primates. *Ann. Carnegie Mus.* **47**, 515–40.
Luckett, W. P. and Maier, W. (1982). Development of deciduous and permanent dentition in *Tarsius* and its phylogenetic significance. *Folia Primatol.* **37**, 1–36.
Madden, C. T. (1980). *Zygolophodon* from Subsaharan Africa, with observations on the systematics of palaeomastodontid proboscideans. *J. Paleontol.* **54**, 57–64.
Maglio, V. J. (1973). Origin and evolution of the Elephantidae. *Trans. Am. Philos. Soc.* **63**, 1–149.
Maglio, V. J. and Hendey, Q. B. (1970). New evidence relating to the supposed stegolophodont ancestry of the Elephantidae. *S. Afr. Archaeol. Bull.* **25**, 85–7.
Mahboubi, M., Ameur, R., Crochet, J. Y., and Jaeger, J. J. (1984). Earliest known proboscidean from early Eocene of North-west Africa, *Nature (London)* **308**, 543–4.
Mahboubi, M., Ameur, R., Crochet, J. Y., and Jaeger, J. J. (1986). El Kohol (Saharan Atlas, Algeria): a new Eocene mammal locality in North-western Africa, *Palaeontographica, Abt. A.* **192**, 15–49.
Matsumoto, H. (1923). A contribution to the knowledge of *Moeritherium*. *Bull. Am. Mus. Nat. Hist.* **48**, 97–139.
Matsumoto, H. (1924). A revision of *Palaeomastodon* dividing it into two genera, and with description of two new species. *Bull. Am. Mus. Nat. Hist.* **50**, 1–58.
McKenna, M. C. (1975). Toward a phylogenetic classification of the Mammalia. In *Phylogeny of the Primates* (eds. W. P. Luckett and F. S. Szalay), pp. 21–45. Plenum Publishing Co., New York.
McKenna, M. C. and Manning, E. (1977). Affinities and palaeobiogeographic significance of the Mongolian genus *Phenacolophus*. *Géobios, Mém. Spéc.* **1**, 61–85.
Mitchell, E. D. (1963). Pachyostosis in desmostylids, *Spec. Pap. Geol. Soc. Am.* **76**, 214.

Miyamoto, M. M. and Goodman, M. (1986). Biomolecular systematics of eutherian mammals. *Syst. Zool.* **35**, 230–40.

Novacek, M. J. (1982). Information for molecular studies from anatomical and fossil evidence on higher eutherian phylogeny. In *Macromolecular Sequences in Systematic and Evolutionary Biology* (ed. M. Goodman), pp. 3–41. Plenum Press, New York.

Novacek, M. J. (1986*a*). The skull of leptictid insectivorans and the higher-level classification of eutherian mammals. *Bull. Am. Mus. Nat. Hist.* **183**, 1–112.

Novacek, M. J. (1986*b*). The primitive eutherian dental formula. *J. Vertebr. Paleontol.* **6**, 191–6.

Novacek, M. J. and Wyss, A. R. (1986). Higher-level relationships of the recent eutherian orders: morphological evidence. *Cladistics* **2**, 257–87.

Novacek, M. J. and Wyss, A. R. (1987). Selected features of the desmostylian skeleton and their phylogenetic implications. *Am. Mus. Novit.* **2870**, 1–8.

Osborn, H. F. (1909). The feeding habits of *Moeritherium* and *Palaeomastodon*. *Nature (London)* **81**, 139–40.

Owen, R. (1848). Description of teeth and portions of jaws of two extinct anthracotheroid quadrupeds (*Hyopotamus vectianus* and *Hyop. bovinus*) discovered by the Marchioness of Hastings in the Eocene deposits on the N. W. coast of the Isle of Wight: with an attempt to develop Cuvier's idea of the classification of pachyderms by the number of their toes. *Q. J. Geol. Soc. London* **4**, 103–41.

Owen, R. (1875). On *Prorastomus sirenoides* (Ow.). *Q. J. Geol. Soc. London* **31**, 559–67.

Perry, J. S. (1974). Implantation, foetal membranes and early placentation of the African elephant *Loxodonta africana*. *Philos. Trans. R. Soc. London, B* **269**, 109–35.

Prager, E. M., Wilson, A. C., Lowenstein, J. M., and Sarich, V. M. (1980). Mammoth albumin. *Science N.Y.* **209**, 287–9.

Rainey, W. C., Lowenstein, J. M., Sarich, V. M., and Magor, D. M. (1984). Sirenian molecular systematics—including the extinct Steller's sea cow (*Hydrodamalis gigas*). *Naturwissenschaften* **71**, 586–8.

Savage, R. J. G. (1976) Review of early Sirenia. *Syst. Zool.* **25**, 344–51.

Schwartz, J. H. and Krishtalka, L. (1976). The lower antemolar of *Litolestes ignotus*, a Late Paleocene erinaceid (Mammalia, Insectivora). *Ann. Carnegie Mus.* **46**, 1–6.

Shikama, T. (1966). Postcranial skeletons of Japanese Desmostylia. *Palaeontol. Soc. Jpn., Spec. Pap.* **12**, 1–202.

Shoshani, J. (1986*a*). *On the phylogenetic relationships among Paenungulata and within Elephantidae as demonstrated by molecular and osteological evidence*, Ph.D. Thesis, 649 pp. Wayne State University, Detroit.

Shoshani, J. (1986*b*). Mammalian phylogeny: comparison of morphological and molecular results. *Mol. Biol. Evol.* **3**, 222–42.

Shoshani, J., Goodman, M., Prychodko, W., and Morrison, K. (1977). Relationships among the 'subungulates' as demonstrated by immunodiffusion analysis of serum protein. *Am. Soc. Mammal., East Lansing, Mich.*, Abstr. 118.

Shoshani, J., Goodman, M., Barnhart, M. I., Prychodko, W., Vereshchagin, N. K., and Mikhelson, V. M. (1981). Blood cells and proteins in the Magadan mammoth calf: immunodiffusion comparisons of *Mammuthus* to extant paenungulates and tissue ultrastructure. In *Magadan Baby Mammoth, Mammuthus primigenius (Blumenbach)* (eds. K. Vereshchagin and V. M. Mikhelson), pp. 191–220. Nauka, Leningrad.

Shoshani, J. *et al.* [76 co-authors] (1982). On the dissection of a female Asian elephant (*Elephas maximus maximus* Linnaeus, 1758) and data from other elephants, *Elephant* **2**, 3–93.

Shoshani, J., Lowenstein, J. M., Waltz, D. A., and Goodman, M. (1985*a*). Proboscidean origins of mastodon and woolly mammoth demonstrated immunologically, *Paleobiology* **11**, 429–37.

Shoshani, J. G., Waltz, D. A., Goodman, M., Lowenstein, J. M., and Prychodko, W. (1985*b*). Protein and anatomical evidence of the phylogenetic position of *Mammuthus primigenius* within the Elephantidae. *Acta Zool. Fenn.* **170**, 237–40.

Sickenberg, O. (1934). Beiträge zur Kenntnis tertiärer Sirenen. I. Die Eozäne Sirenen des Mittelmeergebietes. II. Die Sirenen des Belgischen Tertiärs. *Mem. Mus. R. Hist. Nat. Belg.* **63**, 1–352.

Simpson, G. G. (1932). Fossil Sirenia of Florida and the evolution of the Sirenia. *Bull. Am. Mus. Nat. Hist.* **59**, 419–503.

Simpson, G. G. (1945). The principles of classification and a classification of mammals. *Bull. Am. Mus. Nat. Hist.* **85**, 1–350.

Slijper, E. J. (1979). *Whales*, translated by A. J. Pomerans, 2nd edn. Hutchinson, London.

Szalay, F. S. (1977). Phylogenetic relationships and a classification of the eutherian Mammalia. In *Major Patterns in Vertebrate Evolution* (eds. M. K. Hecht, P. C. Goody and B. M. Hecht), pp. 315–74. Plenum Publishing Co., New York.

Tassy, P. (1979). Relations phylogénétiques du genre *Moeritherium* Andrews, 1901 (Mammalia). *C. R. Acad. Sci, Ser. D* **289**, 85–8.

Tassy, P. (1981). Le crâne de *Moeritherium* (Proboscidea, Mammalia) de l'Éocène de Dor el Talha (Libye) et le problème de la classification phylogénétique du genre dans les Tethytheria McKenna, 1975. *Bull. Mus. Natl. Hist. Nat. Ser. 4* **3-C**, 87–147.

Tassy, P. (1982). Les principales dichotomies dans l'histoire des Proboscidea (Mammalia): une approche phylogénétique. *Géobios, Mém. Spéc.* **6**, 225–45.

Tassy, P. (1983). Les Elephantoidea miocènes du Plateau du Potwar, Groupe de Siwalik, Pakistan, *Ann. Paléontol.* **69**, 99–136, 235–98, 317–54.

Tassy, P. (1985). La place des mastodontes miocènes de l'Ancien Monde dans la phylogénie des Proboscidea (Mammalia): hypothèses et conjectures. Unpublished thèse Doct. ès Sci. *Mem. Sci. Terre Univ. Curie, Paris* **85–34**, 1–862.

Tassy, P. (1986). *Nouveaux Elephantoidea (Mammalia) dans le Miocène du Kenya.*

Essai de réévaluation systématique. Cahiers de Paléontologie Est-africaine. CNRS, Paris.

Tassy, P. (1988). The classification of Proboscidea: how many cladistic classifications? *Cladistics* **4**, (in press).

Tobien, H. (1971). *Moeritherium, Palaeomastodon, Phiomia* aus dem Paläogen Nordafrikas und die Abstammung de Mastodonten (Proboscidea, Mammalia). *Mitt. Geol. Inst. Techn. Univ. Hannover* **10**, 141–63.

Tobien, H. (1976). Zur paläontologischen Geschichte der Mastodonten (Proboscidea, Mammalia). *Mainzer Geowiss. Mitt.* **5**, 143–225.

Tobien, H. (1978*a*). The structure of the mastodont molar (Proboscidea, Mammalia). Part 3: the Oligocene mastodont genera *Palaeomastodon, Phiomia* and the Eo/Oligocene paenungulate *Moeritherium*. *Mainzer Geowiss. Mitt.* **6**, 177–208.

Tobien, H. (1978*b*). An advanced zygodont mastodont (Proboscidea, Mammalia) from the Miocene of the Siwalik of Pakistan. *Mainzer Geowiss. Mitt.* **7**, 239–48.

VanderHoof, V. L. (1937). A study of the Miocene sirenian *Desmostylus*. *Univ. Calif. Publ. Bull. Dept. Geol. Sci.* **24**, 169–262.

Watson, M. (1872). Contributions to the anatomy of the Indian elephant. Part I. The thoracic viscera. *J. Anat. Phys.* **6**, 82–94.

Weitz, B. (1953). Serological relationships of hyrax and elephant. *Nature (London)* **171**, 261.

Wells, N. E. and Gingerich, P. D. (1983). Review of Eocene Anthracobunidae (Mammalia, Proboscidea) with a new genus and species, *Jozaria palustris*, from the Kuldana Formation of Kohat (Pakistan). *Contrib. Mus. Paleontol. Univ. Mich.* **26**, 117–39.

West, R. M. (1980). Middle Eocene large mammal assemblage with Tethyan affinities, Ganda Kas region, Pakistan. *J. Paleontol.* **54**, 508–33.

West, R. M. (1983). South Asian Middle Eocene moeritheres (Mammalia: Tethytheria). *Ann. Carnegie Mus.* **52**, 359–73.

Wiley, E. O. (1979). An annotated Linnaean hierarchy, with comments on natural taxa and competing systems. *Syst. Zool.* **28**, 308–37.

Winge, H. (1942). *The Interrelationships of the Mammalian Genera. Vol. 3. Ungulata, Cetacea*, translated from Danish by E. Deichmann and G. M. Allen. Reitzel Forlag, Copenhagen.

Wislocki, G. B. and Van der Westhuysen, O. P. (1940). The placentation of *Procavia capensis*, with a discussion of the placental affinities of the Hyracoidea, *Contrib. Embryol.* **28**, 65–88.

Zhang, Y. (1978). Two new genera of condylarthran phenacolophids from the Paleocene of Nanxiong Basin, Guangdong. *Vertebr. Palasiat.* **16**, 267–74. [In Chinese, English summary.]

Systematic index

aardvark 47, 207, 219, 222, 223, 225
Achaenodon 239, 252–4, 257, 258
Acinonyx 80
Acreodi 205, 213, 216
adapid 152, 153, 161
Adapiformes 43, 50, 52, 53, 55, 58, 60–4, 67–71
Adapis 150, 153, 170
Adapisorex 117, 120–3, 126, 127, 132
Adapisoricidae 123
Adunator 117, 121, 126, 127, 132, 134
aegialodontid 26, 27
Aelurocyon 108
Aeluroidea 75, 79, 83–91, 101–4
Aelurotherium 104
Afrocricetodontinae 181
Agriochoeridae 210, 252, 254, 257
Ailuridae 83, 92, 96
Ailuropoda 92, 96, 99
Ailurus 85, 91, 92, 94, 95–8, 107
Aletomeryx 277, 279
Allodesmus 108, 109
Amblypoda 209, 220
Amebelodon 291
Amebelodontidae 291, 292
amniote 4, 5
Amphechinus 123, 128
Amphictis 107
Amphicynodon 106
Amphicynodontinae 106
Amphicyonidae 77, 87, 100, 105, 106
Amphidozotherium 128, 132, 133
Amphilemuridae 117, 120, 123, 127, 132
amphilestine 25
Amphimerycidae 239, 240, 242, 251–4, 257, 260, 261
Amphimoschus 275, 277, 280
Amphitragulus 273, 275, 277, 279, 280
Anacodon 212, 227
Anagale 52
Anagalida 39, 49, 51, 52, 65
Anancus 291
Anaptomorphidae 150, 160, 162, 171
Ancodonta 235, 238, 239, 256
Andegameryx 277
Andrewsarchus 212
Anisonchus 212, 227
Ankylodon 125, 132
Anomaluridae 177, 180, 186, 187, 193
Anomodontia 9, 13, 14
Anoplotheriidae 239, 240, 253, 258, 259
Anoplotheriina 239, 241, 252–4, 256, 257
Anoplotherioidea 259, 265, 266
anteater 42, 44
Anthracobune 283, 285, 288, 300–4
Anthracobunidae 302
Anthracothema 248, 252, 254, 255, 257, 258
Anthracotheriidae 235, 239–41, 248, 253, 258, 266
Anthracotherium 252, 254, 257
Anthropoidea 143, 144, 150, 155–63, 167–71
Antiacodon 252, 254, 257, 260
Antilocapra 260–2, 264, 280
Antilocapridae 235, 237, 239, 240, 251, 260, 261, 262, 273, 274, 276–9
apatemyoid 81
ape 143, 156, 169
Aplodontia 184
Aplodontidae 192, 193
Apternodontidae 128, 131
Apternodus 117, 120, 132, 133
Archaeobelodon 291
Archaeomeryx 260, 261, 264, 265, 277
Archaeotherium 249
Archonta 31, 38, 48, 59–62, 65, 80, 151
Arctocyon 205, 207, 210, 212, 228
Arctocyonia 39
Arctocyonidae 201, 202, 204, 205, 207, 208, 212, 215, 216, 227, 236, 242
Arctocyonides 205, 212
Arctoidea 73, 75, 78, 82, 84–7, 91–3, 97–102, 104–10
Arctoidei 86, 94–6, 98
armadillo 42
arsinoithere 201–3, 207, 210, 211, 217, 218
Arsinoitherium 212, 218, 284, 307, 308
Artiodactyla 37, 38, 43, 47, 55, 59, 62, 65, 201, 202, 205, 207–14, 217, 222, 223, 225–7, 235, 236, 238–41, 251, 255, 259, 266, 274, 305, 306, 310
Asioryctes 43, 51, 120, 121, 135, 308
Asioryctidae 39

Astrapotheria 39, 221

Bachitheriidae 276, 277
Bachitherium 260, 261, 264, 265, 273, 274, 277, 279
Baioconodon 212
Balaenoptera 286
Baluchimyinae 180–4, 186
Barbourofelinae 87, 104
Barbourofelis 104
Barbouromeryx 277, 279
Barytherioidea 294
Barytherium 283, 293–9, 303
Bassariscus 79, 94, 95, 107
bat 48, 49, 122, 144, 207
Bathyergidae 177, 180, 185, 186
Bathyergus 185
Batodon 134, 136
Batodonoides 130, 131
Bauriamorpha 10
Baurioidea 10
bear 93, 94, 96, 106
Behemotops 300, 302–4
Biarmosaurus 11
Biarmosuchia 9, 11, 13
Biarmosuchus 11
bird 4, 5
Blastomerycini 263, 274, 275, 277, 279
Blastomeryx 275, 277
boselaphine 261
Bothriodontinae 252, 254, 255, 257, 258
Bouromeryx 277
Bovidae 235, 237, 239, 240, 251, 260–4, 273–80
Brachyericinae 123
Brachyerix 123, 132
bradypodid 44, 45
brithopodid 8, 13
Broiliana 107
Brontotheriidae 212
Bruta 285
Bunodontia 235, 237, 253, 256–8, 265
Bunomerycidae 252–4, 257, 259, 265
Bunomeryx 266
Bunophorus 241
Bunoselenodontia 238–40
Butselia 131, 133

Cainotheriidae 239–41, 252–4, 256, 257, 259, 265
Camelidae 240, 249, 252–9
Cameloidea 259, 266
Canidae 73, 76, 79, 83, 85, 87, 91, 92, 94–6, 99, 101, 102, 104–7
Caniformia 73–5, 79, 84, 86, 87, 92–6, 99–103, 105, 106, 109
Canoidea 84, 86
Canthumeryx 275, 277
Carcinodon 236
Carnivora 33, 37, 38, 43, 46, 48, 52, 55, 59, 61, 66, 73–87, 89–91, 93–6, 98–101, 103, 104, 123, 204, 215, 216
Carodnia 212, 221
Carolozittelia 212
Carsioptychus 212
Caseidae 5, 6, 7
castorid 192, 193
cat 88
Catarrhini 143, 146, 149, 164, 169, 170
Caviomorpha 177, 180, 183, 184, 186
Cebochoeridae 239, 252–4, 256–8, 266
ceboid 168
Cedrocherus 127
Centetodon 117, 120, 122, 130–3, 136
Cephalogale 106
Ceratomorpha 213, 236
Cercopithecoidea 168–70
Cervidae 235, 237, 239–41, 251, 261–4, 275–81
Cervoidea 251, 260, 261, 263, 265, 273, 276–80
Cetacea 37, 38, 43, 55, 59, 62, 65, 205, 207, 212, 213, 216, 222, 223, 286, 306, 310
Cete 212, 213, 215, 216
Chailicyon 104
chalicothere 210, 217
Chalicotheriidae 212
Chapattimyidae 180–3
Chiniquodon 15
Chiniquodontidae 14–18
Chiroptera 31, 37, 38, 43, 44, 48, 49, 55, 58, 62, 65, 80, 151
Choerolophodon 292, 293
Choeropotamidae 239, 241, 252–8, 266
Chorlakkia 241
Chriacus 210, 212, 236
Chrysochloridae 117, 118, 120, 122, 128, 129, 132, 133, 134
Cimolesta 39
Cimolestes 81, 135, 136
Climacoceras 260–2, 264, 275, 277, 278
Climacoceridae 276, 277, 279
Cocomyidae 190
Cocomys 177, 179, 180, 187, 189–91, 193
colugo 144
Conacodon 212
Condylarthra 32, 38, 39, 47, 53, 75, 136, 148, 202, 203, 209, 215, 217, 225, 227, 228, 307
Condylura 122
Conoryctes 227
cony 284
Cranioceras 277
cranioceratine 280
Creodonta 39, 46, 66, 73, 75, 77–82, 87, 100, 204, 205, 215, 216

Criadiatherium 218
cricetid 181, 190
Cricetomyidae 190
crocodile 4, 5
Cryptomeryx 277
Cryptoprocta 89, 90
Cryptoproctinae 90
Ctenodactylidae 177, 179–81, 187, 189–91
Ctenodactyloidea 183
Cuisitherium 248
Cuon 94
Cuvieronius 291
Cylindrodontidae 177, 185, 187–9
Cynelos 106
Cynodontia 1–3, 5, 9–12, 14, 17, 18
Cynofeloidea 95, 102
Cynognathidae 15–17
Cynognathus 14–16
Cynoidea 78, 83, 84, 86, 87, 92, 100

dacryotheriid 248
Dacrytheriinae 239, 241, 252, 254, 257
Daphoenidae 107
Daphoeninae 100, 104, 106
Dartonius 126, 132
dasypodid, 44, 45
Dasypodoidea 42
dasyurid 45
Daubentonia 152
deinothere 288, 293–5, 297–9
Deinotheriidae 296, 297
Deinotherioidea 294
Deinotherium 293
Deltatheridia 26, 27, 299
Deltatherium 81, 220
Dendrohyrax 202
Dendromuridae 190
Dermoptera 31, 37, 38, 43, 46, 48–50, 52, 55, 58, 59, 62, 65, 122, 151, 207, 300
desman 118
Desmana 122
Desmaninae 122, 128, 129, 133
Desmostylia 52–4, 65, 66, 201, 202, 217, 218, 236, 283–6, 290, 299–304, 308
Desmostylus 290, 300, 302, 307
Deuterogonodon 212
Diacocherus 126
Diacodexis 210, 211, 214, 227, 235, 239–41, 248, 252–60, 266
Diacodon 121, 127, 129, 132, 136
Diademodon 154
Diademodontidae 15–18
diapsid 4
Diatomys 181
Dichobune 214, 266
Dichobunidae 235, 239, 241, 253, 256, 260, 266
Dichobuninae 252, 254, 257, 266
Dichobunoidea 239
Dicrocerus 263, 277
Dicynodontia 3, 5, 8, 9, 11–13
didelphid 45, 119, 156, 207, 208
didelphodontine 81
Didelphodus 204
didolodont 220, 221, 227
Didolodus 220
Didymictidae 100
Dimetrodon 8
Dimictis 103
Dimylidae 123, 128, 131–3
Dinnetherium 25
Dinocephalia 3, 8, 9, 11–13
Dinocerata 39, 213, 217, 221, 284
Dipodoidea 177, 188, 189, 192
Dissacus 212, 216
Docodonta 25, 26, 236
Dorcabune 277
Dorcatherium 277
Dormaaliidae 125, 127
Dormaaliinae 117, 125
Downsimys 186
Dremotherium 261, 263, 264, 275, 277
Drepanomeryx 277
dromasaur 10
Dromatheridae 15
Dromomerycidae 275, 279, 280
Dromomerycinae 260, 273, 277, 280
Dromomeryx 277
Dugong 287
dugong 284, 286
dugongid 305
Dusisiren 290
Dvinia 14
Dviniidae 17

Echinops 122
Ectocion 212
Ectoconus 212, 220
Edaphosauria 5
Edaphosauridae 5, 6
Edentata 31, 32, 37, 38, 42–5, 50 55, 60–3, 65, 78, 80, 82, 122, 207, 223, 288, 305, 306, 310
Elaphodus 262
elephant 201–3, 209, 211, 217, 222–5, 284–6, 288, 290, 295, 296, 305, 307, 309
Elephantidae 291, 293
Elephantiformes 291, 292, 296–9
Elephantoidea 286, 288, 291, 292, 295
Elephas 284, 287, 290, 293, 304
Eliomys 192
Embrithopoda 39, 213, 217, 218, 284, 288, 300, 304
Enaliarctinae 108

Enaliarctos 108
Entelodontidae 238, 239, 249, 252–4, 257, 258, 265
Entomolestes 117, 124, 132
Eoconodon 210, 212
Eogliravus 192
Eolestes 127, 132
Eosygodon 291
Eotalpa 133
Eotherapsida 2
Eotheroides 285, 288, 300
Eothyrididae 5, 6
Eotitanosuchia 11
Eotitanosuchus 11, 13
Epimycteri 86
Epitheria 31, 45, 46, 60–2, 65, 286
Equidae 53, 212, 213, 274, 278
equoid 219
Erinaceidae 55, 117–24, 126–9, 135, 300
Erinaceinae 123, 132
Erinaceoidea 122, 123, 131, 134, 136, 148, 151
Erinaceomorpha 47, 117–21, 123, 126–8
Erinaceota 123
Erinaceus 120
Ernotheria 51
Erythrozootes 128
Estemmenosuchus 8
Eubelodon 291
Eucervoidea 276, 280
Eucynodontia 2, 14, 15
Eumeryx 276, 279
Eumys 190
eupantothere 24–6
Eupecora 274, 277, 278
Eupleres 89
euprimate 48, 150 153
Euprotogonia 209
Eurymylidae 49, 51, 177, 189–92, 194
eurymyloid 194
Eutheria 23, 31, 32, 37–9, 42, 45–7, 50, 51, 54, 55, 57–60, 63–6, 76–82, 89, 100, 118, 120–3, 134, 136, 148, 153, 158, 163, 164, 202–4, 208, 210, 214, 222, 223, 227, 242, 305, 306, 308
Eutheriodonta 12
Exaeretodon 18
Exallerix 123

Fallomus 181, 182
Felidae 73, 76, 79, 83, 85, 87, 88, 90, 91, 94–6, 101, 103, 104
Feliformia 73, 74, 84–7, 90, 95, 99, 100–2
Felinae 102
Felis 92
Feloidea 78, 79, 82, 84, 86, 100, 102
Ferae 73–5, 80, 81, 87, 220
Ferungulata 38, 62, 80, 204, 205
Fissipeda 73, 75, 83, 84, 92, 100
flying lemur 48, 49
Fossa 89
Franimorpha 183, 186, 191

galagid 154
Galemys 120, 130
Galericini 121, 123, 124, 127, 132
Galerix 119, 124, 125, 127
Galesaurus 14
Galidiinae 89
Gelocidae 273, 274, 277, 279
Gelocus 235, 261, 264, 265, 276, 278, 279
Geogale 120
Geolabididae 128, 131, 134, 136
Geomyoidea 177, 184, 187–9, 192
Gerbillidae 190
Gesneropithex 120
Giraffa 262
Giraffidae 237, 239, 240, 251, 260–2, 264, 275–80
Giraffinae 277
Giraffoidea 235, 260–3, 265, 273, 275, 276, 278, 279
Glires 31, 49–52, 57, 60–2, 65, 194, 208, 226
Gliridae 192, 193
Gnathabelodon 291
Gobiconodon 25
Gobioceras 277
Gobiohyus 239, 252–5, 257, 258
Gobiomeryx 276, 278
golden mole 118
gomphothere 291, 293
Gomphotherium 290, 291, 293–5
Goniacodon 212
gonostylopoid 221
Gorgonopsia 9–11
gorgonopsid 12, 13
Griphodon 212
Gypsonictops 134, 214

Hadropithecus 152
Haemothermia 4, 5
Hallautherium 25
Hapalodectes 212, 216
Hapalodectini 213
Haplobunodon 249
Haplobunodontidae 241
Haploconus 212
Haplomastodon 291
haplorhine 153, 158, 160, 163, 167, 168
Haplorhini 143, 144, 146, 150, 151, 154–6, 171
Haptodus 8
haramiyid 25
Harpagolestes 212

Harpiocephalus 131
hedgehog 118
Helohyidae 239, 241, 249
Helohyus 239, 252–4, 257, 258
Hemicyoninae 106, 108
Hemiechinus 120
Hemimastodon 292
Hemithlaeus 212
Hendryomeryx 277
Heomys 191, 192
Herpestidae 73, 85, 87–9, 91, 96
Herpestinae 90
Herpestoidei 86
Hesperocyon 79, 104
Heterocemas 277, 281
Heterosoricinae 128
Hexacodus 241
Hippopotamidae 235, 238–40, 252–4, 257, 258
Hippopotamoidea 266
hippopotamus 202, 249
Hispanomeryx 277
Homacodon 248, 252–4, 257, 259, 260
Homacodontinae 235, 253, 256, 259, 260
homaladothere 210
hominid 54
Hominoidea 149, 168–70
Hoplitomerycidae 263, 276, 277, 279, 280
Hoplitomeryx 263, 277, 279, 280
horse 222, 224
hyaena 216
Hyaenidae 73, 83, 85–8, 90, 91, 94, 96, 102
Hyaenodon 82
hyaenodont 204, 209
Hyaenodontidae 75, 81, 82
hyaenodontoid 81
Hydropotes 261–5, 272, 277, 279, 280
Hyemosuchus 277
Hylomeryx 266
Hylomys 120
Hyodonta 238, 239
hyopsodont 201, 207, 208, 210, 215, 216, 220, 221, 227
Hyopsondontidae 53, 212
Hyopsodus 210, 215
Hyperdichobune 260, 266
Hyperdichobuninae 252, 254, 257, 260, 266
Hypertragulidae 242, 251–5, 257, 260, 261, 275–7
Hypertragulus 277
Hypisodus 277
Hypomycteri 86
Hypsodontus 277
hyracodontid 202
Hyracoidea 31, 37, 38, 43, 53–5, 62, 65, 212, 219, 283–5, 288, 300, 304–10
Hyracotherium 219
hyrax 201, 202, 205, 207–11, 217–19, 221–5, 284
Hystricidae 177, 180, 185, 186
Hystricognatha 177, 182–7, 189
Hystricomorpha 177, 179–81, 183, 186–9, 191

Iberomeryx 277
ictidosaur 15, 16, 18
Ictidosuchidae 11
Ictopidium 117, 123, 124
Imagoteria 108
Indomeryx 277
Injanatherium 277
Insectivora 32, 37, 38, 43, 46, 47, 50, 55, 59, 61, 65, 66, 75, 80, 117, 118, 120, 125, 129, 134, 157, 158, 204, 209, 226, 227
Ischyromyoidea 177, 187–93

Jozaria 302

Kennalestes 42, 51, 121, 204
Kennalestidae 39
Kubanotragus 277
Kuehneotheriidae 15, 24, 26

lagomerycid 281
Lagomeryx 275, 277, 281
Lagomorpha 31, 37, 43, 46, 47, 49–51, 55, 59, 60, 62, 63, 177, 181, 194, 207, 305
Lammidhania 303
Lantianius 221, 241
Leipsanolestes 117, 127, 132
lemur 122, 143, 146, 152, 153, 155, 156, 158, 159, 161, 167
lemurid 156
Lemuriformes 143, 150, 153–5, 157–9, 161–3, 167–9, 171
Lemuroidea 149, 153, 164
leporid 50
Leptacodon 117, 120–2, 126, 128, 130–2, 134, 136
Leptadapis 150, 153, 170
Leptictidae 35, 39, 47, 51, 65, 117, 119, 121, 123, 125, 127, 134–6, 220, 310
leptictimorph 220
Leptictis 120
leptictoid 134
Leptochoeridae 238, 252–4, 259, 260, 264, 265
Leptomerycidae 249, 252, 254, 257, 259, 260, 264, 265, 274–7
Leptomeryx 260, 261, 264, 265, 277, 279
Leptotheridium 241
Leptotragulidae 241

Lipotyphla 46, 47, 65, 117–23, 125, 127, 133–6, 226, 227
Litocherinae 117, 127
Litocherus 127, 132
Litocranius 211
Litolestes 117, 126, 127, 132
Litopterna 39, 202, 220, 221
llama 222
Longirostromeryx 277
Lophiomerycidae 276, 277
Lophiomeryx 273, 274, 277
loris 143, 146, 152, 153, 155, 156, 158, 159, 167
lorisid 154, 168
lorisiform 143, 154, 157, 159, 162, 171
Lorisoidea 149, 153–5, 157, 165
Loxodonta 284, 287, 292–5, 305
lutrine 85, 92
Lycaon 94

Machaeomeryx 277
Machairodontinae 102
Macrocranion 117, 122, 125, 126, 132
Macroscelidea 31, 37, 39, 43, 46, 47, 49–52, 55, 61, 62, 65, 118, 134, 135
Mammalia 2–5, 10, 14–19, 23–7, 32, 34–6, 44, 46, 54, 64, 78, 93, 119, 121, 123, 144, 145, 147, 148, 157, 170, 194, 226, 285
Mammaliformes 16
Mammut 295
Mammuthus 293, 295, 292
Mammutidae 288, 292, 293
manatee 223, 284, 286, 288, 309
Manis 44
Marsupialia 23–7, 39, 45, 63, 76, 78 82, 121
Martes 79
Mckennatherium 126
megachiropteran 49, 89
Megalonichoidea 42
Megatherioidea 42
Meles 91
Meniscodon 266
meniscothere 201, 207–9, 211, 217, 227
Meniscotherium 210, 212, 217, 227, 228, 307, 309
Menotyphla 50
Mentoclaenodon 212
mephitine 85, 92
Meridiungulata 66, 220, 221
Meryceros 261
Merycodontinae 260–2, 264, 265
merycoidodont 256
Merycoidodontidae 239, 252, 254, 257
Merycoidodontoidea 253, 256, 259, 265
Merycopotamus 258
Merycotheria 257
Mesaxonia 305–8, 310
Mesocyon 104
Mesomeryx 266
Mesonychidae 201, 202, 205, 207–9, 213, 215, 216, 227, 228
mesonychoid 80
Mesonyx 207, 210, 212
Messelbunodon 252, 254, 256, 257, 260
Metatheria 23
Metriotherium 266
Miacidae 33, 75, 77, 87, 100, 101, 103
Miacoidea 73, 77–9, 87, 100, 101, 103, 104
microchiropteran 49, 158
Microchoeridae 143, 150, 160, 162–4, 168, 169, 171
Microgale 119, 120
Micromeryx 275, 280
Microparamyidae 187, 192
Micropternodontidae 128, 135, 136
Micropternodus 132, 135
Microsus 252–5, 260
Microsyopidae 48, 66, 151, 160
Mimotona 51
Mimotonidae 51, 194
Mimotricentes 212
Minchenella 303
Mioclaenus 227
Miomastodon 291
Miomeryx 277
Mixotheriidae 235, 252–4, 256, 257, 259, 265
Mixotherium 248
moerithere 283, 302
Moeritherioidea 294
Moeritherium 283, 285, 288–90, 293–9, 302, 303, 307, 309, 310
mole 118, 122, 129, 133
Mongolestes 212
Mongolonyx 212
Mongolotherium 218
monkey 143, 156, 169
Monotherium 109
Monotremata 16, 23, 25, 26, 39, 40, 43, 44, 78, 82, 121
moon-rat 118
Morganucodontidae 15, 24, 25
Moschidae 262, 263, 273–80
Moschina 276, 277
Moschus 261–4, 275, 277, 280
Mouillacitherium 260, 266
multituberculate 25, 26
Mungotidae 89
munticine 279, 281
Muntiacus 262
Muridae 93, 190
Muroidea 177, 188–90, 192
Mus 190
Mustela 91
Mustelictis 107

Mustelidae 73, 79, 83, 85, 87, 91–9, 102, 106, 107, 109
Mustelinae 76
Musteloidea 87, 93, 96, 99, 105, 107, 109, 110
Mydaus 95
Myocricetodontinae 190
Myodonta 188, 192
Myomorpha 177, 188, 189
Myosorex 120, 128
Myrmecophaga 78
Myrmecophagidae 42, 45
Mytonomeryx 266

Nandinia 77, 85, 87, 89, 90, 100, 104
Nannopithex 148
Nanotragulus 277
Necrolemur 162
Neoselenodontia 235, 240, 253, 259
Nesophontes 117, 118–21, 122, 128, 130, 131–3
Nesophontidae 118
Neurogymnuridae 123
Neurogymnus 132
Nimravidae 74, 77, 79, 87, 95, 101–4
Notharctus 150
Notiomastodon 291
Notomeryx 276, 278
Notoungulata 39, 47, 53, 220, 221, 226
Numidotherium 283, 294–6, 298, 301, 303, 309
Nyanzameryx 277
Nycticeboides 154
Nyctitheriidae 117, 121, 128, 131–3
Nyctitherium 131, 132

Obdurodon 25
ochotonid 50
Odobenidae 73, 84, 85, 92, 96–8, 108–10
Odobenus 98
Okapia 262
Oligokyphus 19
Omomyidae 150, 160, 163, 164
Omomyoidea 143, 150, 155–8, 160–2, 167–9, 171
Oodectes 204
Ophiacodontia 2, 5
Ophiacodontidae 5, 6
opossum 223
Oreodonta 235, 240, 257, 266
Ornithorhynchus 24, 25
Ormerycidae 252–4, 256–9, 266
Orycteropus 304
Osteria 10
Otariidae 73, 84, 85, 91, 92, 96–8, 108, 110
otter-shrew 118
ox 222
Oxyaenidae 75, 81, 204
oxyaenoid 81
oxyclaenine 216
Oxyclaenus 212

Pachyaena 207
Pachycynodon 106
Pachydermata 226, 236
Paenungulata 31, 32, 47, 52–4, 57, 60, 62, 65, 80, 201, 202, 212, 213, 217, 219, 223, 225, 226, 284, 285, 304–10
Pakilestes 133
Palaeanodonta 65
Palaeoamasia 218
Palaeodonta 235, 238, 239, 256
Palaeogale 87, 101, 102
Palaeohypsodontus 277
Palaeomastodon 291–3
Palaeomerycidae 273, 275–7, 280, 281
Palaeomeryx 251, 260–4, 273, 275, 277, 279, 280
Palaeoparadoxia 280, 301–3
Palaeoryctes 117, 121, 128, 132, 134, 135
Palaeroyctidae 117, 134–6
palaeoryctine 220
palaeoryctoid 80, 81, 121, 134, 136
palaeothere 208, 218
Palaeotheriidae 212
panda 93–6, 99, 107
pangolin 44, 45, 80
pantherine 88
Pantodonta 81, 207, 209, 215, 217–20, 284
Pantolambda 220
Pantolambdodon 219
pantolestid 81, 220
pantolestoid 80
Pantomesaxonia 212, 213, 310
Parablastomeryx 277, 280
Paraceratherium 202
Paragale 108
Paramys 204
Paranyctoides 117, 134, 136
Paratetralophodon 291
Paratriisodon 212
Paromyiformes 150
Parvitragulus 277
Pecora 235, 237–9, 249, 260–5, 273–9
Pedetes 181
Pedetidae 180–2, 186, 193
Pediomeryx 277
Pelycosauria 1–8, 13
Pentacemylus 266
Periptychidae 201, 207–10, 213, 215, 216, 220, 227
Periptychus 207, 210, 212, 227
Perissodactyla 37, 38, 43, 47, 53–5, 59, 62, 65, 80, 201, 202, 205, 207, 208, 210–13,

Perissodactula—cont.
217–19, 222–4, 226, 227, 236, 274, 283–5, 288, 300, 304–10
Phaner 154
Phenacodonta 53, 201, 207, 209–11, 217, 227, 236, 285, 305
Phenacodus 209, 210, 212, 217, 227, 228, 307, 309
phenacolophid 303
Phenacolophus 212, 218, 308
Phiomia 288, 290–2
Phiomyidae 182, 183
Phocidae 73, 84, 85, 91, 92, 96–8, 108–10
Pholidocercus 127, 128
Pholidota 31, 37, 38, 42–6, 55, 59–62, 65, 288, 306, 310
Phthinosuchus 11
Physeter 286
pig 222
Pilgrimella 304
Pilosa 44, 45
Pinnarctidion 108, 109
Pinnipedia 73, 75, 83–5, 87, 90, 92, 95, 96, 98, 99, 108–10
Placentalia 23, 25–7
Plagioctenodon 132
Platybelodon 291
Platyrrhini 143, 146, 149, 164, 169, 170
plesiadapid 169
Plesiadapiformes 48, 66, 143, 145, 149–52, 154, 155, 158, 160–4, 168, 170
Plesiadapoidea 145, 150
Plesictis 107, 108
Plesiodimylus 130
Plesiogale 108
Plesiosorex 130–3
Plesiosoricidae 122, 128, 131, 133
Plesitarsiiformes 155
Pleuroaspidotherium 205, 212
Pliomastodon 291
Poebrotherium 249
Pontifactor 130–2
Potamogale 131
Potamogalinae 121, 122
Potamotherium 85, 109, 110
Potos 95
Preptotheria 299
Primates 31, 37, 38, 43, 46, 48–50, 52, 54, 55, 60, 65, 76, 80, 118, 136, 143–6, 148–54, 157–63, 168–71, 204, 208, 300
Primelephas 292
Prionodontinae 90
Proailurus 104
Probainognathidae 15, 17, 19
Probainognathus 15
Probelesodon 15, 16
Problastomeryx 277
Proboscidea 31, 37, 38, 43, 44, 52–4, 62, 65, 66, 201, 207, 210, 217, 218, 223, 224, 236, 283–8, 290, 291, 293–305, 308, 310
Procavia 207, 284, 304, 305
Procervulus 263, 277, 281
Procranioceras 277
Procynodictis 100
Procynosuchia 2
Procynosuchidae 16
Procynosuchus 14
Procyon 91
Procyonidae 73, 76, 79, 83, 85, 87, 92–8, 106, 107
Prodeinotherium 290, 293
Prodremotherium 276, 278, 279
Progalago 154
Prolapsus 183, 184
Prolibytherium 273, 275, 277, 280
Promartes 108
Pronodens 277
Propalaeoryx 273, 275, 276, 278
Prorastomus 288, 299, 300, 302, 308, 310
Prosarcodon 121, 135
Proscalopidae 132, 133
Prosimii 144, 151, 154, 155, 156, 167, 168
Prosiren 300
Protadelomys 187
Protanancus 291
Proteles 88
Protenrec 128, 131
Protericinae 123
Proterix 132
proterothere 217
Proteutheria 65, 125
Protoceratidae 241, 252–4, 256, 257, 259, 266
Protodichobune 252, 254, 255
Protogonodon 205, 212
Protophiomys 182
Prototheria 23, 24
Prototherium 300
protungulate 151
Protungulatum 136, 148, 204, 210–12, 214
Pseudaelurus 104
Pseudictopidae 49–51
Pseudictops 218
Pseudoceras 277, 279
Pseudomeryx 277
Pseudoparablastomeryx 277
Purgatorius 204, 214
Pyrotheria 39, 202, 213, 221
Pyrotherium 212

rabbit 208
Raliomeryx 277
Raoellidae 239, 252–4, 257, 258
Rattus 190
Rhinoceratoidea 202, 205, 212

rhinoceros 202, 219
Rhynchotherium 291
Robiacina 241
Rodentia 31, 37, 43, 46, 47, 49, 51, 55, 60, 62, 65, 93, 122, 151, 177–9, 181–4, 186, 187, 189–94, 202, 204, 207–9
Ruminantia 93, 211, 235, 237, 238, 240, 242, 251, 253, 255–7, 259, 261, 263, 266, 273–9
Rutitherium 276, 279
Rythroparamyinae 191

Sanotheriidae 241
Sansanosmilus 104
Saturninia 128, 130–2
Sauropsida 4
Scalopus 122
Scandentia 37, 43, 48, 55, 118, 148, 152, 170
Scenopagidae 117, 125–7
Scenopaginae 125
Scenopagus 125–7, 132
Sciuravidae 177, 180, 185, 187–9, 193
Sciuridae 181, 192, 193
sciurognath 181, 183, 186
Sciuromorpha 193
Scraevia 132
sea-cow 217, 284
seal 97
sea-lion 97
Selenodontia 235, 237, 253, 256–60, 265
Semantor 109, 110
Serbelodon 291
Sespedictinae 125, 132
Setifer 120
sheep 222
shrew 118, 120, 129, 130, 131, 133
Shuotherium 25, 26
Silvatheriinae 277
Simiiformes 155
Simiolemuriformes 155
Simmeryx 277
Simocyon 107
Simpsonodon 25
Sinclairomeryx 277
Sinoconodon 24
Sinoconodontidae 15
Sirenia 31, 37, 38, 43, 52–5, 62, 65, 201–3, 209, 210, 217, 218, 236, 283–8, 290, 291, 295, 299, 300–2, 304, 306–8, 310
sivathere 262
sloth 44
Smilodectes 150
Solenodon 117–20, 128–33, 135
Solenodontidae 55, 118
Soricidae 55, 117, 118, 120, 122, 123, 128–34
Soricomorpha 117–20, 123, 128, 130, 131, 133–5
Spalax 184
Speothos 94
Sphenacodontia 2, 5
Sphenacodontidae 5–8
sphenacodontine 8
Stegodibelodon 292
Stegodon 291, 293
Stegodontidae 291–3
Stegolophodon 291, 293
Stegomastodon 291
Stegotetrabelodon 292
Stephanoceras 263
Steropodon 25
Strepsirhini 144, 146, 150, 152, 153, 155, 156, 159, 160, 168, 171
Stromeriella 107
Suidae 238, 239, 241, 252–4, 257, 258
Suiformes 238, 265
Suillotaxus 95
suine 222
Suoidea 240, 241, 248, 253, 258, 265, 305
Sylvisorex 130
symmetrodont 24, 26
Synapsida 1–5, 19
Synoplotherium 212

taeniodont 42, 81, 220, 227
Taligrada 209
Talpa 121
Talpavus 126, 132
Talpidae 55, 117, 118, 120–3, 128–33
Talpinae 133
Tamandua 78
tapir 205, 219, 223
Tapiroidea 212
Tapirulus 241
tarsier 143, 144, 150, 154–61, 163, 167–9, 171
tarsiid 169
Tarsiiformes 143, 146, 149, 155, 159, 160, 163, 164, 167–9, 171
tarsioid 168
Tayassuidae 215, 238–40, 252–4, 257, 258
Teilhardina 171
tenrec 18
Tenrecidae 55, 117, 118, 120–2, 128–34
Tenrecinae 121, 129
Tethytheria 31, 32, 52, 53, 62, 65, 66, 201, 205, 212, 218, 221, 225, 226, 283, 285–90, 299, 300, 302, 203, 305–10
Tetraclaenodon 209, 210, 212, 217, 227
Tetracus 123
Tetralophodon 291
Texodon 241
Therapsida 1–5, 7–13
Theria 24–7, 45–7
Theridomyidae 186, 187

Theriodontia 2, 9, 12–14
Therocephalia 9–13
Therosauria 12, 13
Thrinaxodon 14
Thrinaxodontidae 17
Thryonomyidae 177, 180
Thrynomyoidea 182, 183, 185
Tillodontia 39
titanothere 202
Tokotheria 299, 300
toxodont 202
Tragulidae 239, 252, 254, 257, 260, 261, 274–7, 279
Tragulina 238, 264, 265, 273, 274, 276
traguloid 273, 274, 277, 279
Tragulus 264, 277
Traversodontidae 15–18
tree-shrew 31, 48, 55, 145, 148, 151–3, 170
Tribosphenidae 26
Tricentes 212, 236
Triceromeryx 275, 277, 280
Trichechus 287, 305
triconodont 25
Triisodon 205, 212
Triisodontinae 213, 215, 216
Trirachodon 15
Tritheledontidae 1, 15–18
Tritylodon 1
Tritylodontidae 15–17, 19
Tritylodontoidea 18
Tsaganomyidae 185, 186, 188
Tubulidentata 37, 42, 43, 47, 55, 60–2, 66, 304, 306
tupaiid 46–8, 50–2, 119, 131, 135, 156, 157, 160, 219
Tupaiodon 117, 123–6, 132
Tupaioidea 143, 163, 164
Tupaiodontinae 117, 124, 125, 127
turtle 5
Tylopoda 222, 235, 238–40, 253, 257, 259, 265

uintathere 50, 201–3, 207, 209–11, 217, 218, 220, 221
Uintatheriidae 212
Unguiculata 38, 48, 62
Ungulata 31, 37, 38, 47, 52, 53, 57, 59–62, 65, 66, 75, 80, 201–5, 208–11, 213–27, 236, 288, 309
Uropsilinae 133
Uropsilus 121, 128, 130 131, 133
Urotrichus 129
Ursidae 73, 76, 83, 85, 87, 91, 92, 94–7, 99, 105, 106, 108–10
Ursinae 107
Ursoidea 86, 107
Ursus 91

Varanopseidae 5, 6
Vassacyon 100
Venjukovia 10
viverravid 104
Viverravus 77, 102
Viverridae 73, 79, 83, 85, 87–91, 95, 96, 100–2
Viverrini 90
Vulpavus 100

Walangania 275, 276, 279
walrus 97, 98, 109
Wasatchia 241
Whaitsiidae 12
whale 201–3, 208, 216, 222, 223, 228

Xenarthra 42, 44, 45
Xenungulata 39
Xiphodontidae 239–41, 252–4, 256–9, 266

Yuomyidae 180, 188

Zalamdalestidae 51
Zarafa 275
Zodiolestes 108
Zygolophodon 291

Systematics Association Publications

1. Bibliography of key works for the identification of the British fauna and flora *3rd edition* (1967)
   Edited by G. J. Kerrich, R. D. Meikle and N. Tebble
2. Function and taxonomic importance (1959)
   Edited by A. J. Cain
3. The species concept in palaeontology (1956)
   Edited by P. C. Sylvester-Bradley
4. Taxonomy and geography (1962)
   Edited by D. Nichols
5. Speciation in the sea (1963)
   Edited by J. P. Harding and N. Tebble
6. Phenetic and phylogenetic classification (1964)
   Edited by V. H. Heywood and J. McNeil
7. Aspects of Tethyan biogeography (1967)
   Edited by C. G. Adams and D. V. Ager
8. The soil ecosystem (1969)
   Edited by H. Sheals
9. Organisms and continents through time (1973)†
   Edited by N. F. Hughes

Published by the Association (out of print)

Systematics Association Special Volumes

1. The new systematics (1940)
   Edited by Julian Huxley (Reprinted 1971)
2. Chemotaxonomy and serotaxonomy (1968)*
   Edited by J. G. Hawkes
3. Data processing in biology and geology (1971)*
   Edited by J. L. Cutbill
   Out of print

4. Scanning electron microscopy (1971)*
   Edited by V. H. Heywood
   Out of print
5. Taxonomy and ecology (1973)*
   Edited by V. H. Heywood
6. The changing flora and fauna of Britain (1974)*
   Edited by D. L. Hawksworth
7. Biological identification with computers (1975)*
   Edited by R. J. Pankhurst
8. Lichenology: progress and problems (1976)*
   Edited by D. H. Brown, D. L. Hawksworth and R. H. Bailey
9. Key works to the fauna and flora of the British Isles and north-western Europe (1978)*
   Edited by G. J. Kerrich, D. L. Hawksworth and R. W. Sims
10. Modern approaches to the taxonomy of red and brown algae (1978)*
    Edited by D. E. G. Irvine and J. H. Price
11. Biology and systematics of colonial organisms (1979)*
    Edited by G. Larwood and B. R. Rosen
12. The origin of major invertebrate groups (1979)*
    Edited by M. R. House
13. Advances in Bryozoology (1979)*
    Edited by G. P. Larwood and M. B. Abbot
14. Bryophyte systematics (1979)*
    Edited by G. C. S. Clarke and J. G. Duckett
15. The terrestrial environment and the origin of land vertebrates (1980)*
    Edited by A. L. Panchen
16. Chemosystematics: principles and practice (1980)*
    Edited by F. A. Bisby, J. G. Vaughan and C. A. Wright
17. The shore environment: methods and ecosystems (2 Volumes) (1980)*
    Edited by J. H. Price, D. E. G. Irvine and W. F. Farnham
18. The Ammonoidea (1981)*
    Edited by M. R. House and J. R. Senior
19. Biosystematics of social insects (1981)*
    Edited by P. E. Howse and J. -L. Clément
20. Genome evolution (1982)*
    Edited by G. A. Dover and R. B. Flavell
21. Problems of phylogenetic reconstruction (1982)*
    Edited by K. A. Joysey and A. E. Friday
22. Concepts in nematode systematics (1983)*
    Edited by A. R. Stone, H. M. Platt and L. F. Khalil

23. Evolution, time and space: the emergence of the biosphere (1983)*
Edited by R. W. Sims, J. H. Price and P. E. S. Whalley
24. Protein polymorphism: adaptive and taxonomic significance (1983)*
Edited by G. S. Oxford and D. Rollinson
25. Current concepts in plant taxonomy (1983)*
Edited by V. H. Heywood and D. M. Moore
26. Databases in systematics (1984)*
Edited by R. Allkin and F. A. Bisby
27. Systematics of the green algae (1984)*
Edited by D. E. G. Irvine and D. M. John
28. The origins and relationships of lower invertebrates (1985)‡
Edited by S. Conway Morris, J. D. George, R. Gibson and H. M. Platt
29. Infraspecific classification of wild and cultivated plants (1986)‡
Edited by B. T. Styles
30. Biomineralization in lower plants and animals (1986)‡
Edited by B. S. C. Leadbeater and R. Riding
31. Systematic and taxonomic approaches in palaeobotany (1986)‡
Edited by R. A. Spicer and B. A. Thomas
32. Coevolution and systematics (1986)‡
Edited by A. R. Stone and D. L. Hawksworth
33. Keyworks to the fauna and flora of the British Isles and north-western Europe *5th edition* (1988)‡
Edited by R. W. Sims, P. Freeman and D. L. Hawksworth
34. Extinction and survival in the fossil record (1988)‡
Edited by G. P. Larwood

35A. The phylogeny and classification of the tetrapods. Volume 1: amphibians, reptiles, birds (1988)‡
Edited by M. J. Benton

35B. The phylogeny and classification of the tetrapods. Volume 2: mammals (1988)‡
Edited by M. J. Benton

*Published by Academic Press for the Systematics Association
†Published by the Palaeontological Association in conjunction with the Systematics Association
‡Published by the Oxford University Press for the Systematics Association